T0313886

Organic Synthesis with Carbohydrates

GEERT-JAN BOONS
Complex Carbohydrates Research Center
Athens, Georgia
USA

KARL J. HALE
Department of Chemistry
University College London
UK

Sheffield
Academic Press

Blackwell
Science

Organic Synthesis with Carbohydrates

Postgraduate Chemistry Series

A series designed to provide a broad understanding of selected growth areas of chemistry at postgraduate student and research level. Volumes concentrate on material in advance of a normal undergraduate text, although the relevant background to a subject is included. Key discoveries and trends in current research are highlighted, and volumes are extensively referenced and cross-referenced. Detailed and effective indexes are an important feature of the series. In some universities, the series will also serve as a valuable reference for final year honour students.

Editorial Board

Professor James Coxon (Editor-in-Chief), Department of Chemistry, University of Canterbury, New Zealand.
Professor Pat Bailey, Department of Chemistry, Heriot-Watt University, UK.
Professor Les Field, Department of Chemistry, University of Sydney, Australia.
Professor Dr John Gladysz, Institut für Organische Chemie, Universität Erlangen-Nürnberg, Germany.
Professor Philip Parsons, School of Chemistry, Physics and Environmental Science, University of Sussex, UK.
Professor Peter Stang, Department of Chemistry, University of Utah, USA.

Titles in the Series:

Catalysis in Asymmetric Synthesis
Jonathan M.J. Williams

Protecting Groups in Organic Synthesis
James R. Hanson

Organic Synthesis with Carbohydrates
Geert-Jan Boons and Karl J. Hale

Organic Synthesis with Carbohydrates

GEERT-JAN BOONS
Complex Carbohydrates Research Center
Athens, Georgia
USA

KARL J. HALE
Department of Chemistry
University College London
UK

Sheffield
Academic Press

Blackwell
Science

First published 2000
Copyright © 2000 Sheffield Academic Press

Published by
Sheffield Academic Press Ltd
Mansion House, 19 Kingfield Road
Sheffield S11 9AS, England

ISBN 1-85075-913-8

Published in the U.S.A. and Canada (only) by
Blackwell Science, Inc.
Commerce Place
350 Main Street
Malden, MA 02148-5018, U.S.A.
Orders from the U.S.A. and Canada (only) to Blackwell Science, Inc.

U.S.A. and Canada only:
ISBN 0-6320-4508-6

NOTICE: The authors of this volume have taken care that the information contained herein is
accurate and compatible with the standards generally accepted at the time of publication.
Nevertheless, it is difficult to ensure that all the information given is entirely acccurate for all
circumstances. The publisher and authors do not guarantee the contents of this book and dis-
claim liability, loss, or damage incurred as a consequence, directly or indirectly, of the use and
application of any of the contents of this volume.

Trademark Notice: Product or corporate names may be trademarks or registered trademarks,
and are used only for identification and explanation, without intent to infringe.

Printed on acid-free paper in Great Britain by
Bookcraft Ltd, Midsomer Norton, Bath

British Library Cataloguing-in-Publication Data:
A catalogue record for this book is available from the British Library

Library of Congress Cataloging-in-Publication Data:
A catalog record for this book is available from the Library of Congress

Preface

The carbohydrates or saccharides constitute the most abundant group of compounds found in nature. They are structurally very diverse and are endowed with a wealth of stereochemical properties. Saccharides are available in cyclic and acyclic forms, can have different chain lengths and oxidation and reduction states, and can be substituted with a wide range of functionalities. Furthermore, monosaccharides can be linked together through glycosidic linkages to give oligo- or polysaccharides. Many saccharides are readily and cheaply available and provide an attractive, renewable source of material.

Not surprisingly, these compounds are important starting materials in organic synthesis, and there are thousands of research papers and numerous industrial processes in which carbohydrates feature prominently.

This book provides broad coverage of the use of carbohydrates in organic synthesis, at postgraduate student level. Each chapter describes established and widely used methods and approaches, but also covers recent and promising reports. Many citations to the primary literature are provided. It is hoped, therefore, that this book will also be of use to synthetic organic chemists and carbohydrate chemists in academic and industrial laboratories.

The authors recognise that one book cannot cover all aspects of synthetic carbohydrate chemistry. Part A focuses on monosaccharide chemistry, complex oligosaccharides and glycoconjugate synthesis. For a long time, this area of chemistry was the domain of a small and specialised group of researchers. In the early eighties, it became apparent that oligosaccharides are involved in many important biological processes, such as cell-cell recognition, fertilisation, embryogenesis, neuronal development, viral and bacterial infections and tumour cell metastasis. Consequently, the preparation of complex glycoconjugates became part of mainstream organic chemistry and it is now part of the undergraduate or postgraduate chemistry curriculum in many universities. Chapter one covers important properties of saccharides, such as configuration, conformation, the anomeric effect and equilibrium composition in solution. This basic knowledge is key to many of the discussions that follow. The next two chapters detail the use of protecting groups in carbohydrate chemistry and the preparation of functionalised monosaccharides. Chapters four and five deal with glycosidic bond chemistry, preparation of complex oligosaccharides and the synthesis of glycopeptides.

Part B discusses enantioselective natural product synthesis from monosaccharides. Nowadays, most natural product syntheses are performed in an asymmetric manner. This development is due principally to the realisation that enantiomers may have very different biological properties: one of them may have the desired property, while the other may be potentially harmful, or at least undesirable. Many methods are available for obtaining compounds in an optically pure form. However, each method involves, at a particular stage, a chiral molecule obtained from a natural source, either by using a chiral starting material or chiral auxiliary, or by employing a chiral catalyst. Carbohydrates have been used extensively as chiral starting materials but they have also been utilised as chiral auxiliaries and ligands of chiral catalysts. The examples covered in chapters six to eighteen illustrate the use of carbohydrates in the synthesis of a wide range of natural products. In many cases, the origin of the starting material cannot be recognised in the final product. These chapters demonstrate how the rich stereochemistry of carbohydrates can be used efficiently to install chiral centres into target compounds. To ensure that this material is suitable for teaching, emphasis is placed on retrosynthetic analysis as well as on mechanistic explanations for key and novel reactions.

Geert-Jan Boons and Karl J. Hale

Contents

**PART B: NATURAL PRODUCT SYNTHESIS FROM
                MONOSACCHARIDES**

Part A

Structure and Synthesis of Saccharides and Glycoproteins

1 Mono- and oligosaccharides: structure, configuration and conformation

G.-J. Boons

1.1 Introduction

Carbohydrates constitute the most abundant group of natural products. This fact is exemplified by the process of photosynthesis, which alone produces 4×10^{14} kg of carbohydrates each year. As their name implies, carbohydrates were originally believed to consist solely of carbon and water and thus were commonly designated by the generalised formula $C_x(H_2O)_y$. The present-day definition[1] is that 'the carbohydrates' are a much larger family of compounds, comprising monosaccharides, oligosaccharides and polysaccharides, of which monosaccharides are the simplest compounds, as they cannot be hydrolysed further to smaller constituent units. Furthermore, the family comprises substances derived from monosaccharides by reduction of the anomeric carbonyl group (alditols), oxidation of one or more terminal groups to carboxylic acids or replacement of one or more hydroxyl group(s) by a hydrogen, amino or thiol group or a similar heteroatomic functionality. Carbohydrates can also be covalently linked to other biopolymers, such as lipids (glycolipids) and proteins (glycoproteins).

Carbohydrates are the main source of energy supply in most cells. Furthermore, polysaccharides such as cellulose, pectin and xylan determine the structure of plants. Chitin is a major component of the exoskeleton of insects, crabs and lobsters. Apart from these structural and energy storage roles, saccharides are involved in a wide range of biological processes. In 1952, Watkins disclosed that the major blood group antigens are composed of oligosaccharides.[2] Carbohydrates are now implicated in a wide range of processes[3] such as cell–cell recognition, fertilisation, embryogenesis, neuronal development, hormone activities, the proliferation of cells and their organisation into specific tissues, viral and bacterial infections and tumour cell metastasis. It is not surprising that saccharides are key biological molecules since by virtue of the various glycosidic combinations possible they have potentially a very high information content.[4]

In this chapter, the configurational, conformational and dynamic properties of mono- and oligosaccharides will be discussed and, in general, reference is made to reviews that cover these aspects. These properties, as described in the discussion which follows, are not placed in a historical context.

1.2 Configuration of monosaccharides[5, 6]

Monosaccharides are chiral polyhydroxy carbonyl compounds, which often exist in a cyclic hemiacetal form. Monosaccharides can be divided into two main groups according to whether their acyclic form possesses an aldehyde (aldoses) or keto group (ketoses). These, in turn, are further classified, according to the number of carbon atoms in the monomeric chain (3–10) into trioses, tetroses, pentoses, hexoses, etc. and the types of functionalities that are present. D-Glucose is the most abundant monosaccharide found in nature and has been studied in more detail than any other member of the family. D-Glucose exists in solution as a mixture of isomers. The linear form of glucose is energetically unfavourable relative to the cyclic hemiacetal forms. Ring closure to the pyranose form occurs by nucleophilic attack of the C(5) hydroxyl on the carbonyl carbon atom of the acyclic species (Scheme 1.1). Hemiacetal

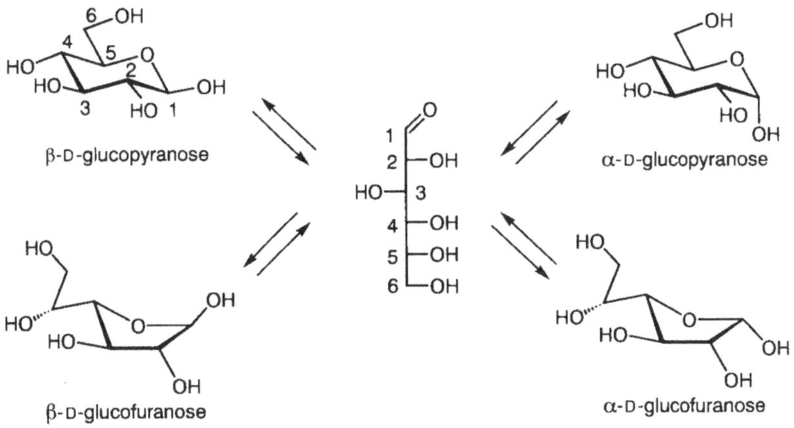

β-D-glucopyranose

α-D-glucopyranose

β-D-glucofuranose

α-D-glucofuranose

Scheme 1.1   Different forms of D-glucose.

ring formation generates a new asymmetric carbon atom at C(1), the anomeric centre, thereby giving rise to diastereoisomeric hemiacetals which are named α and β anomers depending on whether the C(1) substituent resides on the bottom or top of the sugar ring. Cyclisation involving O(4) rather than O(5) results in a five-membered ring structurally akin to furan and is therefore designated as a furanose.

Accordingly, the six-membered pyran-like monosaccharides are termed pyranoses.

All the common hexoses contain four asymmetric centres in their linear form and therefore 2^4 (16) stereoisomers exist which can be grouped into eight pairs of enantiomers. The pairs of enantiomers are classified as D and L sugars. In the D sugars the highest numbered asymmetric hydroxyl group [C(5) in glucose] has the same configuration as the asymmetric centre in D-glyceraldehyde and, likewise, for all L sugars the configuration is that of L-glyceraldehyde (Figure 1.1). The acyclic and pyranose forms of the D-aldoses are depicted in Figures 1.2 and 1.3, respectively.

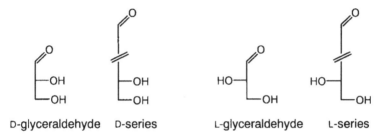

D-glyceraldehyde   D-series   L-glyceraldehyde   L-series

Figure 1.1   D and L sugars.

Monosaccharides have been projected in several ways, the Fischer projection being the oldest (Figure 1.4). In the Fischer projection, the monosaccharides are depicted in an acyclic form and the carbon chain is drawn vertically, with the carbonyl group (or nearest group to the carbonyl) at the top. Each carbon atom is rotated around its vertical axis until all of the C—C bonds lie below a curved imaginary plane. It is only when the projection of this plane is flattened that it can be termed a Fischer projection. In the α anomer the exocyclic oxygen atom at the anomeric centre is formally *cis*, in the Fischer projection, to the oxygen of the highest-numbered chiral centre [C(5) in glucose]; in the β anomer the oxygens are formally *trans*.

Haworth introduced his formula to give a more realistic picture of the cyclic forms of sugars. The rings are derived from the linear form and drawn as lying perpendicular to the paper with the ring oxygen away from the viewer and are observed obliquely from above. The chair conformation gives a much more accurate representation of the molecular shape of most saccharides and is the preferred way of drawing these compounds. It has to be noted that the Mills formula and zig-zag depiction are particularly useful for revealing the stereochemistry of the carbon centres of the sugars.

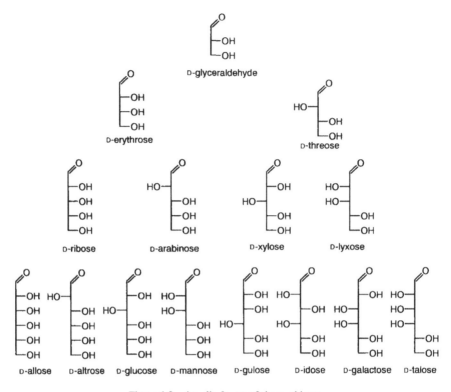

Figure 1.2   Acyclic forms of the D-aldoses.

Apart from the monosaccharides depicted in Figure 1.3, many other types are known. Several natural occurring monosaccharides have more than six carbon atoms and these compounds are named the higher carbon sugars. L-Glycero-D-manno-heptose is such a sugar and is an important constituent of lipopolysaccharides (LPS) of Gram-negative bacteria (Figure 1.5).

Some saccharides are branched and these types are found as constituents of various natural products. For example, D-apiose occurs widely in plant polysaccharides. Antibiotics produced by the micro-organism *Streptomyces* are another rich source of branched chain sugars.

As already mentioned, the ketoses are an important class of sugars. Ketoses or uloses are isomers of the aldoses but with the carbonyl group occurring at a secondary position. In principle, the keto group can be at each position of the sugar chain, but in naturally occurring ketoses the keto group, with a very few exceptions, is normally at the 2-position. D-Fructose is the most abundant ketose and adopts mainly the pyranose form.

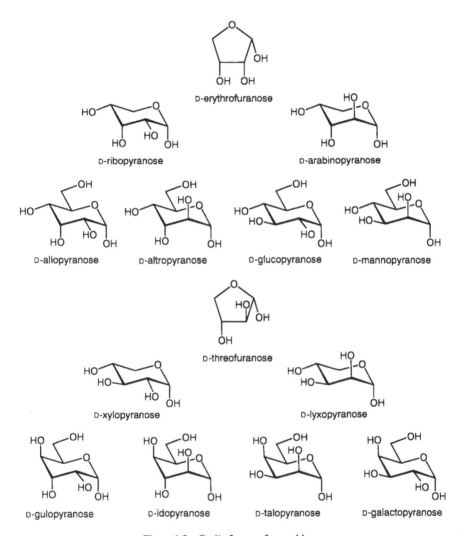

Figure 1.3 Cyclic forms of α-D-aldoses.

The uronic acids are aldoses that contain a carboxylic acid group as the chain-terminating function. They occur in nature as important constituents of many polysaccharides. The ketoaldonic acids are another group of acidic monosaccharides, and notable compounds of this class are 3-deoxy-D-manno-2-octulosonic acid (Kdo) and N-acetyl neuraminic acid (Neu5Ac). Kdo is a constituent of LPS of Gram-negative bacteria and links an antigenic oligosaccharide to Lipid A. N-Acetyl-neuraminic acid is found in many animal and bacterial polysaccharides and is critically involved in a host of biological processes.

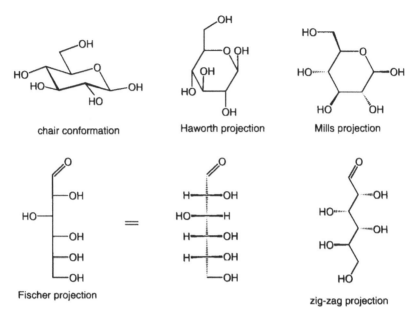

Figure 1.4   Different projections of D-glucopyranose.

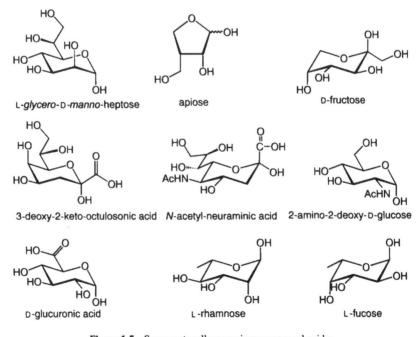

Figure 1.5   Some naturally occuring monosaccharides.

Monosaccharides may possess functionalities other than hydroxyls. Amino sugars are aldoses or ketoses which have a hydroxyl group replaced by an amino functionality. 2-Amino-2-deoxy-glucose is one of the most abundant amino sugars; it is a constituent of the polysaccharide chitin. It also appears in mammalian glycoproteins, linking the sugar chain to the protein. Monosaccharides may also be substituted with sulfates and phosphates. Furthermore, deoxy functions can often be present, and important examples of this class of monosaccharides are L-fucose and L-rhamnose.

1.3 Conformational properties of monosaccharides[7-10]

1.3.1 Ring shapes of pyranoses and furanoses

The concepts of conformational analysis are fundamental to a proper understanding of the relationship between the structure and properties of carbohydrates. Conformational analysis of monosaccharides is based on the assumption that the geometry of the pyranose ring is substantially the same as that of cyclohexane and that of furanoses is the same as that of cyclopentane. The ring oxygen of saccharides causes a slight change in molecular geometry, the carbon–oxygen bond being somewhat shorter than the carbon–carbon bond.

There are a number of recognised conformers for the pyranose ring [11, 12] there being two chairs ($^1C_4$, $^4C_1$), six boats ($^{1,4}B$, $B_{1,4}$, $^{2,5}B$, $B_{2,5}$, $^{0,3}B$, $B_{0,3}$), twelve half chairs ($^0H_1$, $^1H_0$, $^1H_2$, $^2H_1$, $^2H_3$, $^3H_2$, $^4H_3$, $^3H_4$, $^4H_5$, $^5H_4$, $^5H_0$, $^0H_5$) and six skews ($^1S_5$, $^5S_1$, $^2S_0$, $^0S_2$, $^1S_3$, $^3S_1$). To designate each form, the number(s) of the ring atom(s) lying above the plane of the pyranose ring is put as a superscript before the letter designating the conformational form and the number(s) of ring atoms lying below the plane is put after the letter as a subscript (Figure 1.6). The principal conformations of the furanose ring are the envelope (^1E, E_1, ^2E, E_2, ^3E,

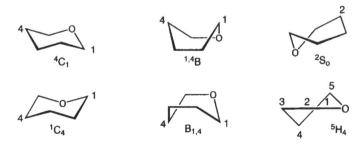

Figure 1.6 Conformers of pyranoses: chair (C), boat (B), skew (S) and half chair (H).

E_3, ^4E, E_4, ^4E, E_4), and the twist form ($^oT_1$, ^1T_o, $^1T_2$, $^2T_1$, $^2T_3$, $^3T_2$, $^3T_4$, $^4T_3$, ^4T_o, $^oT_4$), and they are designated in the same manner as the pyranoses (Figure 1.7).[13]

Most aldohexopyranoses exist in a chair form in which the hydroxymethyl group at C(5) assumes an equatorial position. All the β-D-hexopyranoses exist predominantly in the $^4C_1$ form since the alternative $^1C_4$ conformer involves a large unfavourable *syn*-diaxial interaction between the hydroxymethyl and anomeric group (Figure 1.8). Most of the α-D-hexopyranosides also adopt the $^4C_1$ conformation preferentially. Only α-idopyranoside and α-D-altropyranose show a tendency to exist in the $^1C_4$ conformation, and they coexist with the alternative $^4C_1$ conformations according to ^{1}H-NMR (hydrogen nuclear magnetic resonance) spectroscopy studies.

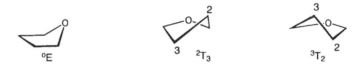

^oE $^2T_3$ $^3T_2$

Figure 1.7 Conformers of furanoses: envelope (E) and twist (T).

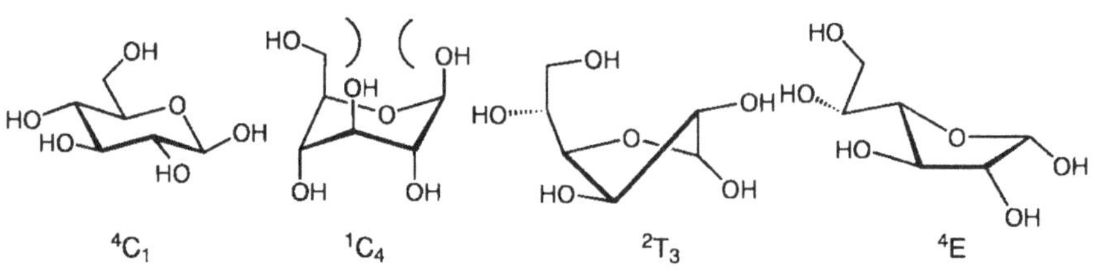

$^4C_1$ $^1C_4$ $^2T_3$ ^4E

Figure 1.8 Some conformations of D-glucopyranose and furanose.

The conformational preferences of the aldopentoses, which have no hydroxymethyl group at C(5), are mainly governed by minimising steric repulsion between the hydroxyl groups. Thus, D-arabinopyranose favours the $^1C_4$ conformer, and α-D-lyxopyranoside and α-D-ribopyranoside are conformational mixtures and the other aldopentoses are predominantly in the $^1C_4$ form.

The preferred conformation of pyranoses in solution can be predicted by empirical approaches.[14] For example, free energies have been successfully estimated by summing quantitative free-energy terms for unfavourable interactions and accounting for the anomeric effects, which are individually depicted in Figure 1.9. The estimated free energies for both chair conformers can be calculated by summation of the various

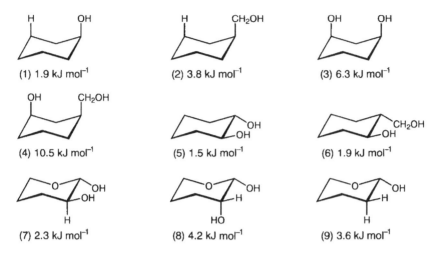

Figure 1.9 Estimated values for nonbonding interactions and anomeric effects in aqueous solution. Interactions (1)–(6) are nonbonding interactions, and interactions (7)–(9) arise from anomeric effects.

steric interactions and taking account of a possible absence of an anomeric affect. The predicted conformational preference was found to be in excellent agreement with experimental data. For example, it has been determined that the $^4C_1$ conformation of β-D-glucopyranose has a conformational energy of 8.7 kJ while that of the $^1C_4$ conformer is 33.6 kJ, which are in agreement with experimental data (Table 1.1). When

Table 1.1 Destabilising values for β-D-glucopyranose in $^4C_1$ and $^1C_4$ conformation

Gauche interactions	Free energy (kJ mol^{-1})	Axial–axial 1–3 interactions	Free energy (kJ mol^{-1})
O-1–O-2	1.5	O-1–O-3	6.3
O-2–O-3	1.5	O-2–O-4	6.3
O-3–O-4	1.5	C-6–O-1	10.5
O-4–O-6	1.9	C-6–O-3	10.5
Anomeric effect	2.3		
Total	8.7	Total	33.6

the free-energy difference between the two chair conformers is less than $3\,kJ\,mol^{-1}$, both conformers will be present in comparable amounts.

Computational methods have been used to predict the anomeric configuration and ring conformation of most aldopyranosides, and generally all are within reasonable agreement with experimental data.[15] Computational studies have also revealed other interesting properties of saccharides. For example, it has been proposed that D-glucose may undergo changes in its ring conformation with a rotation of 10° in the dihedral angles but surprisingly with virtually no changes in energy.[16]

In most cases, the boat and skew conformational isomers are significantly higher in energy and are therefore very sparsely populated conformational states. However, not all monosaccharides take on this conformational behaviour. For example, in solution D-alduronic acid exists as a mixture of a chair and skew conformer. An alduronic acid containing pentasaccharide, that is derived from heparin, has been singled out as having potent antithrombinic activity. It has been proposed that the skew conformation, which the alduronic unit actively adopts, accounts for the biological activity of the pentasaccharide.[17]

Most furanoses prefer the envelope conformation and it appears that a quasi-equatorial exocyclic side chain and a quasi-axial C(1)—O(1) bond (anomeric effect) are equally important stabilising factors (Figure 1.8).

It should be realised that minor conformational isomers may be important reaction intermediates. For example, treatment of 6-O-tosyl-D-glucopyranose with base results in the formation of a 1,6-anhydro derivative. The starting material exists mainly in the $^4C_1$ conformation. However, for reaction to occur the alternative $^1C_4$ conformation has to be adopted (Scheme 1.2). The introduction of protecting groups may alter the preferred conformation of saccharides.

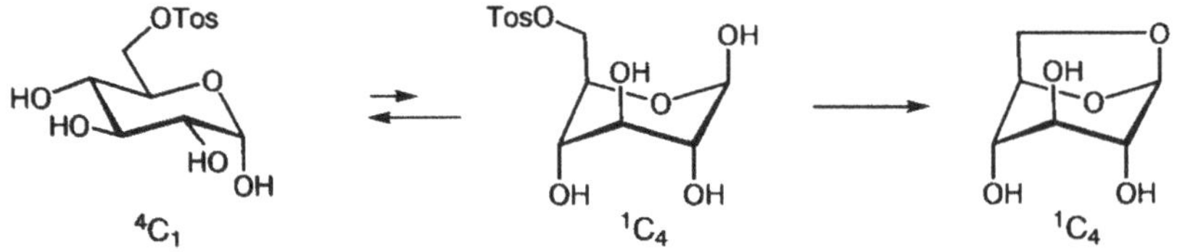

Scheme 1.2 Formation of 1-6-anhydro-D-glucose.

1.3.2 The anomeric effect[18-25]

In general, the stability of a particular conformer can be explained solely by steric factors, and a basic rule for the conformational analysis of cyclohexane derivatives is that the equatorial position is the favoured orientation for a large substituent. The orientation of an electronegative substituent at the anomeric centre of a pyranoside, however, prefers an axial position. For example, in the case of α anomers with a D-gluco configuration, the tendency for axial orientation of the halogen atom is so strong that it is the only observed configuration both in solution and in the solid state. In aqueous solution, unsubstituted glucose exists as 36:64 mixture of the respective α and β anomer. The greater conformational stability of the β isomer with all its substituents in the equatorial orientation seems to be in accord with the conformational behaviour of substituted cyclohexanes. However, the A-value of the hydroxyl group in aqueous solution has been determined at $-1.25\,\text{kcal}\,\text{mol}^{-1}$ and hence an α:β ratio of 11:89 is the predicted value.

The tendency of an electronegative substituent to adopt an axial orientation was first described by Edward[26] and named by Lemieux and Chü[27] 'the anomeric effect'. This orientational effect is observed in many other types of compounds that have the general feature of two heteroatoms linked to a tetrahedral centre; i.e. $C—X—C—Y$, in which $X = N$, O, S, and $Y = Br$, Cl, F, N, O or S, and is termed the generalised anomeric effect.[28, 29]

Over the years, several models have been proposed to explain the anomeric effect, which has been the subject of considerable controversy. It has been proposed that the anomeric effect arises from a destabilising dipole–dipole or electron-pair–electron-pair-repulsion (Figure 1.10). These interactions are greatest in the β anomer, which therefore, is disfavoured. The repulsive dipole–dipole interactions will be reduced in solvents with high dielectric constants.[30] Indeed, the conformational

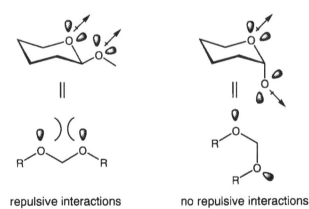

repulsive interactions   no repulsive interactions

Figure 1.10 The anomeric effect: unfavourable dipole–dipole interactions in an equatorially substituted compound.

equilibrium of 2-methoxytetrahydropyran is strongly solvent-dependent, and the highest proportion of the axially substituted conformer is observed in tetrachloromethane and benzene, both solvents having very low dielectric constants (see Table 1.2).[31, 32]

Table 1.2 Solvent dependence of the conformational equilibrium of 2-methoxy-tetrahydropyran

Solvent	Dielectric constant (ε)	Percentage axially substituted conformer
CCl_4	2.2	83
C_6H_{12}	2.3	82
CS_2	2.6	80
$CHCl_3$	4.7	71
$(CH_3)_2CO$	20.7	72
CH_3OH	32.6	69
CH_3CN	37.5	68
H_2O	78.5	52

Detailed examination of the geometry of compounds that experience an anomeric effect reveals that there are characteristic patterns of bond lengths and angles associated with particular conformations. For example, the C—Cl bond of chlorotetrahydropyran, which prefers the axial orientation, is significantly lengthened, and the adjacent C—O bond is shortened.[33] However, this effect is only observed in compounds with the favoured *gauche* conformation about the RO—C—X group. Thus, it is not seen in equatorially substituted compounds. Dipole–dipole interactions fail to account for the differences in bond length and bond angle observed between α and β anomers. To account for these effects, an alternative explanation for the anomeric effect has been proposed.[34] Thus, the axial conformer is stabilised by delocalisation of an electron pair of the oxygen atom to the periplanar C—X bond (e.g. X=Cl) antibonding orbital (Figure 1.11). This interaction, which is not present in the β anomer, explains the shortening of the C—O bond of the α anomer, which has some double bond character. The size of the alkoxy group has little effect on the anomeric preference. For example, in a solution of chloroform, 2-methoxytetrahydropyran (R = Me) and 2-*tert*-butoxytetrahydropyran (R = *t*-Bu) both adopt a chair conformation with the substituent mainly in the axial orientation.[35] On the other hand, the electron-withdrawing ability of the anomeric substituent has a marked

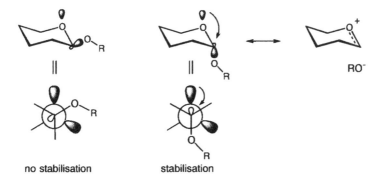

no stabilisation   stabilisation

Figure 1.11 The anomeric effect: interaction of the endocyclic oxygen electron lone pair with the nonbonding orbital in an axially substituted compound.

effect on the axial preference[36] and, in general, a more electronegative anomeric substituent exhibits a stronger preference for an axial orientation. The partial transfer of electron density from a heteroatom to an antibonding σ-orbital is enhanced by the presence of a more electronegative anomeric substituent.

The term 'exoanomeric effect' was introduced to describe an orientational effect of the aglycon part.[31] In this case, the electron density of the lone pair of the exocyclic oxygen atom is transferred to the antibonding orbital of the endocyclic C—O bond (Figure 1.12). Essentially, this effect

E_1 E_2 A_1 A_2

Figure 1.12 Conformations that are stabilised by the exoanomeric effect.

is maximised when the p orbital for an unshared pair of electrons is periplanar to the C(1)–ring-oxygen bond. As can be seen in Figure 1.12, the exoanomeric effect is present in the α as well as in the β anomer. Thus, the α anomer can be stabilised by two anomeric effects (both exo and endo) and the β anomer by only one (exo). Furthermore, two conformations (E_1 and E_2) for the equatorial substituted anomer can be identified that are stabilised by an exoanomeric effect. However, E_2 experiences unfavourable steric interactions between the aglycon and ring moiety and is approximately $0.6 \, \text{kcal mol}^{-1}$ higher in energy than the corresponding E_1 conformer. In the case of the axially substituted anomer also, two conformations are stabilised by an anomeric effect (A_1 and A_2) but A_2 is strongly disfavoured for steric reasons. In the case of the α anomer, the two anomeric effects compete for electron delocalisation towards the anomeric carbon. In the case of a β anomer this competition is absent and hence its exoanomeric effect is stronger.

Another remarkable anomeric effect has been observed which has been named the 'reverse anomeric effect'.[37] By protonation of the imidazole-substituted D-xylo derivative the equilibrium shifts from mainly axial form to mainly equatorial form (Scheme 1.3). There are no changes in the steric requirement between the two compounds and therefore only a stereoelectronic explanation can account for this anomaly. Lemieux has

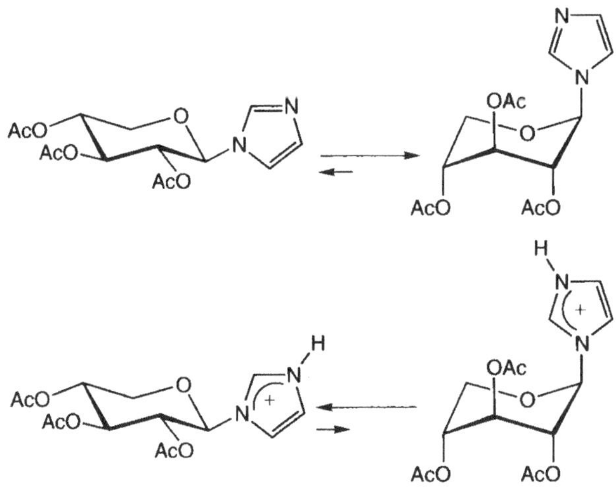

Scheme 1.3   The reverse anomeric effect.

proposed that a strongly electronegative aglycon is unable to stabilise a glycosidic linkage because of the lack of lone-pair electrons. An alternative argument is that the anomeric effect for such a protonated compound is reversed because dipole–dipole interactions no longer reinforce the stereoelectronic preference.

The conformational effects arising from the endoanomeric effect are for furanoses much less profound and as a result relatively little research has been performed in this area. The puckering of the furanose ring of an α and a β anomer usually adjusts the anomeric substituent in a quasi-axial orientation and hence both anomers experience a similar stereoelectronic effect. On the other hand, the conformational preference of the exocyclic C—O bond is controlled by the exoanomeric effect in the usual way.

1.3.3 The equilibrium composition of monosaccharides in solution[38, 14b]

In solution, the α and β forms of D-glucose have a characteristic optical rotation that changes with time until a constant value is reached. This change in optical rotation is called mutarotation and is indicative of an anomeric equilibration occurring in solution.

For some monosaccharides, the rate of mutarotation ($K = k_1 + k_2$) is found to obey a simple first-order rate law in which $-d[\alpha]/dt = k_1[\alpha] - k_2[\beta]$ (Scheme 1.4). Glucose, mannose, lyxose and xylose exhibit

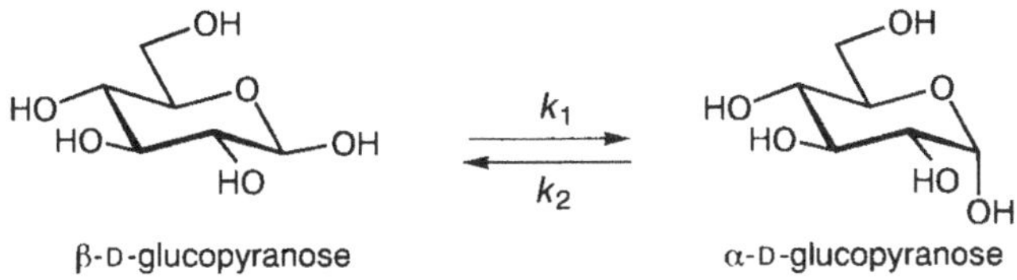

Scheme 1.4 Mutarotation of glucose.

this behaviour. The equilibrium mixture consists predominantly of the α and β pyranoses, when mutarotation can be described by this equation whether measured starting from the α or β anomer. Other sugars such as arabinose, ribose, galactose and talose show a much more complex mutarotation consisting of a fast change of optical rotation followed by a slow change. The fast change in optical rotation is attributed to a pyranose–furanose equilibration, and the slow part to anomerisation.

In general, a six-membered pyranose form is preferred over a five-membered furanose form because of the lower ring strain, and these cyclic forms are very much favoured over the acyclic aldehyde or ketone forms. As can be seen in Table 1.3, at equilibrium, the anomeric ratios of pyranoses differ considerably between aldoses. These observations are a direct consequence of differences in anomeric and steric effects between monosaccharides. The amount of the pyranose and furanose present in aqueous solution varies considerably for the different monosaccharides. Some sugars, such as D-glucose, have undetectable amounts of furanose according ^{1}H-NMR spectroscopic measurements whereas others, such as D-altrose, have 30% furanose content under identical conditions.

The main steric interactions in a five-membered ring are between 1,2-*cis* substituents. For example, D-glucofuranose experiences an unfavourable interaction between the 3-hydroxyl group and the carbon side chain at C(4), which explains its small quantity in solution. On the other hand, this steric interaction is absent in galactofuranose, and, at equilibrium, the latter isomer is present in significant quantity (Figure 1.13).

Table 1.3 Composition of some aldoses at equilibrium in aqueous solution

	Pyranose (%)			Furanose (%)		
Aldose	α	β	total	α	β	total
Glucose	38	62	100	0.1	0.2	0.3
Mannose	65.5	34.5	100	–	–	–
Gulose	0.1	78	78	<0.1	22	22
Idose	39	36	75	11	14	25
Galactose	29	64	93	3	4	7
Talose	40	29	69	20	11	31
Ribose	21	59	80	6	14	20
Xylose	36.5	63	99.5	–	–	<0.5
Lyxose	70	28	98	1.5	0.5	2
Altrose	27	43	70	17	13	30

3-Deoxy-D-glucose which also lacks this unfavourable steric interaction has 28% of the furanose form in aqueous solution.

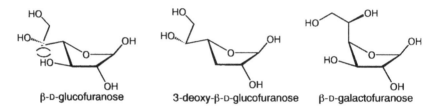

β-D-glucofuranose   3-deoxy-β-D-glucofuranose   β-D-galactofuranose

Figure 1.13 Conformations of β-D-glucofuranose, 3-deoxy-β-D-glucofuranose and β-D-galactofuranose.

The orientation of a C(2) substituent has a remarkable effect on the anomeric equilibrium. In general, an axial alkoxy group at C(2) increases and an equatorial alkoxy group decreases the anomeric effect. For example, in aqueous solution, D-mannose contains at equilibrium as much as 65.5% of the α anomer whereas only 38% of this form is present for D-glucose. Reeves argued[39] that for D-mannose the β anomer is destabilised by the proximity of the endocyclic oxygen and the C(1) and C(2) oxygen atoms, resulting in unfavourable dipole–dipole interactions (Figure 1.14). This effect, which was named the Δ2 effect, has also been explained in stereoelectronic terms. It has been proposed[40] that the anomeric effect for α-D-mannose is significantly stronger because of lowering of the antibonding orbital of the C(1)—O(1) bond as a result of secondary orbital overlap between the antibonding orbitals of C(1)—O(1) and C(2)—O(2).

The presence of particular substituents and the nature of the solvent appear to have an effect on the equilibrium composition of particular monosaccharides. As already discussed, the anomeric effect is stronger in

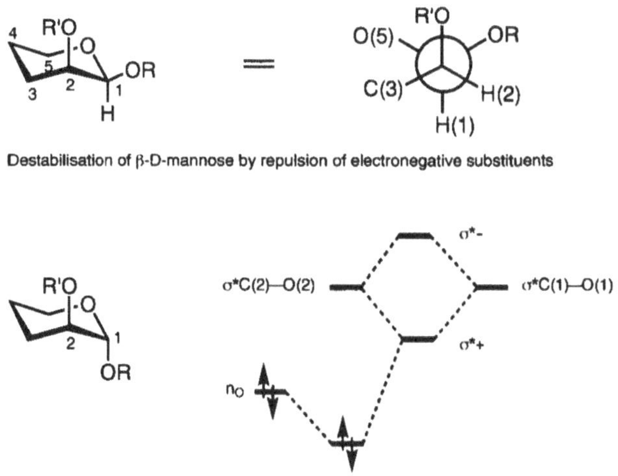

Destabilisation of β-D-mannose by repulsion of electronegative substituents

Increase of endoanomeric effect by antibonding–antibonding orbital overlap

Figure 1.14 The Δ2 effect.

apolar solvents and therefore in these solvents the α anomer is present in higher proportions. However, pyranose:furanose ratios also depend strongly on the nature of the solvent. For example, in dimethylsulfoxide (DMSO), arabinose contains as much as 33% of the furanose form whereas in water only 3% of the same form is present. This observation may reflect differences in solvation for the various hydroxyls. In water, the hydroxyl group is both a hydrogen bond acceptor and donor. In DMSO, this functionality is only a hydrogen bond donor. Thus, the apparent bulk of hydroxyl groups in water is much larger and 1–2 *cis* interactions are stronger and less favoured.

Different levels of substitution exert varying degrees of α anomerisation in saccharides. For example, in an aqueous solution at equilibrium, D-mannose contains 65.5% of the α anomer. However, 2-*O*-methyl mannose contains 75% and 2,3-di-*O*-methyl mannose 86% of the α isomer. The methylation probably increases steric interactions in the β anomer making this anomer less favoured. Furthermore, the electron-donating methyl substituent at C(2) oxygen atoms makes the C(2)—O(2) antibonding orbital a better acceptor resulting in a stronger Δ2 effect for the α anomer.

The equilibrium composition also depends on the temperature, and, in general, increasing the temperature results in a decrease of the β anomer population whereas the proportion of α anomer does not change, although the content of furanoses does increase.

1.4   Conformational properties of oligosaccharides[40-49]

Oligosaccharides are compounds in which monosaccharides are joined by glycosidic linkages. Their saccharidic lengths are described by a prefix: for example, disaccharides and trisaccharides are composed of two and three monosaccharide units, respectively. The borderline between oligo- and polysaccharides cannot be drawn strictly; however, the term 'oligosaccharide' is commonly used to refer to well-defined structures as opposed to a polymer of unspecified length and composition. In most oligosaccharides glycosidic linkages are formed between the anomeric centre of one saccharide and a hydroxyl of another saccharide. However, some saccharides are linked through their anomeric centres and are named trehaloses. Oligosaccharides can be linear as well as branched.

It is now widely accepted that the conformational properties of glycosidic linkages are a major factor determining the overall shape of oligosaccharides. The relative spatial disposition between two glycosidi-cally-linked monosaccharides can be described by two torsional angles, ϕ and ψ, or additionally by a third torsional angle, ω, in the case of a 1–6 glycosidic linkage (Figure 1.15). In the case of a 1–4 linkage, ϕ is defined as the rotation around the C(1)—O(1) bond [rotation for the H(1)—C(1)—O(1)—C(4) fragment] and ψ is defined as the rotation around O(1)—C(4) [ψ rotation for C(1)—O(1)—C(4)—H(4)]. The additional degree of freedom in a 1–6 linkage is defined by the free rotation of the hydroxymethyl group around the C(5)—C(6) bond [rotation for O(6)—C(6)—C(5)—O(5)]. For relatively simple oligo-saccharides, the exoanomeric effect is an important contributor to the

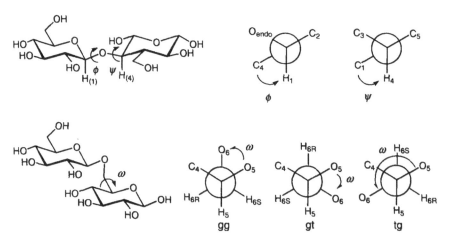

Figure 1.15   Conformations of disaccharides.

preferred conformation of a glycosidic linkage. Thus, in the case of an α glycoside the maximum exoanomeric effect is obtained when $\phi = -60°$ and for β glucosides when $\phi = +60°$.

In early conformational studies, oligosaccharides were regarded as rigid bodies having fixed conformations around their glycosidic linkages. Often good agreement was found between conformations predicted by experimental data (NMR or X-ray crystallography) and the global minimum conformation found by computational studies. However, analysis of more complex structures resulted in poor agreement between these approaches and it is now well-established that most glycosidic linkages exist with a degree of conformational variability.[39–45] Therefore, oligosaccharides cannot be described in terms of a rigid global minimum-energy conformation. However, it appears that the conformational space adopted by an oligosaccharide is restricted around several low-energy minima. When an oligosaccharide complexes with a protein, one of these low-energy minima may be selected as the binding conformation. Several examples are known in which the binding conformation differs substantially from the global energy conformation. For example, high-resolution X-ray crystallographic studies of the Fc region of human immunoglobulin G shows that the non saccharide attached to this protein adopts a different conformation from when it is in solution.[50] The torsional angles of the core pentasaccharide differ only slightly from those predicted by NMR and computational methods. In contrast, the dihedral angles of the β-GlcNAc-Man on the 1–6-arm and the β-Gal(1–4)GlcNAc moiety on the 1–3 arm are significantly different from the predicted solution values.

1.5   Acid-catalysed glycoside bond formation and cleavage

In this section, acid-catalysed glycosidic bond synthesis and cleavage will be covered. This knowledge is critical to many discussions that will follow in subsequent chapters. In Chapter 4, modern approaches to the introduction of glycosidic linkages are discussed.

The Fischer glycosidation is a valuable method for the direct formation of glycosides from unprotected carbohydrates and simple alcohols.[51] The reaction requires strong acids such as hydrochloric acid, trifluoromethanesulfonic acid or $FeCl_3$. As a rule, furanosides predominate in the early stages of a Fischer glycosidation whereas at equilibrium the pyranoses are the dominant product (Scheme 1.5). At equilibrium, the percentage composition of the different forms of methyl glycosides depends on steric and electronic factors and follows similar rules as the process of mutarotation.

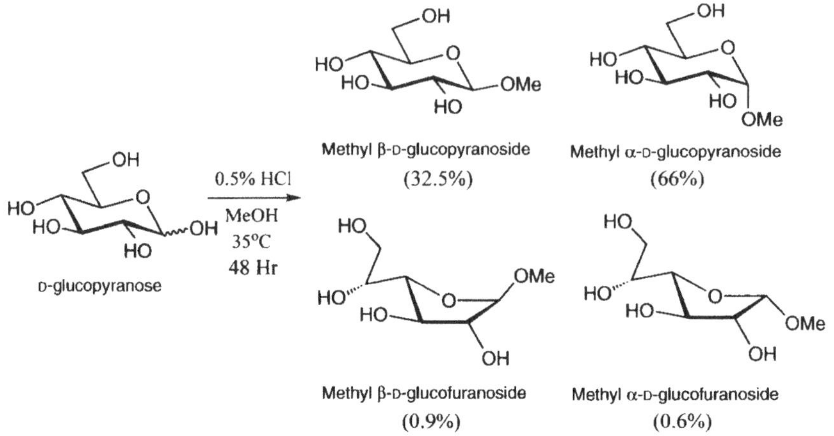

Scheme 1.5 Fischer glycosidation.

The reaction mechanism of the Fischer glycosidation is complex and not fully understood. In the first step, the aldehydo intermediate reacts with an alcohol to give a hemiacetal intermediate (Scheme 1.6). As discussed in Section 1.2, the aldose forms are in equilibrium with the linear aldehydo form. Only a small fraction is in the aldehydo form, however, acids accelerate the equilibration. The aldehydo intermediate will react most quickly with alcohols. This reaction results in a mixture of diastereoisomers of hemiacetals, which may either revert to the aldehydo form or undergo ring closure. In the first instance, ring closure will lead predominantly to furanoside formation (five-membered ring formation). It is well established that five-membered ring closure is much faster than six-membered ring closure. This part of the reaction is kinetically controlled. Next, the kinetically controlled mixture of products will equilibrate to a thermodynamic mixture of products. This equilibration involves ring expansion to the thermodynamically more stable pyranose forms. Probably, the transformation proceeds by protonation of the endocyclic oxygen followed by nucleophilic attack of methanol at the anomeric centre leading to a dimethyl acetal. Acid-mediated cyclisation of the acetal will give either a furanoside or pyranoside. The pyranosides are the thermodynamically more stable compounds because they experience fewer nonbonding steric interactions and have more substantial anomeric stabilisations. In the first instance, the cyclisation of the dialkyl acetals will give equal quantities of α and β anomers. The final stages of the Fischer reaction involve an anomeric an equilibration of the pyranosides. Relatively strong acidic conditions favour thermodynamic control of the reaction. Most monosaccharides follow the described reaction mechanism. However, in the initial stages, the Fischer reaction of D-mannose, D-lyxose and L-arabinose gives substantial pyranose formation concurrently with furanose formation.

Glycosidic linkages can be cleaved by treatment with aqueous acid. The reaction can be described simply as the fast protonation of the exocyclic oxygen, rather than of the ring oxygen atom, followed by slow cleavage of the C(1)—O(1) bond to give an oxocarbenium ion. The latter intermediate is quenched by reaction with water to give a lactol. It has been proposed that important stereoelectronic effects govern reactions at the anomeric centre. Many examples can be found in which an axially positioned anomeric leaving group is more reactive than an equatorial leaving group. These observations have been explained by the assumption that departure of an axial anomeric leaving group is assisted by a lone pair of the endocyclic oxygen (Scheme 1.7).[52] Thus, this theory is a simple

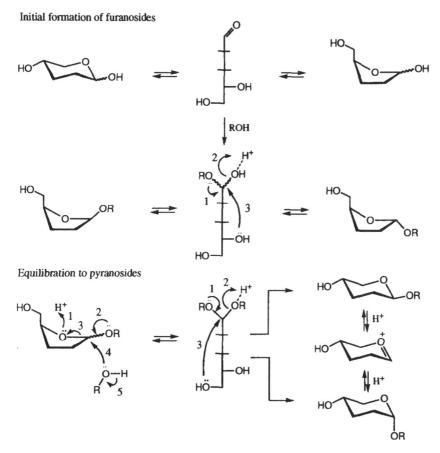

Initial formation of furanosides

Equilibration to pyranosides

Scheme 1.6   The mechanism of the Fischer reaction.

extension of the n–σ* overlap model of the static anomeric effect to transition and ground states. Such assistance cannot be provided when the anomeric leaving group is in an equatorial orientation. This effect has been named the 'kinetic anomeric effect'. Relevant pioneering studies date back to the early 1930s when Isbell[53] discovered that β glycosides are more readily oxidised than are α anomers. Later, Eliel and Nader demonstrated[54] that the reaction of axially substituted 2-alkoxy-1,3-dioxanes with Grignard reagents proceed readily to give acetals (Scheme 1.7). Equatorially substituted derivatives, however, were unreactive when reacted under similar conditions. Based on stereoelectronic considera-

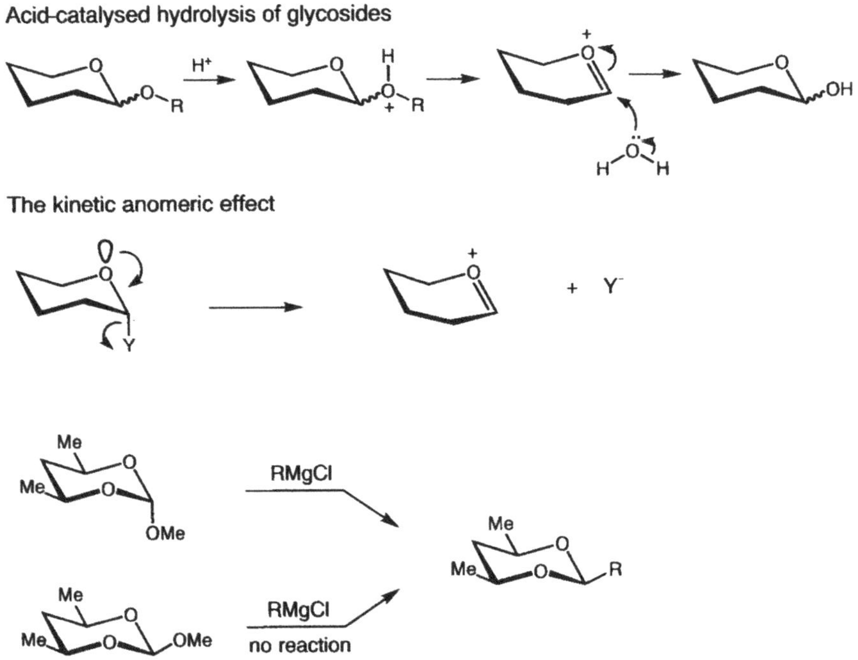

Acid-catalysed hydrolysis of glycosides

The kinetic anomeric effect

Scheme 1.7 Reactions at the anomeric centre. Note: Y = leaving group.

tions, it would be expected that cleavage of axial glycosides would proceed faster than hydrolysis of equatorial glycosides. However, experimental results show the opposite; for example, β-D-glycopyranosides hydrolyse about two to three times faster than α-D-glycopyranosides.[55] Apparently, acid-catalysed hydrolysis of glycosides in water is not governed by a kinetic anomeric effect. A number of hypotheses have been proposed to explain the experimental results. One hypothesis is based on the principle of least molecular motion.[56] An alternative explanation is known as the *syn*-periplanar lone-pair hypothesis.[57] Computational and experimental results indicate that departure of an equatorial anomeric group is stabilised by *syn*-periplanar lone pair interactions in the energetically accessible half-chair conformation, and this interaction is equivalent to the *anti*-periplanar interactions for the departure of an axial leaving group. It has further been suggested that the time of the transition state may be an important determinant of the reactivity of an anomeric centre.[58] Isotope effects indicate that an unassisted departure of an anomeric leaving group to an oxocarbenium ion is characterised by a very late transition state. The Hammond postulate states that in such a case the transition-state structure closely resembles that of the oxocarbenium ion. Departure of an α or β anomeric leaving group will lead to the same oxocarbenium ion and therefore will have very similar transition-state energies. The α anomer is significantly more stable than the β anomer because it is stabilised by an endoanomeric effect and therefore has a higher activation energy and is less reactive. Figure 1.16 summarises this reaction profile and, as can be seen, differences in reactivity of α and β anomers are explained by differences in ground-state energies. The model predicts that the strength of the anomeric effect is directly related to the ratio of rates of hydrolysis of α and β anomeric leaving groups. Experimental evidence supports this prediction. For example, loss of fluoride from β-D-glucosyl fluoride in water is 40 times faster than the same reaction for the α anomer. The anomeric fluoride is much more

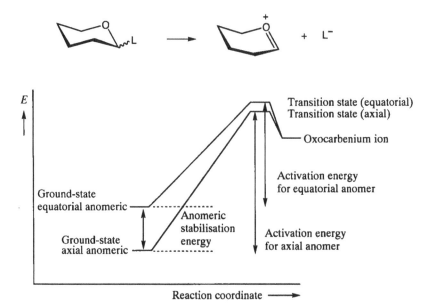

Figure 1.16   Reaction profiles of hydrolysis of axial and equatorial glycosides. Note: L = leaving group.

electronegative than oxygen and therefore has a much stronger anomeric effect. Thus, the α anomer has a significantly higher anomeric stabilisation energy, and therefore, is less reactive.

References

1. A. McNaught, 1996, *Pure Appl. Chem.*, 1919.
2. W.M. Watkins and W.I.J. Morgan, 1952, *Nature*, 169, 825.
3. (a) A. Varki, 1993, *Glycobiology*, 3, 97; (b) R.A. Dwek, 1996, *Chem. Rev.*, 96, 683.
4. R.R. Schmidt, 1986, *Angew. Chem. Int. Ed. Engl.*, 25, 212.
5. J.A. Mills, 1955, *Adv. Carbohydr. Chem. Biochem.*, 10, 1.
6. B. Capon and W.G. Overend, 1960, *Adv. Carbohydr. Chem. Biochem.*, 15, 11.
7. P.L. Durett and D. Horton, 1971, *Adv. Carbohydr. Chem. Biochem.*, 26, 49.
8. (a) D.H.R. Barton, 1950, *Experientia*, 6, 316; (b) D.H.R. Barton, 1970, *Science*, 169, 539.
9. J.F. Stoddart, 1972, *Stereochemistry of Carbohydrates*, Wiley-Interscience, New York.
10. E.L. Eliel and S.H. Willen, 1994, *Stereochemistry of Organic Compounds*, Wiley, New York.
11. J.C. Schwarz, 1973, *J. Chem. Soc., Chem. Commun.*, 505.
12. IUPAC–IUB Joint Commission on Biochemical Nomenclature (JCBN), 1981, *Pure Appl. Chem.*, 53, 1901.
13. N. Cyr and A.S. Perlin, 1979, *Can. J. Chem.*, 57, 2504.
14. (a) S.J. Angyal, 1968, *Aust. J. Chem.*, 21, 2737; (b) S.J. Angyal, 1969, *Angew. Chem. Int. Ed. Engl.*, 8, 157.
15. (a) R. Virudachalam and V.R.S. Rao, 1976, *Carbohydr. Res.*, 51, 135; (b) K. Kildeby, S. Melberg and K. Rasmussen, 1977, *Acta Chem. Scand.*, A31, 1; (c) K. Rasmussen and S. Melberg, 1982, *Acta Chem. Scand.*, A36, 323.
16. N.V. Joshi and V.S.R. Rao, 1979, *Biopolymers*, 18, 2993.
17. N. Sakairi, J.E.M. van Basten, G.A. van der Marel, C.A.A. van Boeckel and J.H. van Boom, 1996, *Chem. Eur.*, 2, 1007.
18. R.U. Lemieux, 1971, *Pure Appl. Chem.*, 25, 527.
19. D.G. Gorenstein, 1987, *Chem. Rev.*, 87, 1047.
20. I. Tvarosaka and T. Bleha, 1989, *Adv. Carbohydr. Chem. Biochem.*, 47, 45.
21. E. Juaristi and G. Cuevas, 1992, *Tetrahedron*, 5019.
22. P. Deslongchamps, 1983, *Stereoelectronic Effects in Organic Chemistry*, Pergamon Press, Oxford.
23. T. Kirby, 1983, *The Anomeric Effect and Related Stereoelectronic Effects at Oxygen*, Springer, Berlin.
24. G.R.J. Thatcher (ed.), 1992, *The Anomeric Effect and Associated Stereoelectronic Effects*, ACS Symp. Series.
25. E. Juaristi and G. Cuevas, 1995, *The Anomeric Effect*, CRC Press, Boca Raton, FL.
26. J.T. Edward, 1955, *Chem. Ind. (London)*, 1102.
27. R.U. Lemieux and N.J. Chü, 1958, *Chemical Abstracts, Am. Chem. Soc.*, 133, 31N.
28. R.U. Lemieux, 1964, in *Molecular Rearrangements* (P. de Mayo, ed.), Wiley Interscience, New York, 709.
29. M.C. Krol, C.J.M. Huige and C. Altona, 1990, *J. Comp. Chem.*, 11, 765.
30. E.L. Eliel and C.A. Giza, 1968, *J. Org. Chem.*, 33, 3754.
31. (a) R.U. Lemieux, A.A. Pavia, J.C. Martin and K.A. Watanabe, 1969, *Can. J. Chem.*, 47, 4427; (b) R.U. Lemieux, A.A. Pavia and J.C. Martin, 1987, *Can. J. Chem.*, 65, 213.
32. M.D. Walkinshaw, 1987, *J. Chem. Soc., Perkin Trans. II*, 1903.
33. E. Juaristi, 1989, *Accs. Chem. Res.*, 22, 357.

34. (a) C. Romers, C. Altona, H.R. Buys and E. Havinga, 1969, in *Topics in Stereochemistry*, Vol. 4 (E.L. Eliel and N.L. Allinger, eds.), Wiley Interscience, New York, 39; (b) S. Wolfe, M.-H. Whangbo and D.J. Mitchell, 1979, *Carbohydr. Res.*, 69, 1.

35. (a) E.L. Eliel and C.A. Giza, 1968, *J. Org. Chem.*, 33, 3754; (b) R.W. Warrent, C.N. Caughlan, J.H. Hargis, K.C. Yee and W.G. Bentrude, 1978, *J. Org. Chem.*, 43, 4266.

36. I. Tvaroska and T. Bleha, 1989, *Advances in Carbohydr. Chem. Biochem.*, 45-123.

37. R.U. Lemieux and A.R. Morgan, 1965, *Can. J. Chem.*, 43, 2205.

38. (a) A.W. Pigman and H.S. Isbell, 1969, *Adv. Carbohydr. Chem.*, 23, 11; (b) A.W. Pigman and H.S. Isbell, 1969, *Adv. Carbohydr. Chem. Biochem*, 24, 13.

39. (a) R.E. Reeves, 1950, *J. Am. Chem. Soc*, 72, 1499; (b) R.E. Reeves, 1958, *Annu. Rev. Biochem.*, 27, 15.

40. H. Steinlin, L. Camarda and A. Vasella, 1979, *Helv. Chim. Acta*, 62, 378.

41. K. Bock, 1983, *Pure Appl. Chem.*, 55, 605.

42. B. Meyer, 1990, *Topics in Current Chemistry*, 154, 143.

43. J.P. Carver, 1993, *Pure Appl. Chem.*, 65, 763.

44. (a) S.W. Homans, 1993, *Glycobiology*, 3, 551; (b) T.J. Rutherford, J. Partridge, C.T. Weller and S.W. Homans, 1993, *Biochemistry*, 32, 12715.

45. D.A. Cumming, R.N. Shan, J.J. Krepinksy, A.A. Gray and J.P. Carver, 1987, *Biochemistry*, 26, 6655.

46. D.A. Cumming and J.P. Carver, 1987, *Biochemistry*, 26, 6664.

47. A. Imberty, V. Tran and S. Perez, 1989, *J. Comp. Chem.*, 11, 205.

48. S.W. Homans, 1990, *Biochemistry*, 29, 9110.

49. L. Poppe and H. van Halbeek, 1992, *J. Am. Chem. Soc.*, 114, 1092.

50. J. Deisenhofer, 1981, *Biochemistry*, 20, 2361.

51. (a) E. Fischer, 1893, *Ber.*, 26; (b) B. Capon, 1969, *Chem. Rev.*, 69, 407; (c) P. Konradsson, P. Robers, B. Fraser-Reid, 1991, *Recl. Trav. Chim. Pays-Bas*, 110, 23, (d) V. Ferriéres, J.-N, Bertho, D. Plusquellec, 1995, *Tetrahedron Lett.*, 36, 2749.

52. P. Deslongchamps, 1983, *Stereoelectronic Effects in Organic Chemistry*, Pergamon Press, Oxford, 30-35.

53. (a) H.S. Isbell, 1932, *J. Research Nat. Bureau Standards*, 8, 615; (b) H.S. Isbell, 1961, *Chem. Ind. (London)*, 593.

54. E.L. Eliel and F.W. Nader, *J. Am. Chem. Soc.*, 1970, 92, 584.

55. (a) L.J. Haynes and F.H. Newth, 1955, *Adv. Carbohydr. Res.*, 10, 207; (b) J.V. O'Conner, R. Barker, 1979, *Carbohydr. Res.*, 73, 227.

56. M.L. Sinott, 1988, *Adv. Phys. Org. Chem.*, 24, 113.

57. (a) A.J. Ratcliffe, D.R. Mootoo, C.W. Andrews and B.J. Fraser-Reid, 1989, *J. Am. Chem. Soc.*, 111, 7661; (b) C.W. Andrews, B.J. Fraser-Reid and J.P. Bowen, 1991, *J. Am. Chem. Soc.*, 113, 8293.

58. M.L. Sinott, 1992, in *The Anomeric Effect and Associated Stereoelectronic Effects*, (G.R.J. Thatcher, ed.), ACS Symposium Series.

2 Protecting groups

G.-J. Boons

2.1 Introduction

Carbohydrates are polyfunctional compounds having several hydroxyls often in combination with other functionalities such as amino and carbonyl groups. During a synthetic sequence, most of these functionalities must be blocked and liberated when a selective chemical manipulation needs to be performed. The choice of a set of protecting groups is one of the decisive factors in the successful synthesis of a complex target compound, and the following issues need to be considered.[1] The reaction conditions required to introduce a protecting group should be compatible with the other functionalities in the compound. Often this point does not cause problems because many protecting groups can be introduced by different methods. A protecting group must be stable under all the conditions used during subsequent synthetic steps and must capable of being cleaved under mild conditions in a highly selective manner and high yield. Furthermore, a protecting group should be cheap and easily available and it should also be borne in mind that some protecting groups may affect the reactivity of other functionalities. For example, electron-withdrawing ester functionalities reduce the nucleophilicity of neighbouring hydroxyls. Furthermore, bulky protecting groups can sterically block other functionalities. Protecting groups can be distinguished as either persistent or temporary. Persistent or permanent protecting groups are used to block functionalities that do not need functionalisation and therefore are present throughout the synthesis. Ideally, all permanent protecting groups will be removed at the end of a synthetic sequence in one chemical operation. Some functional groups need to be protected in such a manner that they can be made available for derivatisation at some point in a synthesis. These functionalities are usually protected with temporary protecting groups. As a rule, a set of orthogonal protecting groups is required when several positions need functionalisation. Orthogonal protecting groups are capable of being removed in any order, with reagents and conditions that do not affect the other protecting groups. Unfortunately, the boundaries between orthogonal sets and the gradation of lability within orthogonal sets are not always well-defined, leading to diminished selectivities.

Different functionalities require different protecting groups and the most commonly applied hydroxyl and amino protecting groups are shown in Figure 2.1. Furthermore, in carbohydrate chemistry, a clear distinction should be made between the protection of the anomeric centre and other hydroxyls.

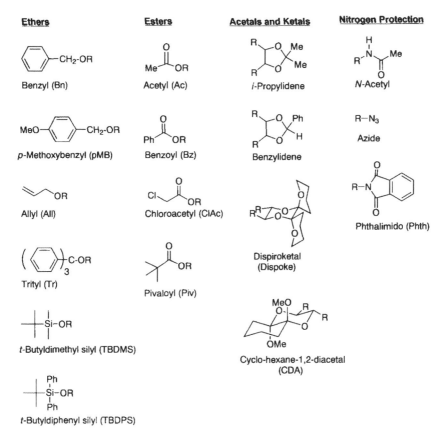

Figure 2.1 Protecting groups commonly used in carbohydrate chemistry.

2.2 Ether protecting groups

2.2.1 Benzyl ethers

In carbohydrate chemistry, benzyl ethers are often applied as permanent protecting groups. They are robust and are stable to a wide range of basic and acidic conditions. They can also withstand hydride reducing agents and mild oxidants. Benzyl ethers can be introduced under basic, neutral and acidic conditions.

The most common procedure used for the preparation of benzyl ethers is O-alkylation of a sodium or potassium alkoxide with benzyl bromide. The alkoxides are usually generated with NaH or KH and the reaction is often performed in a dipolar aprotic solvent such as dimethylformamide (DMF). The latter type of solvents solvate the sodium or potassium ions, which results in an increase in nucleophilicity of the alkoxide. Benzylation reactions can also be accelerated by the addition of a catalytic amount of Bu_4NI.[2] Under these conditions it is suspected that the iodide displaces the bromide of benzyl bromide to generate benzyl iodide *in situ*, which is a much more reactive alkylating agent. Often, anionic benzylation proceeds to a very high yield and allows the simultaneous benzylation of several hydroxyls (Scheme 2.1a). However, the strongly basic conditions required make this method incompatible with base-sensitive functionalities. For example, benzylation with NaH and benzyl bromide of hydroxyls of sugar derivatives that have an N-acetamido group can sometimes result in N-benzylation of the N-acetamido functionality. The application of milder basic conditions can sidestep this problem and the use of $Ba(OH)_2 \cdot 8H_2O$ and BaO as the base often gives satisfactory results.[3] In this reaction, the hydration water of $Ba(OH)_2$ is converted into $Ba(OH)_2$ by reaction with BaO.

When functionalities such as esters are present, neutral or acidic conditions are required for benzylation. The combination of Ag_2O and benzyl bromide in DMF effects benzylation under virtually neutral conditions (Scheme 2.1b).[4] In this reaction, silver oxide complexes with the bromide of benzyl bromide to generate an electrophilic benzyl cation, which then alkylates hydroxyls. In some cases, O-acetyl migration or cleavage occurs. Generally, these reactions require anhydrous conditions and freshly prepared silver oxide.

Benzylation with benzyl trichloroacetimidate and a catalytic amount of triflic acid (TfOH) is a mild and efficient procedure (Scheme 2.1c).[5] The acid protonates the nitrogen of the imidate moiety converting it into a very good leaving group. Nucleophilic attack by an alcohol introduces a benzyl ether. The procedure is often compatible with base- and acid-sensitive functionalities with esters, O-isopropylidene and O-benzylidene acetals. Benzyl trichloroacetimidate is commercially available but can easily be prepared by reaction of benzyl alcohol with trichloroacetonitrile in the presence of a mild base.

Cyclic dibutylstannylene derivatives are convenient intermediates for the regioselective benzylation of polyols.[6] These derivatives can easily be prepared by reaction with Bu_2SnO or $Bu_2Sn(OMe)_2$ with removal of water or methanol, respectively. They can be alkylated in benzene,

(a) Anionic benzylation of carbohydrate hydroxyls

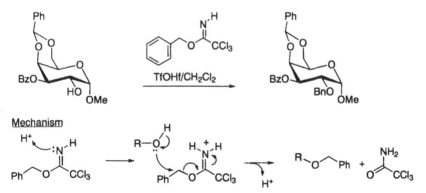

Scheme 2.1   Benzylations of carbohydrate hydroxyls.

toluene or DMF in the presence of added nucleophiles such as tetrabutylammonium halides, caesium fluoride or *N*-methylimidazole to give good yields of monosubstituted products (Scheme 2.2a).

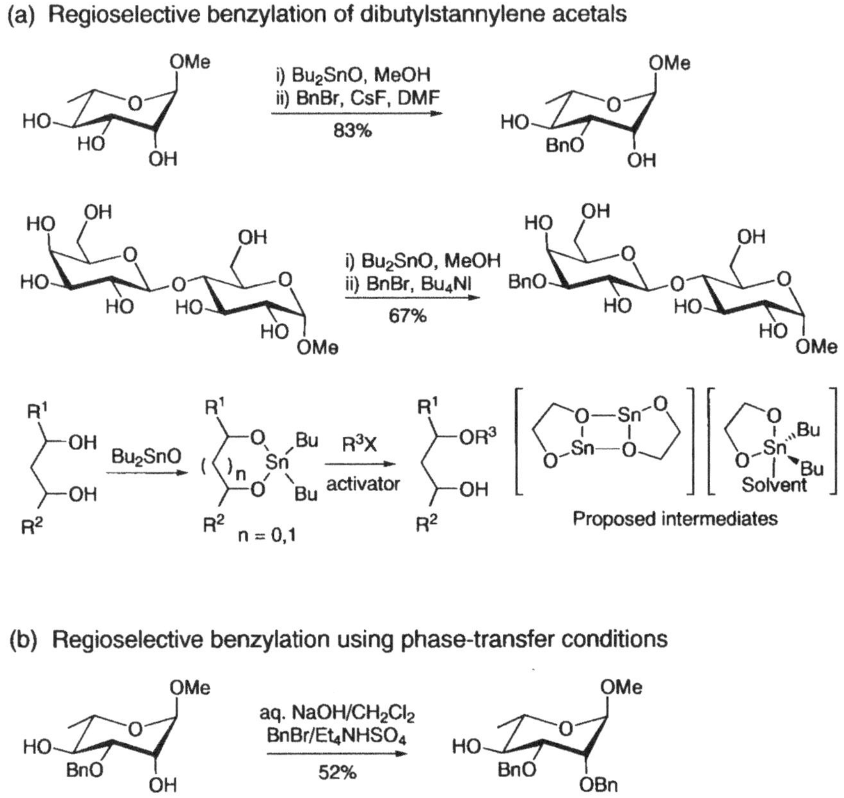

(a) Regioselective benzylation of dibutylstannylene acetals

(b) Regioselective benzylation using phase-transfer conditions

Scheme 2.2 Regioselective benzylations of carbohydrate hydroxyls.

The stannylation of a diol generally enhances the nucleophilicity of one of the hydoxyls. The regioselectivity of these reactions is not well understood but it has been proposed that the stannylene acetals exist as dimers in which the tin atoms are at the centre of a trigonal bipyramid with the butyl groups occupying equatorial positions.[7] The more electronegative of the two oxygen atoms is coordinated with only one atom whereas the less electronegative oxygen atom is coordinated to two tin atoms (Scheme 2.2a). Thus, the observed regioselectivity is a consequence of a cascade of effects beginning with the selection of a particular pair of hydroxyls for stannylene formation followed by orientation of the more electronegative oxygen in an apical position which is intrinsically more reactive. However, it should be realised that the selectivities observed not only depend on the structure of the substrate but also are affected by the reaction conditions. Generally speaking, dibutyl tin acetals derived from mixed primary and secondary diols are selectively alkylated at the primary positions. A tin acetal derived from a *cis*-1,2-cyclohexanoid diol is selectively alkylated at the equatorial position but regioselectivities are generally poor for equatorial–equatorial diols. The preferred formation of a five–membered tin acetal over a six-membered primary or secondary counterpart is borne out in the most impressive selective benzylation of methyl lactoside. In this case the equatorial 3'OH is preferentially derivatised over the primary hydroxyl groups.

Phase-transfer techniques are simple procedures for regioselective alkylation of carbohydrates.[8] In many cases, a mixture of aqueous sodium hydroxide, tetrabutylammonium bromide as a phase-transfer catalyst and benzyl bromide in dichloromethane gives highly regioselective reactions (Scheme 2.2b). In this type of reaction, a partially protected saccharide is deprotonated in the aqueous phase. The resulting alkoxide then complexes with the phase-transfer catalyst and is transferred to the organic phase. Thereafter, the alkoxide reacts with the benzyl bromide that is also present in the organic phase. Further benzylation is much slower since the more lipophilic benzylated saccharide is less likely to return to the aqueous phase. A combination of steric and electronic factors determine the regioselective outcome of a phase-transfer benzylation. Generally, a primary hydroxyl group is preferentially derivatised over a secondary hydroxyl. Because of its higher acidity, the C(2) hydroxyl displays the highest reactivity of secondary hydroxyls.

Catalytic hydrogenolysis using Pd—C, Pd(OH)$_2$ or Pd(OAc)$_2$ is the most commonly employed method for the removal of benzyl ethers, and yields are often quantitative. Cyclohexene, cyclohexadiene, formic acid and ammonium formate can also be used as hydrogen sources rather than hydrogen. Benzyl ethers can also be removed by Birch reduction with lithium or sodium dissolved in liquid ammonia, but this procedure is not often applied in carbohydrate chemistry.

Regioselective debenzylation can be achieved by treatment with Lewis acids such as ferric chloride and SnCl$_4$ or under acetolysis conditions with acetic anhydride and sulfuric acid, and several examples are depicted in Scheme 2.3.[9] Acetolysis results in cleavage of the most acid-sensitive benzyl group. In general, primary benzyl ethers can be selectively acetolysed in the presence of secondary benzyl ethers. The regioselectivity of the reaction can be explained as follows: sulfuric acid protonates acetic anhydride followed by the formation of an acetyl ion and acetic acid. The acetyl ion reacts with the sterically most accessible oxygen which is at

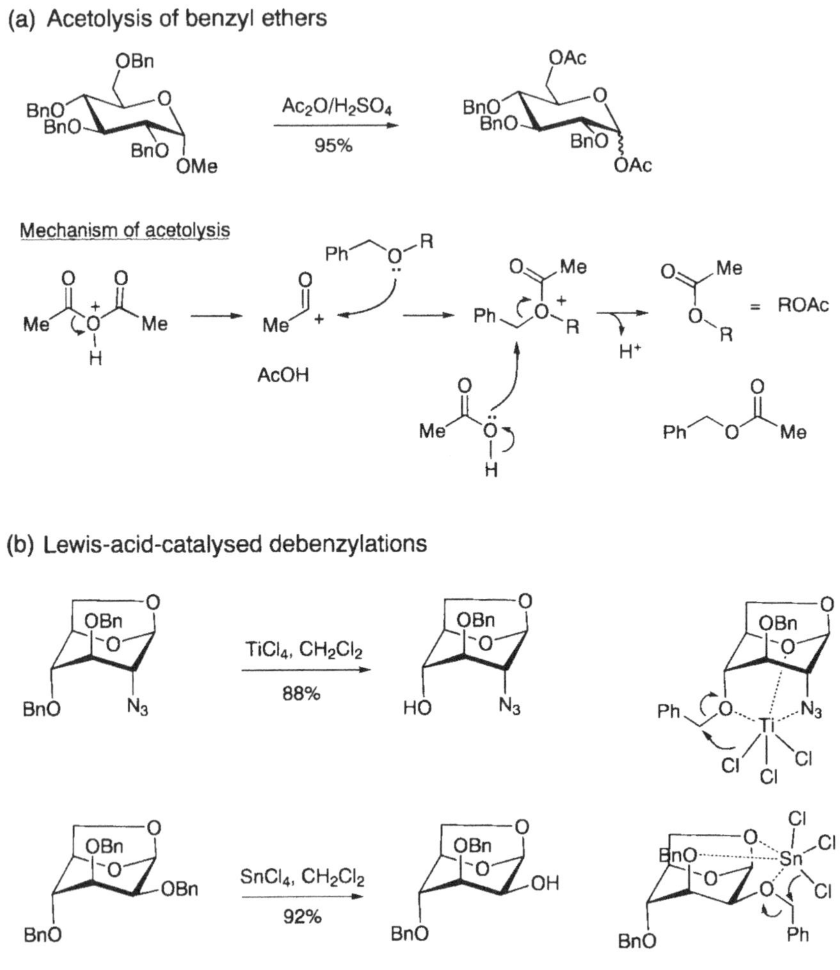

(a) Acetolysis of benzyl ethers

Mechanism of acetolysis

(b) Lewis-acid-catalysed debenzylations

Scheme 2.3   Regioselective debenzylations.

C(6) to give an oxonium ion. Nucleophilic attack, probably by acetic acid, on the benzylic carbon gives a O-acetyl sugar derivative and benzyl acetate. Glycosides are also acetolysed under these reaction conditions.

Mono O-debenzylation can be achieved with Lewis-acid-mediated conditions such as with $TiCl_4$ or $SnCl_4$. Good yields are obtained when the substrate has three continuous *cis*-orientated metal-chelating groups (Scheme 2.3b). Thus, in case of a 1,6-anhydro-mannose derivative chelation takes place between the exocyclic anomeric oxygen and the 2-O- and 3-O-benzyl groups. On the other hand, 1,6-anhydroglucose derivatives are chelated between endocyclic oxygen and the functionalities on

C(2) and C(4). As a result different regioselectivities are observed when
O-benzylated 1,6-anhydro glucose or mannose derivatives are treated
with Lewis acids.

2.2.2 p-*Methoxybenzyl ethers*

Of the substituted benzyl ether protecting groups now available, the *p*-
methoxybenzyl ether (PMB) is the most commonly utilised temporary
protecting group. The electron-rich PMB ether is much more acid labile
than a benzyl ether and can be selectively cleaved by aqueous mineral
acids or camphor sulfonic acid (CSA) in methanol. They are introduced
in a similar fashion as benzyl ethers using *p*-methoxybenzyl chloride in
the presence of NaH, Ba(OH)$_2$, Ag$_2$O or a stannylene acetal. It should be
realised that this moisture-sensitive chloride degrades on standing. PMB
trichloroacetimidate has been used, but this reagent is very reactive and
only a very small amount of acid is required for its activation.[10]

PMB ethers can be cleaved oxidatively with 2,3-dichloro-5,6-dicyano-
1,4-benzoquinone (DDQ)[11] in dichloromethane/water or with cerium
ammonium nitrate (CAN) in acetonitrile/water.[12] Many other protecting
groups such as esters, isopropylidene acetals, benzyl ethers, allyl ethers
and *t*-butyldiphenyl silyl (TBDMS) ethers are stable to these conditions
(Scheme 2.4). The cleavage reaction with DDQ is initiated with a single-

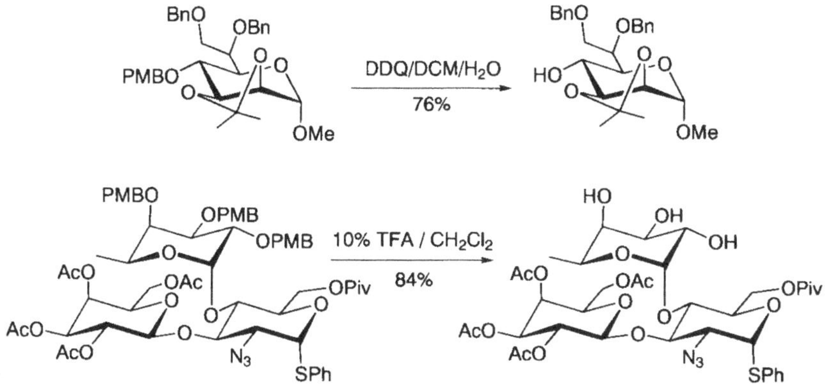

Scheme 2.4 Cleavage of *p*-methoxybenzyl ethers.

electron transfer from the oxygen of the PMB ether to DDQ to generate
an oxonium radical cation and a phenoxy radical anion (Scheme 2.5).
Next, a hydrogen radical is transferred to the phenoxy radical anion to
give 2,3-dicyano-4,5-dichloro-hydroquinone and an oxocarbenium ion.

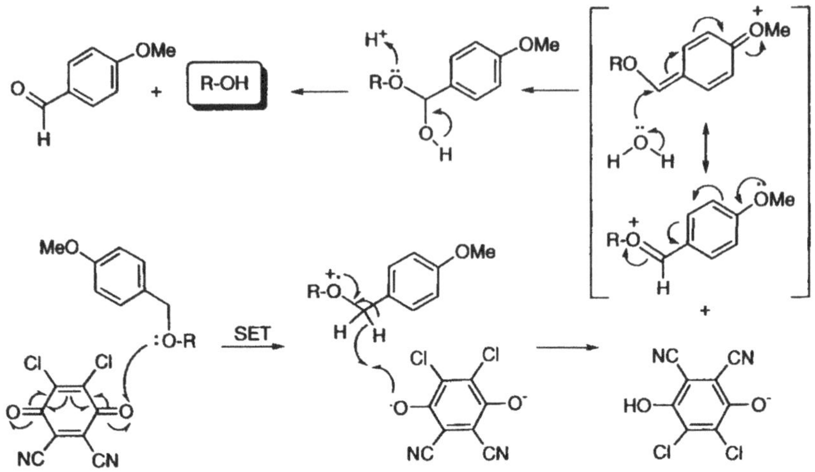

Scheme 2.5   Mechanism of cleavage of *p*-methoxybenzyl ethers with 2,3-dichloro-5,6-dicyano-1,4-benzoquinone. Note: SET = single-electron transfer.

The latter intermediate will hydrolyse in water to give the required alcohol and *p*-methoxybenzaldehyde. Thus, overall a hydride is abstracted from the PMB group to form a resonance-stabilised oxonium ion that gets attacked by water. This reaction leads to a hemiacetal, which breaks down to the parent alcohol and benzaldehyde. The driving force of the reaction is the strong electron-accepting properties of DDQ by virtue of its electron-withdrawing substituent and aromatisation. Furthermore, the methoxy substituent of the PMB group stabilises the radical and positively charged intermediates.

Catalytic hydrogenolysis or treatment with trifluoroacetic acid (TFA) in water and dichloromethane (DCM)[13] will also remove a PMB group. This protecting group is, however, reasonably stable to cold aqueous acetic acid.

2.2.3   *Allyl ethers*

Allyl ethers are often employed as temporary protecting groups. They are stable to a wide range of reaction conditions including to moderately strong acids and bases. They are, however, attacked by strong electrophiles and reduced by catalytic hydrogenation.[14] Allyl ethers can be introduced by procedures similar to those used for benzylation. Thus, reaction of an alkoxide with allyl bromide gives the corresponding allyl ether in high yield, but the strong basic conditions employed to generate the alkoxide limit the scope of this reaction. For base-sensitive substrates, *O*-allylation is best achieved by using allyl bromide and BaO/Ba(OH)$_2$, or Ag$_2$O, or by treatment with allyl trichloroacetimidate and a catalytic amount of TfOH. Regioselective *O*-allylation has been achieved by the stannylene and phase-transfer methods.

An interesting two-step procedure has been described[15] which in some cases gives better yields than a one step approach (Scheme 2.6). The

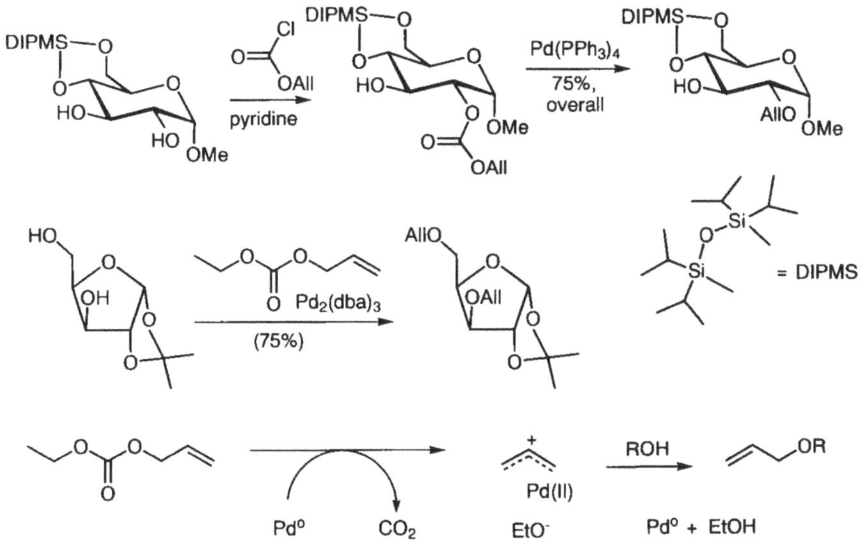

Scheme 2.6 Allylation via an allyl carbonyl derivative followed by palladium-mediated CO$_2$ extrusion.

method involves the reaction of an alcohol with allylchloroformate to give the corresponding allyloxycarbonyl (Aloc) derivative which is then converted into an allyl ether by palladium-catalysed extrusion of CO$_2$.[16] Introduction of an allyl group by *in situ* CO$_2$ extrusion is also possible. This procedure involves a proton exchange between the intermediary ethoxy-palladium π allyl complex and the alcohol substrate.

According to Rudinger's terminology, the allyl group is a 'safety-catch' protecting group. This means that an initially quite stable protecting group is converted into a more labile one as a prelude to the final cleavage step. The most commonly employed method for removing *O*-allyl ether is through transition-metal-catalysed isomerisation to a labile prop-1-enyl ether. This compound is then easily cleaved by mild acid, basic aqueous potassium permanganate, ozonolysis or HgCl$_2$/HgO in acetone–water (Scheme 2.7a). Typical isomerisation catalysts include: Pd—C in MeOH,

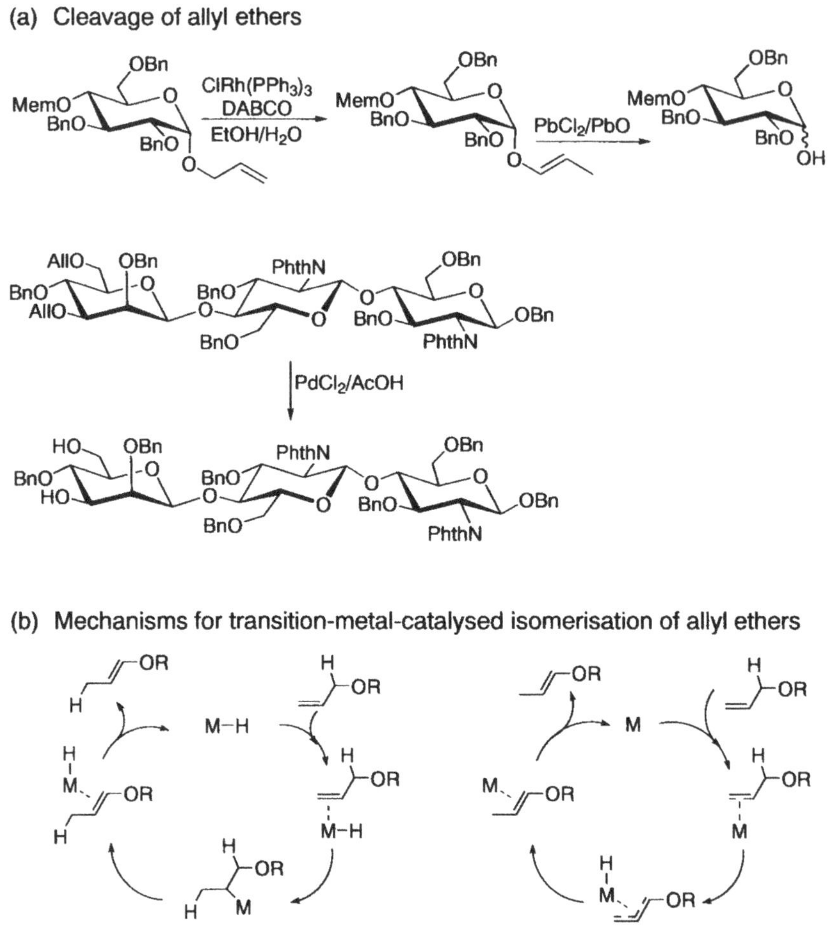

(a) Cleavage of allyl ethers

(b) Mechanisms for transition-metal-catalysed isomerisation of allyl ethers

Scheme 2.7 Transition-metal-catalysed isomerisation of allyl ethers to vinyl ethers, followed by cleavage.

$[Ph_3P]_3RhCl/Dabco$ in MeOH, $[Ph_3P]_3RhH$ or $Ir(COD)[PMePh_2)_2PF_6$ activated with H_2 in THF.[17]

Depending on the nature of the catalyst, the transition-metal-catalysed isomerisation will proceed by an addition–elimination mechanism or by the formation of a π allyl complex followed by a 1,3-hydrogen shift (Scheme 2.7b). The equilibrium of the isomerisation lies strongly in favour of the propenyl ether because of resonance stabilisation between the oxygen lone pair and the π orbital of the double bond. Isomerisation can also be performed under strongly basic conditions by using potassium *t*-butoxide in dimethylsulfoxide (DMSO).[18]

2.2.4 Triphenylmethyl ethers

The bulky triphenylmethyl or trityl (Tr) group is often employed for the regioselective protection of a primary sugar alcohol. It is conveniently introduced into partially protected sugar derivatives by treatment with trityl chloride (TrCl) in pyridine (Scheme 2.8).[19] The protection can be slow but the addition of 4-(dimethyl)aminopyride (DMAP) or 1,8-diazobicyclo[5.4.0]undec-7-ene (DBU) can accelerate the reaction.

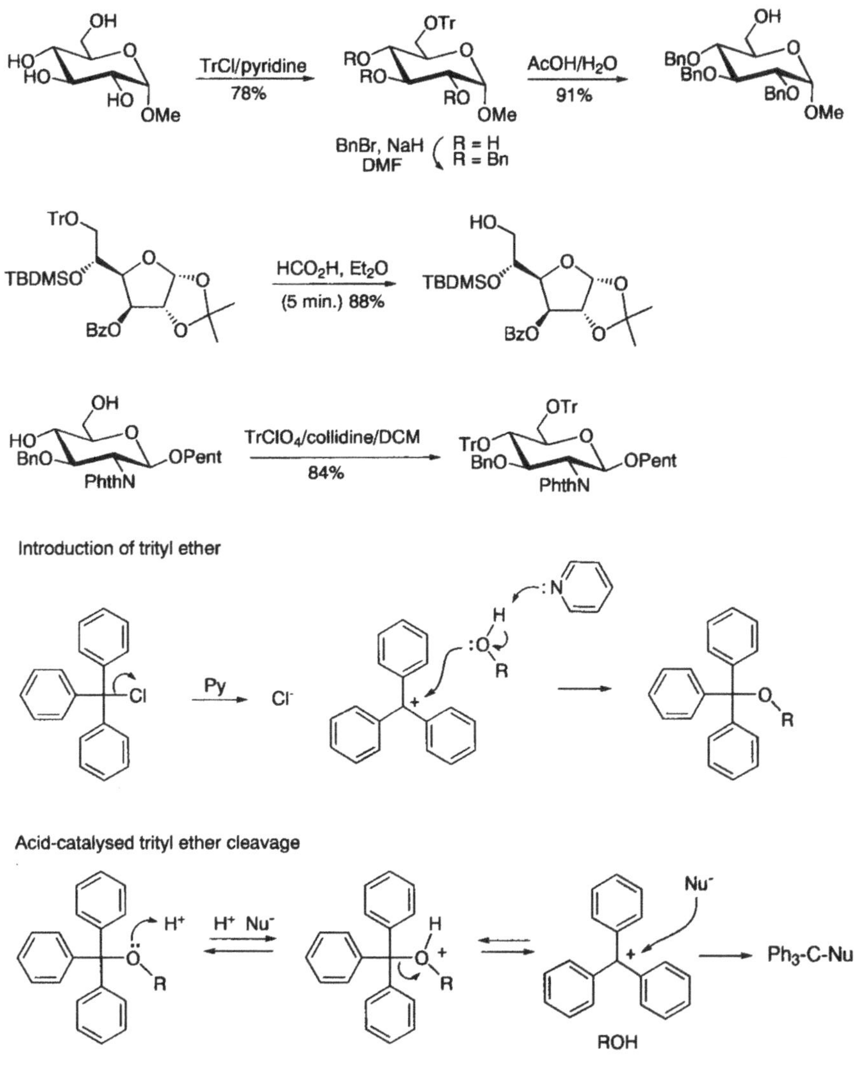

Introduction of trityl ether

Acid-catalysed trityl ether cleavage

Scheme 2.8 Tritylation of sugar hydroxyls.

Secondary trityl ethers can best be prepared by reaction with triphenylmethyl perchlorate, but these derivatives tend to be rather labile.[20]

Trityl ethers are easily cleaved by mild protic acids such as aqueous acetic or trifluoroacetic acid owing to the stability of the triphenylmethyl carbocation. They are also labile in the presence of Lewis acids such as $ZnBr_2$—MeOH, $FeCl_3$ or $BF_3 \cdot Et_2O$.[21] Trityl ethers can be cleaved selectively in the presence of TBDMS ether and isopropylidene acetals by brief exposure to formic acid.[22] Catalytic hydrogenation has also been used to effect O-detritylation.

The introduction and cleavage of the trityl ether proceeds through a very well-stabilised triphenylmethyl carbocation. In the case of trityl ether bond formation, the reaction is performed under anhydrous conditions and the carbocation, which is formed by an S_N1 mechanism, reacts with an alcohol. In the case of cleavage, the triphenylmethyl carbocation ion is formed by treatment with acid, which is then trapped by water or a nucleophilic solvent to give trityl alcohol or other derivatives, respectively. Trityl ethers have also been used to protect thiols.

2.2.5   Silyl ethers

The synthetic potential of silyl ethers as protecting groups for hydroxyls is based on the fact that they can be easily introduced and cleaved under mild conditions and their relative stability can be tuned by varying the substituents on silicon. In carbohydrate chemistry, the *tert*-butyl-dimethylsilyl (TBDMS), *tert*-butyldiphenylsilyl (TBDPS) and triethyl-silyl (TES) ethers are the most often applied silicon-based protecting groups (Scheme 2.9).[23]

The TBDMS and TBDPS groups are normally introduced by treating the sugar alcohol with TBDMSCl or TBDPSCl in DMF in the presence of imidazole. DMF can often advantageously be replaced by pyridine/4-DMAP. If pyridine is used without DMAP, O-silylation is much slower and primary hydroxyls can selectively be silylated.[24] The latter selectivity can also be achieved by employing TBDMSCl in the presence of $AgNO_3$.[25] These reaction conditions are very useful when the substrate contains base labile moieties. In some case, secondary hydroxyls can be selectively protected by reaction with a TBDMSCl and imidazole in DMF, and an example is shown in Scheme 2.9a.[26]

TBDMS and TBDPS triflate are very powerful silylating agents that have been used for the protection of sterically hindered hydroxyls.[27]

Silyl ethers are cleaved under basic and acidic conditions as well as by nucleophilic attack by fluoride ions. The driving force of the latter

(a) *t*-Butyldimethylsilyl ethers (TBDMS)

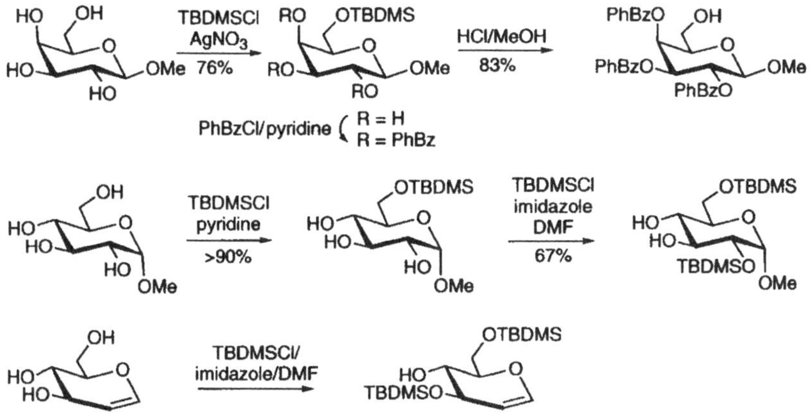

PhBzCl/pyridine $\begin{cases} R = H \\ R = PhBz \end{cases}$

(b) *t*-Butyldiphenylsilyl ethers (TBDPS)

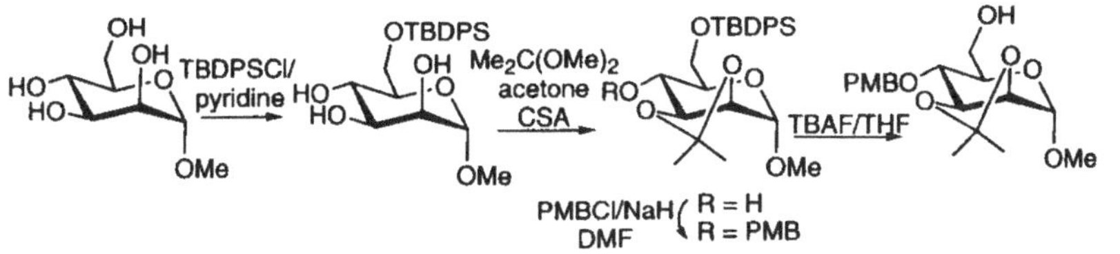

PMBCl/NaH $\begin{cases} R = H \\ R = PMB \end{cases}$
DMF

(c) Triethylsilyl ethers (TES)

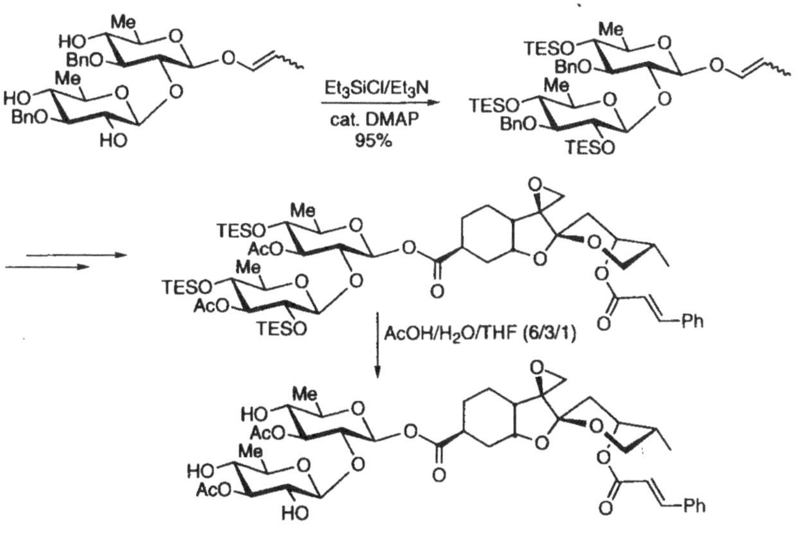

Scheme 2.9 Silylation and desilylation of sugar hydroxyls.

reaction is the high bond energy of the C—F bond ($142 \, \text{kcal mol}^{-1}$). In general, the more bulky the substituents on silicon, the higher the stability of silyl ethers to both acidic and basic hydrolysis. Furthermore, electron-withdrawing functionalities increase the stability to acidic hydrolysis and decrease the stability to basic hydrolysis. Thus, the substitution of a methyl with a phenyl group results in stabilisation of a silicon ether to acidic conditions by steric and electronic effects. However, the same substitution has little effect on the rate of base treatment since the electronic and steric effects oppose each other.

TBDMS and TBDPS are equally stable towards basic solvolysis. Towards acid hydrolysis, the O-TBDPS ether is 250 times more stable than the corresponding O-TBDMS ether. The order of cleavage with basic fluoride reagents such as tetra-n-butylammonium fluoride (TBAF) is similar to that found for basic hydrolysis. In the case of slightly acidic fluoride-based reagents such as HF/acetonitrile, the sequence of stability and the rate of cleavage tend to be more like those associated with acidic hydrolysis. Generally, TBDMS and TBDPS ethers are compatible with benzylation conditions using NaH and BnBr in DMF. Furthermore, the O-TBDPS group is stable towards acidic conditions used to introduce benzylidene and isopropylidene acetals but the TBDMS group may not survive these conditions. Both protecting groups can be removed by treatment with TBAF in tetrahydrofuran (THF). The TBDMS ether can also be cleaved under mild acidic conditions such as aqueous acetic acid and pyridinium tosylate in methanol.

Triethyl silyl (TES) ethers are used as alcohol protecting groups in cases where deprotection of the alcohol needs to be performed under mild conditions in the presence of sensitive functionalities. They are introduced by the reaction of an alcohol with triethylsilyl chloride (TESCl) and triethylamine in DMF. TES ethers can withstand the following conditions: (1) hydrogenolysis with Pd—C in ethyl acetate, although hydrogenation in methanol will cleave TES ethers; (2) ozonolysis of double bonds in dichloromethane at $-78°C$; (3) acetylation with acid anhydrides and acid chlorides in pyridine and (4) reduction with $NaBH_4$ in methanol or THF at $-20°C$. TES ethers can be cleaved in high yields by mild acid hydrolysis in acetic acid/THF/H_2O at room temperature. Deprotection can also be accomplished with TBAF in THF, however; these conditions cannot be used for substrates that are susceptible to base or nucleophilic attack. As can be seen in the example presented in Scheme 2.9c[28], several TES ethers could be hydrolysed with mild acid without affecting the epoxide, the spiroketal and ester moiety. TES ethers are much more acid labile than are TBDMS or TBDPS ethers and often can be cleaved without affecting these silicon ethers.

2.3 Acetal protecting groups

Benzylidene and isopropylidene acetals are often used for the selective protection 1,2-*cis* or 1,3-*cis/trans* diols of sugar derivatives. They are stable to strong basic conditions but quite fragile towards acid. Recently, dispirodiketal and cyclohexane-1,2-diacetal groups have been introduced to protect selectively 1,2-*trans* diols of carbohydrates.

2.3.1 Benzylidene acetals

Methyl hexopyranosides react with benzaldehyde in the presence of zinc chloride to give selectively 4,6-*O*-benzylidine acetals as a highly crystal-line solid. A milder and often preferred method involves *trans*-acetalisation with benzaldehyde dimethyl acetal in the presence of a catalytic amount of acid [CSA or *p*-toluene sulfonic acid (*p*-TsOH)] (Scheme 2.10a).[29] These conditions usually give thermodynamic products, and generally six-membered rings (1,3-diols) are preferred over five-membered rings (1,2-diols). Furthermore, introduction of benzylidene acetals result in a new asymmetric carbon; however, in cases of 1,3 diols, only one diastereoisomer is formed in which the phenyl substituent is in an equatorial orientation. Protection of 1,2-diols as benzylidene acetals usually give mixtures of diastereoisomers (endo/exo mixtures).

Trans-acetalisations are often performed in DMF at elevated temperatures under reduced pressure or a stream of nitrogen. These conditions result in removal of the methanol that is formed in the reaction and, as a result, the benzylidenation is driven to completion. It has been reported that the use of chloroform as a solvent gives higher yields and, in this case, pyridinium *p*-toluene sulfonate (PPTS) may be used as a catalyst.

Benzylidene acetals can be removed by mild aqueous acid hydrolysis (80% AcOH or TFA/DCM/H_2O) or by catalytic hydrogenolysis over Pd(OH)$_2$ or Pd—C.[30]

A number of reagents have been used to cleave these acetals regioselectively to obtain mono *O*-benzylated diol derivatives. For example, treatment of a 4,6-*O*-benzylidene derivative with Dibal-H or LiAlH$_4$–AlCl$_3$ liberates selectively the less hindered C(6) hydroxyl.[31] In these reactions, the reagents act as a Lewis acid and complex with the less hindered primary alcohol which results in weakening of the O(6)—CHPh bond. This bond cleaves preferentially to give a stabilised oxocarbenium ion that results in a benzyl ether at C(4) and selective deprotection of C(6) (Scheme 2.10b). By using NaBH$_3$CN–HCl or BH$_3$·NMe$_3$–AlCl$_3$ as the reducing reagent, the regioselectivity can be reversed. In these cases, the

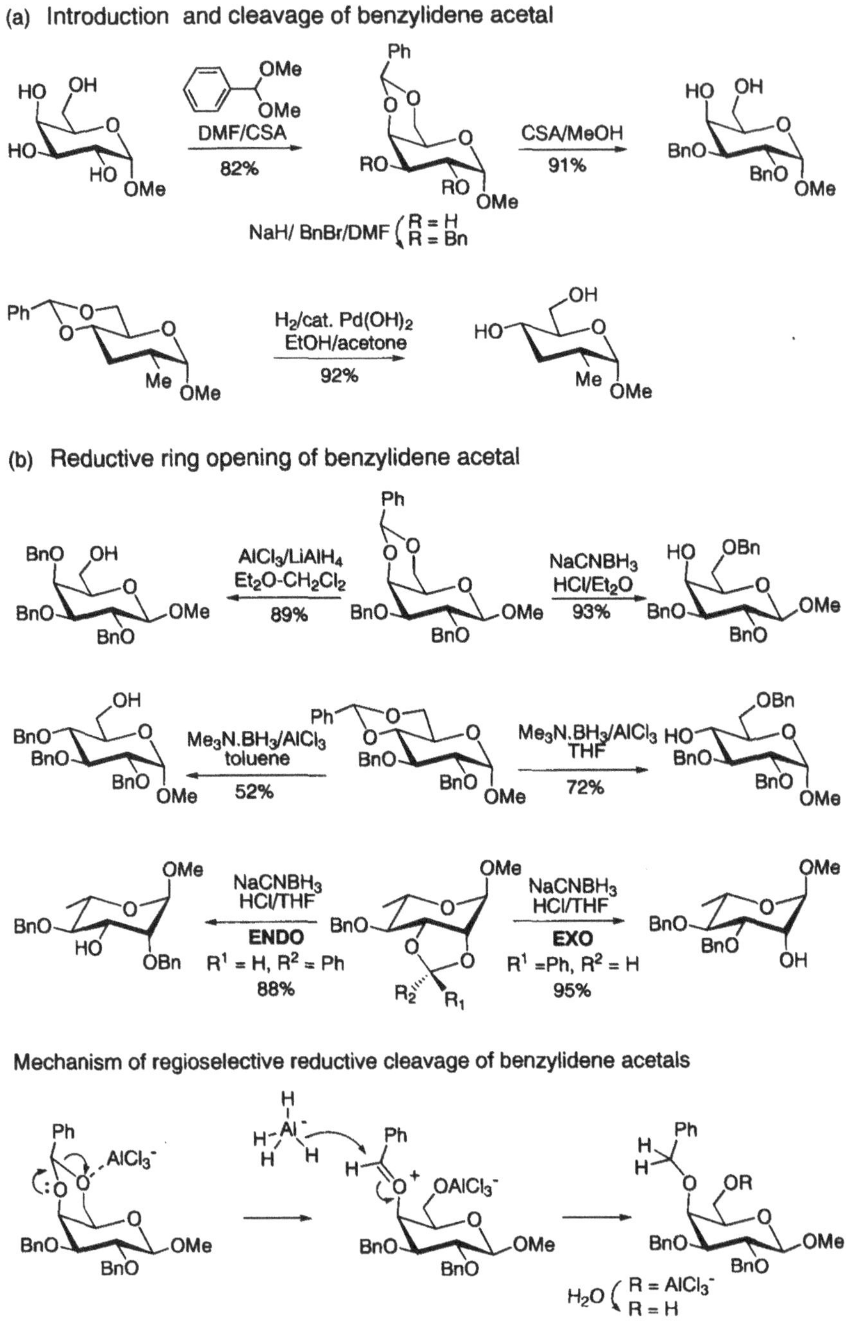

(a) Introduction and cleavage of benzylidene acetal

(b) Reductive ring opening of benzylidene acetal

Mechanism of regioselective reductive cleavage of benzylidene acetals

Scheme 2.10 *O*-benzylidene formation and cleavage.

reducing reagent is much less bulky and therefore the regioselectivity of the reaction is governed by electronic factors. Thus, the Lewis acid complexes at the more basic O(4), with hydride delivery resulting in the formation of a C(4) hydroxyl/C(6) O-benzyl derivative. It is postulated that O(4) is more basic than O(6) because it is further removed from the electron-withdrawing influence of the anomeric centre.

For a five-membered dioxolane ring, the regioselectivity depends on the stereochemistry of the benzylidene acetalic carbon atom. For example, treatment of the methyl endo 2,3-O-benzylidene protected rhamnoside with $LiAlH_4$–$AlCl_3$ gave a 2-O-benzyl derivative, but the same reaction with the 2-exo isomer yielded mainly the 3-O-benzyl derivative.[32]

Benzylidene acetals can be oxidatively cleaved by N-bromosuccinimide (NBS) to give 6-bromo-4-benzoyl-hexopyranosides (see Chapter 3). This reaction has also been performed on 1,2-O-benzylidene derivatives and it appears that the cleavage reactions result in the formation of a derivative that has an axial benzoyl ester.[33]

2.3.2 Isopropylidene acetals

An isopropylidene acetal can be introduced under thermodynamic conditions by treatment of a diol with dry acetone in the presence of an acid catalyst (PPTS, CSA catalysed by H_2SO_4, or I_2).[29] In some cases $CuSO_4$ is added as a dehydrating agent to trap water that is formed in this reaction (Scheme 2.11). Alternatively, an isopropylidene group can be introduced by an acetal exchange reaction employing 2,2-dimethoxy-propane and a catalytic amount of CSA. The latter reaction can be performed under mildly acidic conditions and is often the method of choice when acid-sensitive substrates need to be protected. In general, five-membered 1,3-dioxolanes are formed preferentially over six-membered 1,3-dioxane ring systems. The six-membered ring is destabilised because one of the methyl groups adopts an axial orientation at C(2) of the chair conformation of a 1,3-dioxane and therefore this orientation is energetically disfavoured. The formation of cis-fused five-membered ring isopropylidene derivatives is a strong driving force. This feature is illustrated by the formation of the di-O-isopropylidene protected furanose form of glucose and mannose when treated with acetone in the presence of a catalytic amount of sulfuric acid.

Isopropylidene derivatives can also be obtained under kinetic conditions and in this case 2-methoxypropene in the presence of a catalytic amount of a mild acid is employed. Under these conditions, the initial bond is formed at a primary hydroxyl followed by ring closure.

An interesting property of saccharides that are protected by two isopropylidenes is that often one can be selectively hydrolysed, and these

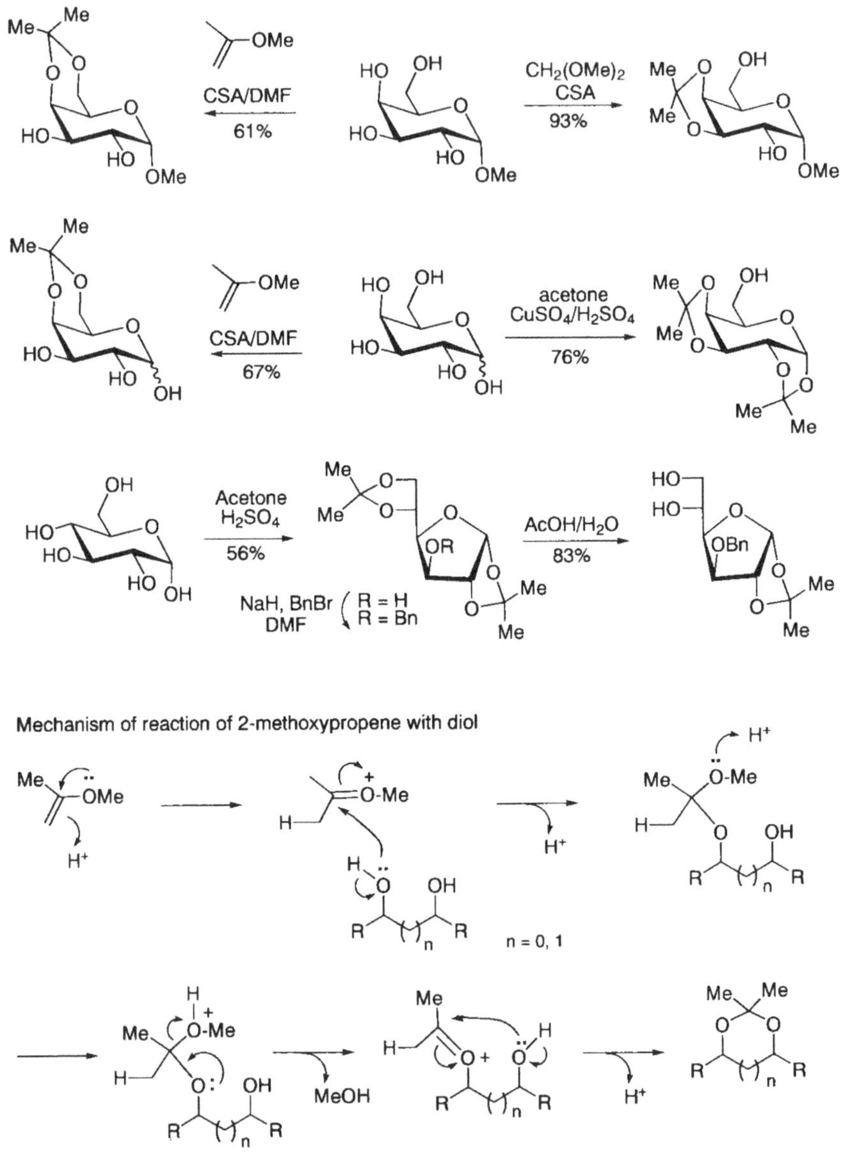

Mechanism of reaction of 2-methoxypropene with diol

Scheme 2.11   Regioselective formation of isopropylidene acetals.

transformations are of great synthetic value.[34] Often exocyclic isopropy-lidene acetals can be selectively cleaved in the presence of endocyclic isopropylidene acetals. For example, the 5,6-O-isopropylidene acetal of 1,2:5,6-di-O-isopropylidene-D-glucose can be cleaved selectively by treatment with aqueous acetic acid at room temperature. Furthermore, a five-membered 1,3-dioxolane is more stable than a six-membered 1,3-dioxane ring, and 1,2-isopropylidenes are more stable than other isopropylidenes.

2.3.3 Dispirodiketal and cyclohexane-1,2-diacetal groups

Benzylidene and isopropylidene acetals of *trans*-1,2-diols are very labile as a result of ring strain and are not often used for synthetic applications. Fortunately, the protection of these diols can be accomplished with the recently developed dispiroketal (dispoke)[35] and cyclohexane-1,2-diacetal (CDA) groups.[36]

Dispirodiketals can be introduced by reaction of a 1,2-*trans*-diol with 3,3',4,4'-tetrahydro-6,6'-bi-2H-pyran in the presence of a catalytic

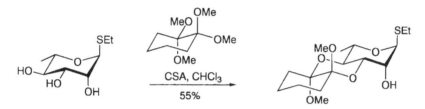

Scheme 2.12   Protection of a *trans*-diol as a cyclohexane-1,2-diacetal.

amount of CSA in chloroform (Scheme 2.12). The two newly formed spiroketal centres are chiral but only one diastereoisomer is formed in which both spiroketal centres are in a configuration that has a maximum stabilisation by anomeric effects. As can be seen in the example presented in Scheme 2.12, a 1,2-*trans* di-equatorial diol is protected in preference over a 1,2-*cis* diol. The latter can be explained as follows: dispiroketal formation results in the formation of a six-membered ring. Two six-membered rings joined together in a 1,2 fashion are named decalins. It is well established that *trans*-decalins are more stable than corresponding *cis*-fused ring systems. As a result, 1,2-diequatorial diols are preferentially protected as dispiroketals. It should be realised that protection of a 1,2-diol as a benzylidene or isopropylidene results in the formation of a new five-membered ring. A 1,2-diequatorial-fused five-membered ring is highly strained and as a result difficult to introduce.

The cyclohexane-1,2-diacetal group can be introduced by transketalisation with 1,1,2,2-tetramethoxycyclohexane. The Dispoke and CDA group can be removed by transketalisation with ethylene glycol in the presence of CSA or with aqueous TFA. The CDA group is more acid labile than the Dispoke group and is preferred for complex oligosaccharide synthesis.

2.4 Ester protecting groups

A large number of ester protecting groups have been described but in carbohydrate chemistry only acetyl (Ac), benzoyl (Bz), pivaloyl (Piv) and chloroacetyl (ClAc) esters are often employed for the protection of hydroxyls. They are affected by strong nucleophiles such as Grignard and organolithium reagents and by metal hydride reducing agents.

In general, esters are prepared by the reaction of an alcohol with an acetic anhydride or acid chloride in pyridine. The reaction can be accelerated by the addition of 4-(dimethylamino)pyridine (DMAP). Several methods have been described for regioselective acylation (Scheme 2.13). The bulky pivaloyl group can be introduced selectively at a primary position simply by reaction of a polyol with PivCl in pyridine.[37] Primary hydroxyls can also be selectively benzoylated but in this case special acylating reagents need to be used such as benzoyl cyanide/triethyl amine[38] or 1-acyloxy-1*H*-benzotriazoles.[39] Partial benzoylation may be achieved when low equivalents of benzoyl chloride in pyridine are used at low temperatures. For example, benzoylation of methyl β-D-galactoside with three equivalents of benzoyl chloride in pyridene results in protection of the C(2), C(3) and C(6) hydroxyl. The axial C(4) hydroxyl has the lowest reactivity because of steric effects and is not acylated. However, perbenzoylation can be achieved when an excess of benzoyl chloride is used and the reaction performed at room temperature.

Regioselective acylations can also be achieved by the stannylene procedure but, in general, selectivities are not as high as for alkylations. For example, treatment of methyl β-D-galactoside with dibutyltin oxide followed by reaction with benzoyl chloride results in the formation of a mixture of methyl 6-*O*-benzoyl-β-D-galactoside (53%) and methyl 3,6-*O*-di-benzoyl-β-D-galactoside (21%).[40]

The axial hydroxyl of a 1,2-*cis*-diol can be selectively acetylated by a two-step procedure, which entails the formation of an orthoester by transesterification with triethylorthoacetate in DMF (or acetonitrile) catalysed by *p*-toluenesulfonic acid followed by ring opening by treatment with aqueous acetic acid. When trimethylorthoacetate is used, a corresponding acetyl group will be obtained, and when trimethylorthobenzoate is applied a benzoate will be introduced (Scheme 2.13).[41,42]

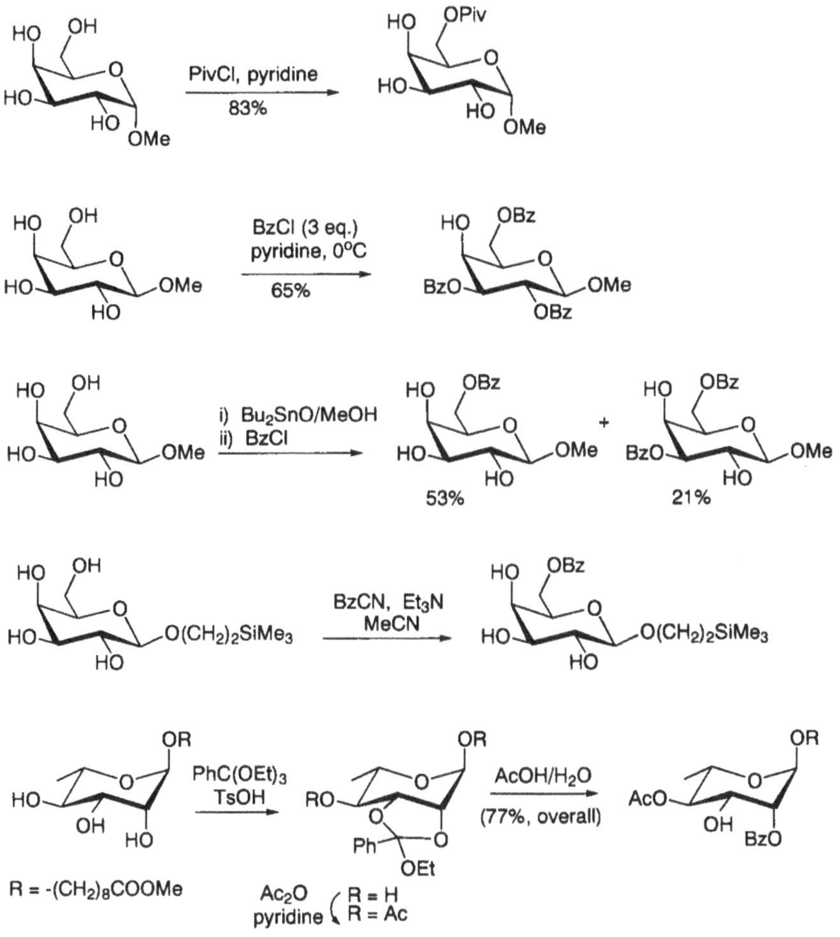

Scheme 2.13   Regioselective acylations.

Lipases have been employed for the regioselective acylation of sugar derivatives. These reactions exploit the phenomenon that the hydrolytic nature of lipases can be reversed under appropriate anhydrous conditions to achieve ester formation. Owing to the high polarity of saccharides, only aprotic polar solvents such as DMF and pyridine can be used. Unfortunately, only a few enzymes remain active in this environment. The choice of acylating reagent is also critical to the success of enzyme catalysed transesterifications. Anhydrides are usually too reactive in dipolar aprotic solvents and give nonenzyme catalysed reactions resulting in low regioselectivities. On the other hand, no spontaneous reactions take place when less reactive 2-haloethyl esters are used. For example, the primary position of the nonreducing end of several disaccharides can selectively be acylated using the enzyme subtilisin in DMF (Scheme 2.14). As outlined in Table 2.1, the regioselectivity depends greatly on configurational features of the substrate.[43]

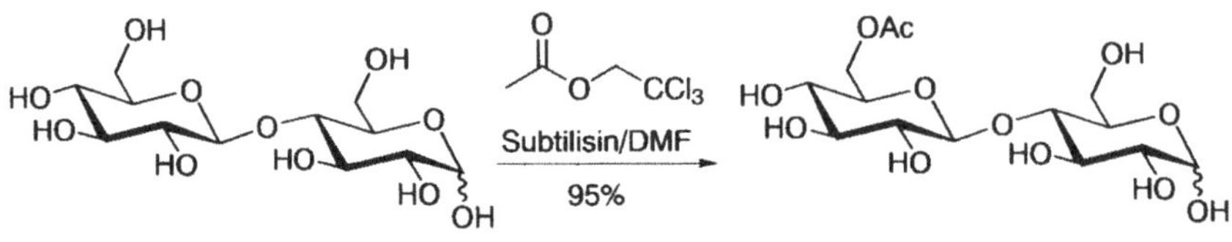

Scheme 2.14   Regioselective acylation using lipases.

Table 2.1   Lipase-catalysed esterification of 4,6-O-benzylidene glycopyranosides

Substrate	2-O-Acetyl (%)	3-O-Acetyl (%)
Methyl 4,6-O-benzylidene-α-D-glucopyranoside	100	0
Methyl 4,6-O-benzylidene-β-D-glucopyranoside	6	94
Methyl 4,6-O-benzylidene-α-D-galactopyranoside	0	0
Methyl 4,6-O-benzylidene-β-D-galactopyranoside	2	98
Methyl 4,6-O-benzylidene-α-D-mannopyranoside	97	3
Methyl 4,6-O-benzylidene-β-D-mannopyranoside	2	98

Esters are usually cleaved by bases such as NaOMe, KOH or NH_3 in methanol but they can also be hydrolysed by acid-catalysed solvolysis (MeOH/HCl).[44] However, they are relatively stable to acid treatment in the absence of water or alcohols. The relative order of base stability of the commonly employed ester groups is as follows:

$$t\text{-BuCO} > \text{PhCO} > \text{MeCO} > \text{ClCH}_2\text{CO}.$$

The t-butyl moiety of the pivaloyl group sterically shields the carbonyl moiety from nucleophilic attack and confers stability to such esters. Cleavage of the pivaloyl protecting group requires strong basic conditions such as KOH in methanol, aqueous methylamine or aqueous methanolic tetramethyl ammonium hydroxide. An acetyl or benzoyl group can easily be cleaved by treatment with NaOMe in methanol. Furthermore, methanolic ammonia selectively removes acetates in the presence of pivaloates. The electron-withdrawing α-Cl substituent of the chloroacetyl group lowers the pK_a of the corresponding acid and therefore is an excellent leaving group. This protecting group can be cleaved in the presence of acetyl groups using NH_3 in methanol or toluene, by using aqueous pyridine or hydrazine acetate in methanol (Scheme 2.15a).

Several mild reagents have been reported to cleave the chloroacetyl group, including 2-mercaptoethylamine ($H_2NCH_2CH_2SH$), thiourea

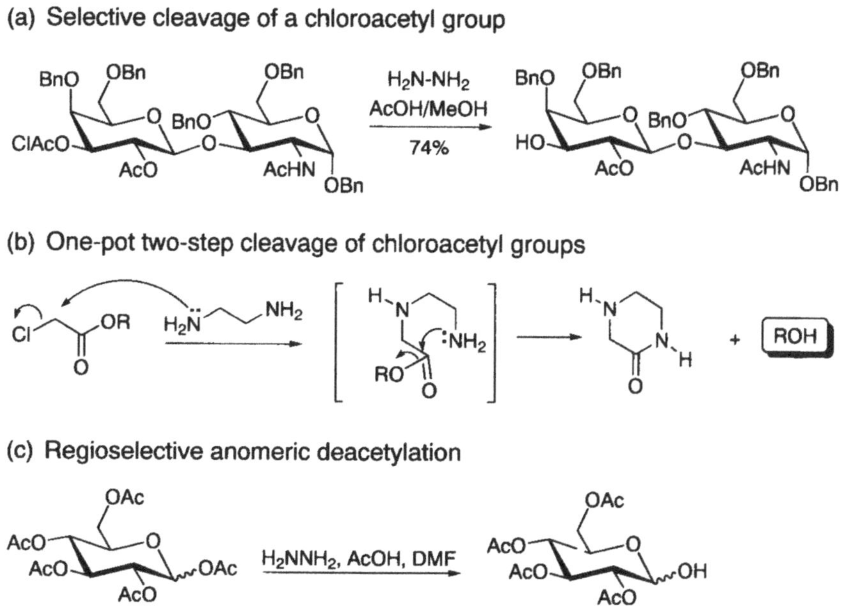

(a) Selective cleavage of a chloroacetyl group

(b) One-pot two-step cleavage of chloroacetyl groups

(c) Regioselective anomeric deacetylation

Scheme 2.15 Regioselective deacetylation.

(H_2NCSNH_2) and hydrazinedithiocarbonate ($H_2NNHC(=S)SH$).[45] The latter reagents cleave the chloroacetyl group by a two-step one-pot reaction. For example, treatment with mercaptoethylamine results in substitution of the chloride followed by intramolecular cyclisation and cleavage (Scheme 2.15b). This reaction exploits the fact that intramolecular reactions are much faster than intermolecular reactions.

An anomeric acetyl group can be removed selectively by treatment with hydrazine acetate or ammonium carbonate in DMF (Scheme 2.15c).[46] The latter transformations are of great importance for the preparation of glycosyl donors such as anomeric trichloroacetimidates.

2.5 Anomeric protecting groups

In general, the anomeric centre is the first position to be protected during a series of protecting group manipulations. The anomeric centre can simply be protected as an alkyl, allyl or benzyl glycoside. These glycosides can be prepared by a classical Fischer reaction in which a monosaccharide is suspended in an alcohol and treated with a catalytic amount of acid (see Chapter 1.5). The use of a cationic exchange resin (H^+ form) gives cleaner products and higher yields.[47] In many cases, the Fischer reactions give mixtures of anomers, which in some cases can be separated by crystallisation. Cleavage of alkyl glycosides can be accomplished under hydrolytic conditions, and often aqueous HCl in dioxane is employed. Benzyl and allyl glycosides can be removed under standard conditions as described above. Intramolecular glycoside formation to give 1,6-anhydro derivatives is an important method for simultaneous temporary protection of the anomeric centre and the C(6) position (Scheme 2.16).[48] These

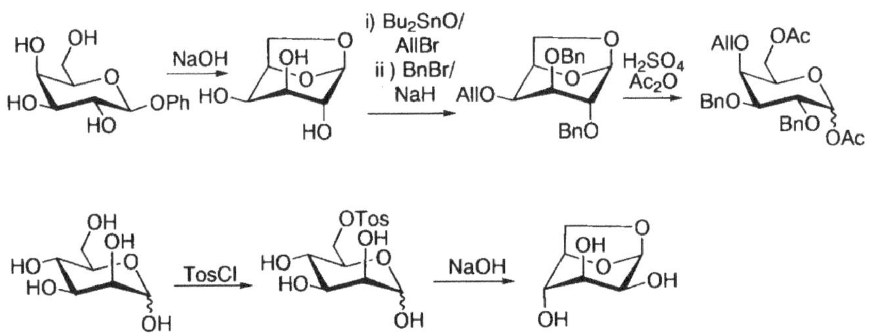

Scheme 2.16 Preparation and selective protection of 1,6-anhydropyranoses.

derivatives are commonly prepared by treatment of a phenyl β-glycoside with base, and the reaction probably proceeds through the formation of a 1,2-anhydro derivative, which is opened by nucleophilic attack by the C(6) hydroxyl. 1,6-Anhydromannose cannot be prepared by this method but selective tosylation of the 6-hydroxyl followed by base treatment provides this derivative in relatively high yield. It is important to note that during the formation of 1,6-anhydro derivatives, the conformation of the six-membered ring changes from $^4C_1$ to $^1C_4$. This change in ring conformation alters the relative reactivity of the hydroxyls. For example, the unreactive axial C(4) hydroxyl of the galactose derivative adopts an equatorial orientation in a 1,6-anhydro derivative and is, therefore, significantly more reactive. The 1,6-anhydro bridge can be cleaved by acetolysis with sulfuric acid and acetic anhydride.

The 2-(trimethylsilyl)ethyl group (TMSEt) has found widespread use as anomeric protecting group.[49] It can be introduced by a glycosylation (see Chapter 4) and cleaved by treatment with $BF_3 \cdot Et_2O$. The mechanistic considerations of the latter cleavage are as follows: a hard base such as a fluoride ion can react with the hard centre of silicon and thereby cause the TMSEt to fragment into ethylene, a TMS derivative and a saccharide derivative carrying a Lewis acid at its anomeric centre. The latter species can react with an acid anhydride to give the corresponding 1-*O*-acyl sugar derivative (Scheme 2.17).

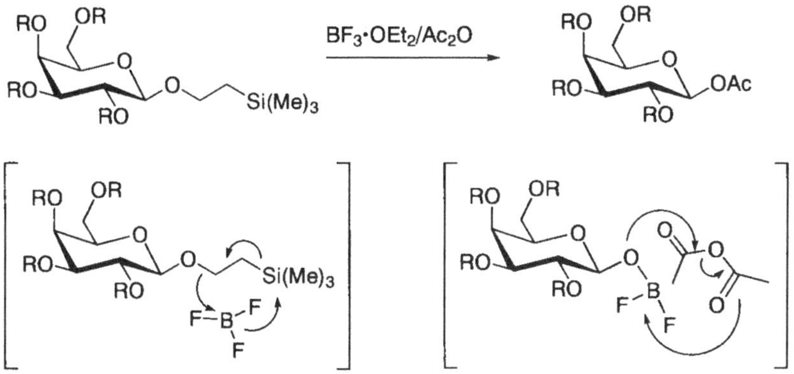

Scheme 2.17   Transformation of 2-(trimethylsilyl)ethyl (TMSEt) glycosides into 1-*O*-acyl sugars.

Another useful anomeric protecting group is the 4-methoxyphenyl group, which is stable under a wide range of reaction conditions but can be removed oxidatively with cerium ammonium nitrate.[50] 4-Methoxyphenyl glycosides can also be converted into the corresponding glycosyl chlorides and bromides and into thiophenyl glycosides. Thioglycosides are useful substrates in oligosaccharide synthesis. The anomeric thio group can be introduced at an early stage of the synthesis, is stable under many protecting group manipulations, yet can act as an efficient anomeric leaving group when activated with thiophilic reagents. The preparation of thioglycosides and its use in glycosylations will be discussed in Chapter 4.

2.6 Amino protecting groups

Amino sugars are widely distributed in living organisms and occur as constituents of glycoproteins, glycolipids and proteoglycans and as parts of various antibiotics. The most commonly used *N*-protecting groups are the *N*-acetyl and phthalimido groups. The azido moiety is often used as an amino-masking functionality since it can easily be reduced to an amino group. Phthalimido and azido groups are often used for the protection of C(2) amino groups because they are compatible with most glycosylation protocols.

2.6.1 Phthalimides

Phthalimides are introduced by the reaction of an amino group with phthalic anhydride in the presence of a base such as K_2CO_3, triethylamine or pyridine (Scheme 2.18).[51] They are stable to a wide range of reaction

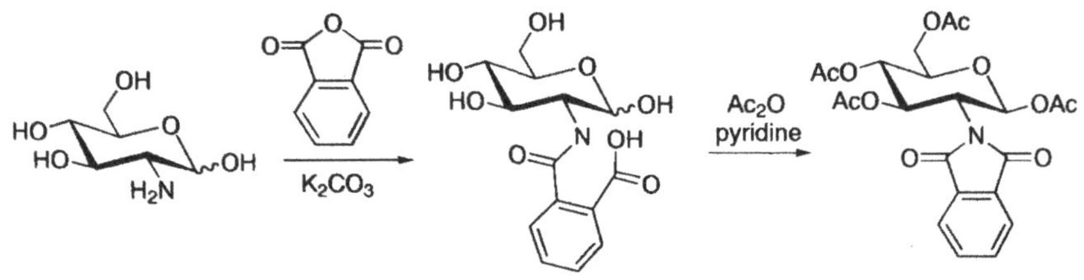

Scheme 2.18   Introduction of the phthalimido group.

conditions, including acids such as HBr/AcOH, oxidative reagents such as H_2O_2, O_3 and Jones oxidation and mild transesterification conditions to cleave acetyl protecting groups. Care has to be taken during benzylations with NaH or BnBr in DMF, and the latter transformation can more reliably be performed in THF as the solvent. The phthalimido moiety can be cleaved by treatment with large excesses of hydrazine, $NaBH_4$, butylamine, hydroxylamine, ethylenediamine or alkyldiamines immobilised on polystyrene beads.[52] These procedures can affect base-sensitive functionalities. Several modified phthalimides have been reported that can be cleaved under milder conditions.[53] For example, the tetrachlorophthaloyl (TCP) group can be introduced under similar conditions to the phthaloyl group but can be cleaved simply by treatment with ethylene diamine in MeCN/THF/EtOH at 60°C. The TCP group is compatible with several glycosylation conditions but is cleaved under benzylation conditions using NaH and BnBr. The functionality is stable under acidic benzylation conditions.

2.6.2 Azides

Formally, the azido group is not an amino protecting group but can easily be converted into an amino functionality by reduction, and the following reagents have been successfully employed for this transformation: catalytic hydrogenation, H_2S/pyridine/Et_3N, Na_2S/Et_3N, 1,3-propanedithiol/Et_3N, Ph_3P and $NaBH_4$.[54] Several methods can be employed for the preparation of azido sugars and some of these will be discussed in Chapter 3. The azido group is stable to a wide range of reaction conditions, including relatively strong acids and bases.

References

1. (a) T.W. Greene and P.G.M. Wuts, 1991, *Protecting Groups in Organic Synthesis*, 2nd edn, John Wiley, New York; (b) P.J. Kocieński, *Protecting Groups*, 1994, Georg Thiem, Stuttgard, New York; (c) A.H. Haines, 1976, *Adv. Carbohydr. Chem. Biochem.*, 33, 11; (d) A.H. Haines, 1981, *Adv. Carbohydr. Chem. Biochem.*, 39, 13; (e) J. Gelas, 1981, *Adv. Carbohydr. Chem. Biochem.*, 39, 71; (f) M. Schelhass and H. Waldmann, 1996, *Angew. Chem. Int. Ed. Engl.*, 35, 2057.
2. (a) S. Czernecki, C. Georgoulis and C. Provelenghiou, 1976, *Tetrahedron Lett.*, 3535; (b) K. Kania, I. Sakamoto, S. Ogawa and T. Suami, 1987, *Bull. Chem. Soc. Jpn.*, 60, 1529.
3. J.C. Jacquinet, J.M. Petit and P. Sinaÿ, 1974, *Carbohydr. Res.*, 38, 305.
4. (a) R. Kuhn, I. Löw and H. Trishmann, 1957, *Chem. Ber.*, 90, 203; (b) I. Croon and B. Lindberg, 1959, *Acta Chem. Scan.*, 13, 593; (c) J. Thiem and M. Wiesner, 1988, *Synthesis*, 125.
5. H.P. Wessel, T. Iversen and D.R. Bundle, 1985, *J. Chem Soc. Perkin. Trans. I*, 2247.
6. (a) S. David and S. Hanessian, 1985, *Tetrahedron*, 41, 643; (b) S. David, A. Thiéffry and A. Veyrieres, 1981, *J. Chem. Soc. Perkin Trans. I*, 1796; (c) S.J. Danishefski and R. Hungate, 1986, *J. Am. Chem. Soc.*, 108, 2486; (d) N. Nagashima and M. Ohno, 1985, *Chem. Pharm. Bull.*, 33, 2243; (e) P. Kovac and K.J. Edgar, 1992, *J. Org. Chem.*, 57, 2455; (f) G.-J. Boons, G.H. Castle, J.A. Clase, P. Grice, S.V. Ley and C. Pinel, 1993, *Synlett*, 913.
7. S. David, C. Pascard and M. Cesaria, 1979, *J. Nouveau Chim*, 3, 63.
8. (a) V. Pozsgay, 1979, *Carbohydr. Res.*, 69, 284; (b) P.J. Garegg, T. Iversen and S. Oscarson, 1976, *Carbohydr. Res.*, 50, C12; (c) S.S. Rana, J.J. Barlow and K.L. Matta, 1980, *Carbohydr. Res.*, 85, 313.
9. (a) R. Allerton and H.G. Fletcher, 1954, *J. Am. Chem. Soc*, 76, 1757; (b) K.P.R. Kartha, F. Dasgupta, P.P. Singh and H.C. Scrivastava, 1986, *J. Carbohydr. Chem.*, 5, 437; (c) A.B. Smith III, K.J. Hale, H.A. Vaccaro and R.A. Rivero, 1991, *J. Am Chem. Soc.*, 113, 2112; (d) H. Hori, Y. Nishida, H. Ohrui and H. Megoro, 1989, *J. Org. Chem.*, 54, 1346.
10. N. Nakajima, K. Horita, R. Abe and O. Yonemitsu, 1988, *Tetrahedron Lett.*, 29, 4139.
11. (a) R. Johansson and B. Samuelsson, 1984, *J. Chem. Soc. Perkin Trans. I*, 2371; (b) N. Nakajima, R. Abe and O. Yonemitsu, 1988, *Chem. Pharm. Bull.*, 36, 4244.
12. (a) Y. Oikawa, T. Yoshioka and O. Yonemitsu, 1982, *Tetrahedron Lett.*, 23, 885; (b) K. Horita, T. Yoshioka, T. Tanaka, Y. Oikawa and O. Yonemitsu, 1986, *Tetrahedron*, 42, 3021; (c) Y. Oikawa, T. Tanaka, K. Horito, T. Yoshioka and O. Yonemitsu, 1984, *Tetrahedron Lett.*, 25, 5393; (d) P.J. Garegg, L. Olsson and S. Oscarson, 1993, *J. Carbohydr. Chem.*, 12, 955.
13. L. Yan and D. Kahne, 1995, *Synlett*, 5, 523.
14. F. Guibé, 1997, *Tetrahedron*, 53, 13509.
15. (a) F. Guibé and Y. Saint M'leux, 1981, *Tetrahedron Lett.*, 22, 3591; (b) J.J. Oltvoort, M. Kloosterman and J.H. van Boom, 1983, *Recl. Trav. Chim. Pays-Bas* 102, 501.
16. (a) R. Lakhmiri, P. Lhoste and D. Sinou, 1989, *Tetrahedron Lett.*, 30, 4669; (b) R. Lakhmiri, P. Lhoste and D. Sinou, 1990, *Synth. Commun.*, 20, 1551.
17. (a) T. Ogawa, T. Kitajima and T. Nukada, 1983, *Carbohydr. Res.*, 123, C5; (b) M. Imoto, N. Kusunose, S. Kusomoto and T. Dhiba, 1988, *Tetrahedron Lett.*, 29, 2227; (c) P.A. Gent and R. Gigg, 1974, *J. Chem. Soc. Perkin Trans. I*, 1835.

18. J. Gigg and R. Gigg, 1966, *J. Chem. Soc. C.*, 82.
19. (a) K.L. Agerwal, A. Yamazaki, P.L. Cahion and H.G. Khorana, 1972, *Angew. Chem.*, *Int. Ed. Engl.*, 11, 451; (b) J.G. Buchanan and J.C.P. Schwarz, 1962, *J. Chem. Soc.*, 4770.
20. (a) Y.E. Tsvetkov, P.I. Kitov, L.V. Backinowsky and N.K. Kochetkov, 1993, *Tetrahedron Lett.*, 34, 7977; (b) A. Demchenko and G.J. Boons, 1997, *Tetrahedron Lett.*, 38, 1629.
21. (a) B. Helferich, 1848, *Adv. Carbohydr. Chem. Biochem.*, 3, 79; (b) R.U. Lemieux and J.P. Barrette, 1958, *J. Am. Chem. Soc.*, 80, 2243; (c) C.W. Baker and R.L. Whistler, 1974, *Carbohydr. Res.*, 33, 372; (d) K. Dax, W. Wolflehner and H. Weidmann, 1978, *Carbohydr. Res.*, 65, 132; (e) N.K. Kotchetkov, B.A. Dmitriev, N.E. Bairamova and A.V. Nikolaev, 1978, *Izv. Akad. Nauk. SSSR, Ser Khim.*, 652; (f) H. Prinz, L. Six, K.-P. Ryess and M. Lielländer, 1985, *Liebigs Ann. Chem.*, 217; (g) K.S. Khim, Y.H. Song, B.H. Lee and C.S. Hahn, 1986, *J. Org. Chem.*, 51, 404.
22. M. Bessodes, D. Komiotis, K. Antonakis, 1986, *Tetrahedron Lett.*, 27, 579.
23. (a) E.J. Corey and A. Venkateswarlu, 1972, *J. Am. Chem. Soc.*, 94, 6190; (b) S. Hanessian and P. Lavallee, 1975, *Can. J. Chem.*, 53, 2975; (c) S.C. Chaudary and O. Hernandez, 1979, *Tetrahedron Lett.*, 99; (d) W. Kinzy and R.R. Schmidt, 1987, *Tetrahedron Lett.*, 28, 1981.
24. F. Frank and R.D. Guthrie 1977, *Aust. J. Chem.*, 30, 639.
25. E.M. Nashed and C.P.J. Glaudemans, 1987, *J. Org. Chem.*, 52, 5255.
26. H. van Hannelore, H. Brandstetter and E. Zbiral, 1978, *Helv. Chim. Acta*, 61, 1832.
27. E.J. Corey, H. Cho, C. Rücker and D.H. Hua, 1981, *Tetrahedron Lett.*, 3455.
28. A.B. Smith, R.A. Rivero, K.J. Hale and H.A. Vaccaro, 1991, *J. Am. Chem. Soc.*, 113, 2092.
29. (a) R.F. Brady, 1971, *Adv. Carbohydr. Chem. Biochem.*, 26, 197; (b) A.N. de Belder, 1977, *Adv. Carbohydr. Chem. Biochem.*, 34, 179.
30. (a) S.Peat and L.F. Wiggins, 1938, *J. Chem. Soc.*, 1088; (b) A.B. Smith III and K.J. Hale, 1989, *Tetrahedron Lett.*, 30, 1037.
31. (a) S.S. Bhattacharjee and P.A.J. Gorin, 1969, *Can. J. Chem.*, 47, 1195; (b) A. Lipták, J. Imre, J. Hangi, P. Nánási and A. Neszmélyi, 1982, *Tetrahedron*, 38, 3721; (c) P.J. Garegg, 1984, *Pure Appl. Chem.*, 56, 845; (d) T. Mikami, H. Asano and O. Mitsunobu, 1987, *Chem. Lett.*, 2033; (e) M.P. de Ninno, J.B. Etienne and K.C. Duplantier, 1995, *Tetrahedron Lett.*, 36, 669.
32. P.J. Garegg, H. Hultberg and S. Willan, 1982, *Carbohydr. Res.*, 108, 97.
33. S. Hanessian, 1968, *Adv. Chem. Ser.*, 74, 159.
34. O.T. Schmidt, 1963, *Methods Carbohydr. Chem.*, 2, 318.
35. S.V. Ley, M. Woods and A. Zanottigerosa, 1992, *Synthesis*, 52.
36. S.V. Ley, W.M. Priepke and S.L. Warriner, 1994, *Angew. Chem. Int. Ed. Engl.*, 33, 2290.
37. S.J. Angyal and M.E. Evans, 1972, *Austr. J. Chem.*, 25, 1495.
38. S. Rio, J.-M. Beau and J.-C. Jacquinet, 1994, *Carbohydr. Res.*, 255, 103.
39. I.F. Pelyvás, T.K. Lindhorst, H. Streicher and J. Thiem, 1991, *Synthesis*, 1015.
40. Y. Tsuda, M.E. Haque and K. Yoshimoto, 1983, *Chem. Pharm. Bull.*, 31, 1612.
41. (a) R.U. Lemeuix and H. Driguez, 1975, *J. Am. Chem. Soc.*, 97, 4069; (b) S. Hanessian and R. Roy, 1985, *Can. J. Chem.*, 63, 163; (c) P.J. Garegg and H. Hultberg, 1979, *Carbohydr. Res.*, 72, 276; (d) V. Potzgay, 1992, *Carbohydr. Res.*, 235, 295.
42. (a) S. Oscarson and A.-K. Tidén, 1993, *Carbohydr. Res.*, 247, 323; (b) S. Josephson and D.R. Bundle, 1980, *J. Chem. Soc. Perkin Trans. I*, 301.
43. (a) S. Riva, 1996, in *Enzymatic Reactions in Organic Media* (A.M.P. Koskinen and A.M. Klibanov, eds.), Blackie/Chapman & Hall, London Academic & Professional, p. 140; (b) L. Panza, M. Luisetti and E. Crociati, 1993, *J. Carbohydr. Chem.*, 12, 125; (c) S. Riva, J. Chopineau and A.P.G. Kieboom, 1988, *J. Am. Chem. Soc.*, 110, 584.
44. N.E. Byramova, M.V. Ovchinnikov, L.V. Backinowsky and N.K. Kochetkov, 1983, *Carbohydr. Res.*, 124, C8.

45. (a) C.A.A. van Boeckel and T. Beetz, 1983, *Tetrahedron Lett.*, 24, 3775; (b) T. Ziegler, 1990, *Liebigs Ann.*, 1125; (c) A.F. Cook and D.T. Maichuk, 1970, *J. Org. Chem.*, 35, 1940; (d) P.M. Aberg, L. Blomberg and T. Norberg, 1994, *J. Carbohydr. Res.*, 13, 141.
46. (a) G. Excoffier, D. Gagnaire and J.-P. Utille, 1975, *Carbohydr. Res.*, 39, 368; (b) A. Nudelman, J. Herzig, H.E. Gottlieb, E. Kerinan and J. Sterling, 1987, *Carbohydr. Res.*, 162, 145; (c) M. Mikamo, 1989, *Carbohydr. Res.*, 191, 150; (d) M.K. Gurjar and U.K. Saha, 1992, *Tetrahedron*, 48, 4039; (e) T. Nakano, Y. Ito and T. Ogawa, 1993, *Carbohydr. Res.*, 243, 43.
47. G.N. Bollenback, 1963, in *Methods in Carbohydrate Chemistry*, Vol. II (R.L. Wistler, M.L. Wolfrom and J.N. BeMiller, eds.), Academic Press, New York, pp. 326-328.
48. (a) M. Cerny and J. Stanek, Jr, 1977, *Adv. Carbohydr. Chem. Biochem.*, 34, 23; (b) M. A. Zottola, R. Alonso, G.D. Vito and B. Fraser-Reid, 1989, *J. Org. Chem.* 54, 6123.
49. K. Jansson, S. Ahlfors, T. Frejd, J. Kihlberg and G. Magnusson, 1988, *J. Org. Chem.*, 53, 5629.
50. (a) T. Slagheck, Y. Nakahara and T. Ogawa, 1992, *Tetrahedron Lett.*, 33, 4971; (b) Z. Zhang and G. Magnusson, 1996, *Carbohydr. Res.*, 925, 41.
51. (a) R.U. Lemieux, T. Takeda and B.Y. Chung, 1976, *A.C.S. Symp. Ser.*, 39, 90; (b) S. Akiya and T. Osawa, 1960, *Chem. Pharm. Bull.*, 8, 583; (c) B.R. Baker, J.P. Joseph, R.E. Schaub and J.H. Williams, 1954, *J. Org. Chem.*, 19, 1786.
52. (a) D.R. Bundle and S. Josephson, 1979, *Can. J. Chem.*, 57, 662; (b) P.L. Durette, E.P. Meitzner and T.Y. Shen, 1979, *Tetrahedron Lett.*, 42, 4013; (c) R. Madsen, U.E. Udodong, C. Roberts, D.R. Mootoo, P. Konradsson and B. Fraser-Reid, 1995, *J. Am. Chem. Soc.*, 117, 1554; (d) J.O. Osby, M.G. Martin and B. Ganem, 1984, *Tetrahedron Lett.*, 25, 2093; (e) P. Stangier and O. Hindsgaul, 1996, *Synlett*, 179.
53. J. Debenham, R. Rodebaugh and B. Fraser-Reid, 1997, *Liebigs Ann./Recueil*, 791.
54. (a) T. Adachi, Y. Yamada, I. Inoue and M. Saneyoshi, 1977, *Synthesis*, 45; (b) R.U. Lemieux, S.Z. Abbas, M.H. Burzynska and R.M. Tatcliffe, 1982, *Can. J. Chem.*, 60, 63; (c) M. Valteir, N. Knouzi and R. Carrie, 1983, *Tetrahedron Lett.*, 24, 763; (d) B.A. Belinka and A. Hassner, 1979, *J. Org. Chem.*, 44, 4712; (e) F. Rolla, 1982, *J. Org. Chem.*, 47, 4327.

3 Functionalised saccharides

G.-J. Boons

3.1 General introduction

The saccharides represent a structurally diverse group of compounds, which are often derivatised with a variety of functional groups. The chemical modification of saccharides often involves the installation of a functional group that at an appropriate time can be converted into another functionality. In this chapter, synthetic methodologies for the preparation of halogenated, unsaturated and deoxygenated sugar derivatives will be discussed and furthermore methods for the introduction of amino, sulfate and phosphate moieties will be covered.

3.2 Deoxyhalogeno sugars

3.2.1 Introduction

Deoxyhalogeno sugars are carbohydrate derivatives in which one or more hydroxyls are replaced by halogen atom(s).[1] Iodo, bromo, and chloro sugar derivatives have been widely used in substitution and elimination reactions. The ease of displacement decreases in the order I > Br > Cl. In general, fluoro sugars are too stable to be used in the above-mentioned reactions but they are often used as probes for studying carbohydrate–protein interaction.[2] Furthermore, sugar derivatives having an ^{18}F label are applied for medical imaging.[3] Direct replacement of a hydroxyl, displacement reactions, epoxide opening, addition reactions to unsaturated derivatives, radical halogenations and oxidative cleavage of a benzylidene acetal with N-bromosuccinimide (NBS) have been used for the introduction of halogen into a saccharide. In this section examples of these methods will be discussed.

3.2.2 Direct halogenation of alcohols

A variety of reagents are available for the direct replacement of a hydroxyl by a halogeno substituent. Many of these approaches are based on the reaction of an activated triphenyl phosphine derivative with a sugar alcohol to give an alkoxyphosphonium ion, which in turn is substituted with a halide ion to give a halogeno sugar derivative. The driving force of this reaction is the formation of a strong $P=O$ bond. A generalised mechanism is depicted in Scheme 3.1. As can be seen, this reaction proceeds with inversion of configuration when performed at a secondary position.

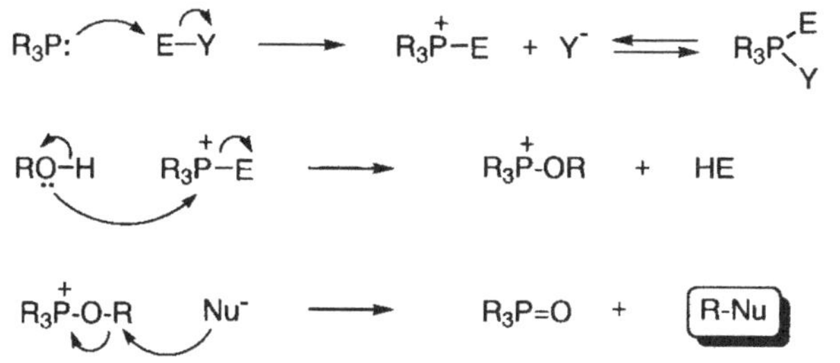

Scheme 3.1 Activation of alcohols by phosphines. Note: Nu = nucleophile.

The reaction of triphenylphosphine with carbon tetrahalides gives the reagent Ph_3P^+—XCX^- that reacts with primary and secondary sugar hydroxyls to give halogeno sugars.[4] However, regioselective halogenations at the primary position of unprotected pyranosides and furanosides can be achieved when the reaction is performed in pyridine (Scheme 3.2a). This type of reagent has been used for the preparation of chlorides, bromides and iodides. Similar transformations have been performed with triphenylphosphine-N-halosuccinimides.[5]

Isolated primary and secondary hydroxy groups of carbohydrate derivatives can be transformed into iodo groups with inversion of configuration by treatment with either triphenylphosphine, iodine and imidazole, or with triphenylphosphine and 2,4,5-triiodoimidazole at elevated temperatures (Scheme 3.2b).[6] At lower reaction temperatures, primary hydroxyls can be selectively replaced by iodo groups. In the absence of imidazole, triphenylphosphine and iodine form an adduct, which is virtually insoluble. If, however, imidazole is added, a partially soluble complex is formed which rapidly combines with an alcohol to give an activated species. When 2,4,5-tribromoimidazole is used, various bromo sugars can be prepared. This reagent has been used for the regioselective bromination of a C(2), C(3) diol. The C(2) hydroxyl is significantly less nucleophilic than the C(3) hydroxyl because of its proximity to the electron-withdrawing anomeric centre.

Mitsunobu conditions [triphenylphosphine/diethyl azodicarboxylate (DEAD)][7] and triphenylphosphite methiodide [$(PhO)_3P^+MeI^-$] or dihalides [$(PhO)_3P^+XX^-$][8] have been successfully applied to the synthesis of halogeno sugars.

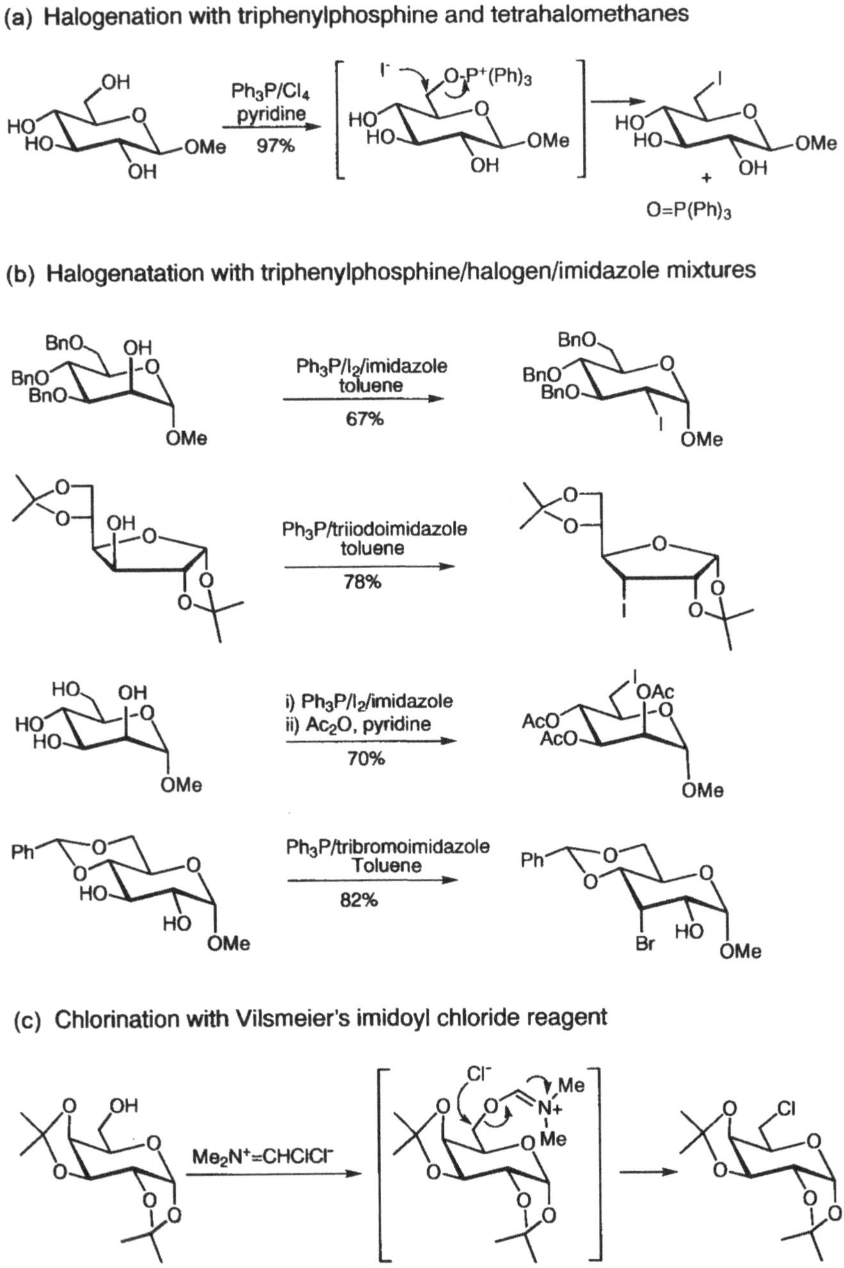

(a) Halogenation with triphenylphosphine and tetrahalomethanes

(b) Halogenatation with triphenylphosphine/halogen/imidazole mixtures

(c) Chlorination with Vilsmeier's imidoyl chloride reagent

Scheme 3.2   Halogenation of saccharides.

(d) Chlorination with thionyl chloride

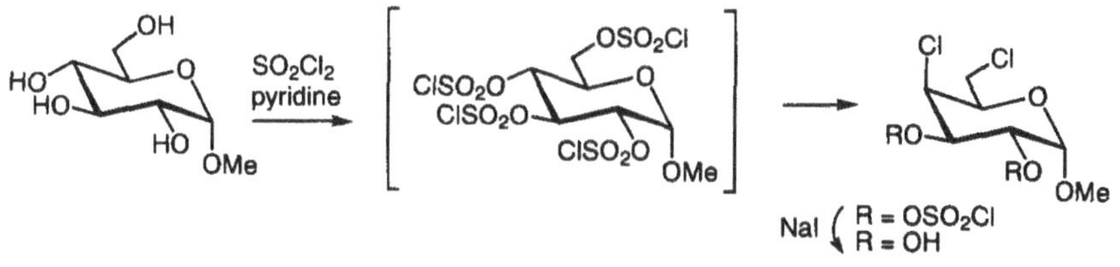

(e) Fluorination with diethylaminosulfur trifluoride (DAST)

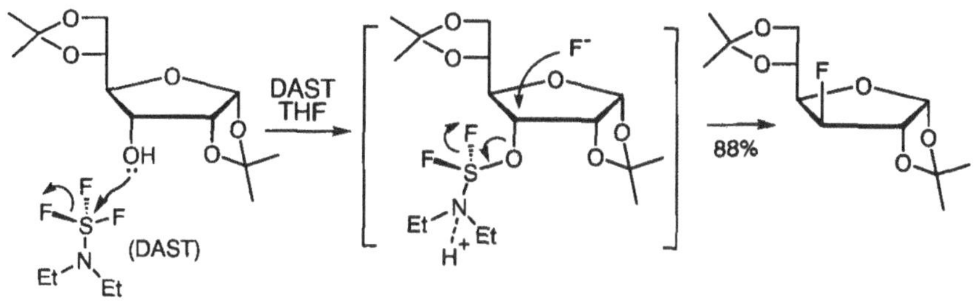

Scheme 3.2   (Continued).

Many methods for the preparation of halogeno sugars rely on the activation of triphenylphosphine or trialkylphosphite. However, a different method is available for the direct conversion of hydroxyls into chlorides. *N,N*-Dimethylformamide (DMF) reacts with chlorides of inorganic acids (thionyl chloride, phosphorus trichloride and phosphoryl chloride) to give the active salt: (chloromethylene)dimethyliminium chloride.[9] This salt, which is also called 'Vilsmeier's imidoyl chloride reagent', reacts with alcohols to give an activated intermediate that can undergo nucleophilic substitutions releasing chloride to give a chlorinated sugar derivative (Scheme 3.2c).

Another interesting reaction is the treatment of partially protected saccharides with sulfuryl chloride (Scheme 3.2d).[10] These reactions proceed by the initial formation of chlorosulfates, which are displaced by the liberated chloride ions. The substitution reaction occurs only at those centres where the steric and polar factors are favourable for an S_N2 reaction. The chlorosulfates that have not been substituted with a chloride can easily be cleaved by treatment with sodium iodide.

The above-discussed methods are mainly used for the synthesis of chloro, bromo and iodo sugars. A different approach has to be taken when a deoxyfluoro sugar derivative is required, and the most commonly applied reagent for direct fluorination is diethylaminosulfur trifluoride

(Et$_2$NSF$_3$, DAST). In this reaction, an alcohol displaces a fluoride of DAST resulting in an activated intermediate, which in turn is displaced by the liberated fluoride (Scheme 3.2e).[11] A *gem*-difluoride is formed when a ketone or aldehyde is treated with DAST.

3.2.3 Displacement reactions

Halogeno saccharides can be prepared by the displacement of a sulfonate ester with a halogen ion. Primary fluorides, bromides, chlorides and iodides can be prepared from mesylates (ROSO$_2$CH$_3$) and tosylates (ROSO$_2$PhMe) (Scheme 3.3a).[12] The latter derivatives can be obtained cheaply by treatment of alcohols with a sulfonyl chloride in pyridine or triethylamine.

The success of a displacement critically depends on the position of the leaving group and the configuration of the sugar ring. The direct replacement of a sulfonate ester normally takes place by an S$_N$2 mechanism, as shown by the inversion of configuration, which accompanies such reactions when performed at asymmetric centres. The geometry of the S$_N$2 transition state involves two highly polar bonds, one in the process of formation and the other in degeneration. The formation of these polar bonds is greatly affected by the presence of neighbouring polar substituents.[12d] The permanent dipole of an electronegative substituent may hinder the development of a transition state when an anionic nucleophile is used. Maximum unfavourable dipolar interactions result when these dipoles are parallel (Figure 3.1). In addition, steric effects may hinder incoming nucleophiles, retarding displacement reactions. Primary sulfonates of hexopyranosides are readily displaced by nucleophiles provided that the C(4) oxygen is in an equatorial orientation. Incoming nucleophiles are hindered when the C(4) substituent is in an axial position (6-*O*-sulfonates of galactosides). In this case, the transition state will experience minimal dipolar interactions when the dipoles associated with the incoming nucleophile is 90° with respect to the permanent dipole of the endocyclic oxygen. However, in such an arrangement, the transition state will experience considerable steric overcrowding. Displacements from other conformations are disfavoured by unfavourable dipolar interactions.

Tosylate and mesylate displacements at C(2) of α-glycosides are very slow owing to unfavourable dipolar interactions in the S$_N$2 transition state. Both polar bonds of the transition state are inclined at an angle of about 30° to permanent dipoles of the C(1)—O(1) and C(1)—O(5) bonds. Displacement of C(2) sulfonates of β-glycosides is much more facile because, in this case, the transition state experiences only one unfavourable dipolar interaction of the C(1)—O(1) bond.[12d, e]

(a) Displacements of tosylates and mesylates

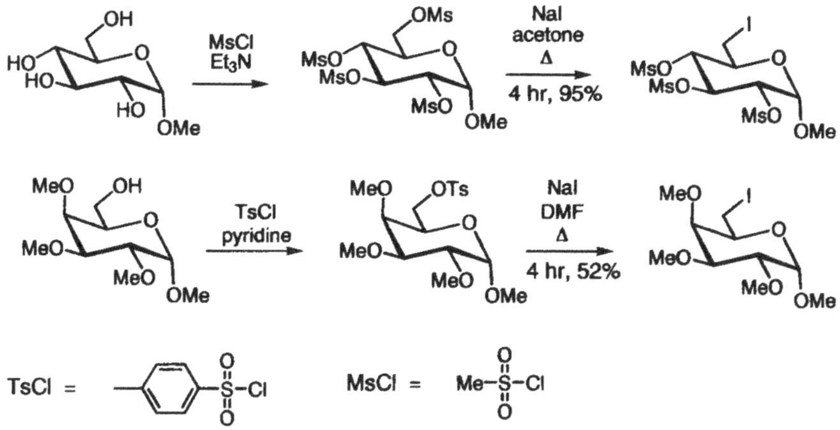

(b) Displacements of triflates

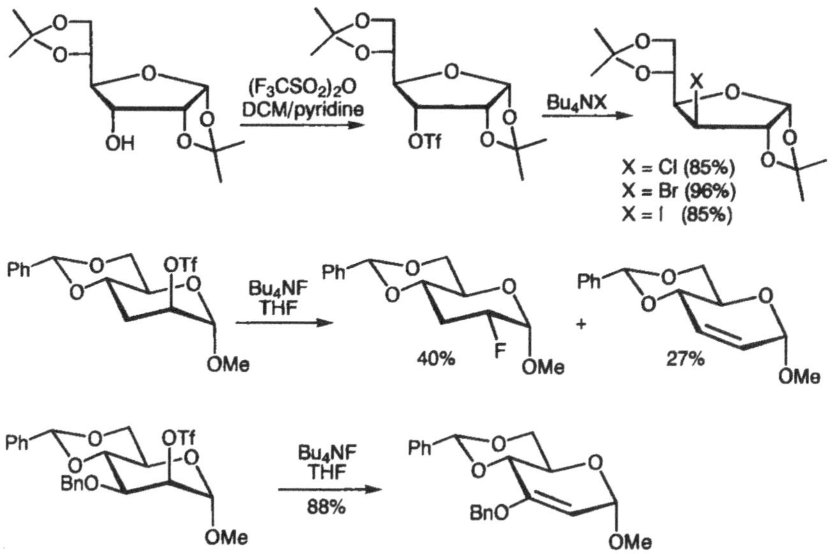

Scheme 3.3 Preparation of halides by sulfonate displacement.

The development of an S_N2 transition state at $C(3)$ or $C(4)$ is not affected by dipoles associated with the anomeric centre but is mainly influenced by steric and polar factors from other groups in the sugar ring. For example, the displacement of equatorially oriented tosylates or

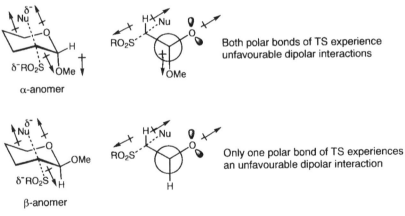

TS destabilised by unfavourable
dipolar interactions

Interference of C(4) axial group

Axial C(4) group interferes with
incoming nucleophile in the preferred
rotamer for nucleophilic attack that
experiences no dipolar interactions

Sulfonate displacements at C(2)

α-anomer

Both polar bonds of TS experience
unfavourable dipolar interactions

β-anomer

Only one polar bond of TS experiences
an unfavourable dipolar interaction

β-*Trans*-axial effects

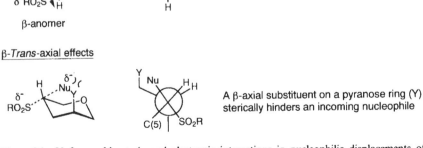

A β-axial substituent on a pyranose ring (Y)
sterically hinders an incoming nucleophile

Figure 3.1 Unfavourable steric and electronic interactions in nucleophilic displacements of
sulfonates. Note: Ts = transition state.

mesylates that have a β-*trans*-axial substituent is greatly retarded by a steric clash between the incoming nucleophile and the β-axial substituent.

Electropositive leaving groups such as phosphonium and ammonium ions [ROP$^+$(Ph)$_3$, RN$^+$R$_3$] can have favourable dipolar interactions in the transition state and therefore reaction rates can be enhanced. Similarly, the use of a neutral nucleophile such as ammonia or hydrazine will reverse the polarity of the forming bond in the transition state, giving rise to favourable polar interactions with neighbouring electronegative substituents.

Triflate displacements are popular for the preparation of secondary halogenated carbohydrates and often give satisfactory results when mesylates and tosylates fail to give products (Scheme 3.3b).[13] In general, triflates react 10^3–10^4 faster than corresponding mesylates. Triflated saccharide derivatives (ROSO$_2$CF$_3$) can easily be prepared by the reaction of a sugar hydroxyl with triflic anhydride in the presence of pyridine or other bases. It has to be realised that an elimination reaction may be a competing side reaction.[14] This reaction is favoured when weak, basic, nucleophiles are used. Among halide ions, fluorides are the strongest bases and the weakest nucleophiles and, as a result, triflate displacements with fluorides are often accompanied by elimination. Elimination reactions are more likely when the leaving group is in a *trans*-diaxial relationship with an α-hydrogen atom at a sterically hindered position and in such a case the main product may be the unsaturated elimination compound.

To circumvent elimination reactions, fluorides have been introduced by nucleophilic ring opening of epoxides.[15] The regioselective opening of aldosyl epoxides proceeds *trans*-diaxially.[16] Epoxides can also be opened with chloride, bromide and iodide ions. The chemistry of epoxides is discussed in detail in Section 3.6.

3.2.4 *Miscellaneous methods*

An efficient procedure for the formation of primary bromides is the reaction of 4,6-*O*-benzylidene hexopyranosides with *N*-bromosuccinimide (NBS) in the presence of barium carbonate. This reaction leads to the corresponding 4-*O*-benzoyl-6-bromide-6-deoxy-glycoside (Scheme 3.4a).[17] Probably, the reaction proceeds by the radical bromination of the benzylic carbon atom followed by rearrangement to the 6-deoxy-6-bromo derivative. The application of this method is very efficient since the benzylidene functionality can act as a protecting group but can be oxidatively cleaved to give a 6-deoxy-6-bromo derivative.

Radical bromination with NBS can result in the replacement of a sugar ring hydrogen by bromine.[18] These reactions can be highly regioselective

(a) Free radiacal halogenation

Mechanistic considerations

Benzylic bromination

Br· + H–R ⟶ HBr + R·

HBr + [N-Br succinimide] ⟶ [NH succinimide] + Br₂

Br₂ + R· ⟶ RBr + Br·

Aliphatic bromination

Br· + H–R ⟶ HBr + R·

R· + [N-Br succinimide] ⟶ [N· succinimide] + RBr

[N· succinimide] + H–R ⟶ [NH succinimide] + R·

(b) Electrophilic halogenation

XeF₂, BF₃·Et₂O

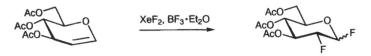

Scheme 3.4   Radical and electrophilic halogenation.

and, in general, the C(5) of peracetylated sugars can be selectively brominated. However, the anomeric centre can be brominated in high yield when it is substituted with a strongly electronegative substituent such as a fluoride or chloride.

Bromination of a benzylidene acetal or aliphatic hydrogens of sugar rings proceed via different reaction mechanisms. In the case of benzylic halogenation, the reaction is initiated by a small amount of bromine radicals. The main propagation steps are hydrogen abstraction at the benzylic position by the bromine radical and bromination of the resulting benzylic radical with bromine. Molecular bromine is maintained at low concentrations by an ionic reaction of NBS with hydrogen bromide. Thus, the role of NBS is to provide a source of bromine and to use up the HBr that is formed. The bromine radical is highly electrophilic and reacts at positions of high electronegativity. The hydrogen at the benzylic position is abstracted because the aromatic ring is somewhat electron-donating and provides resonance stabilisation of the produced radical. In addition, the benzylic carbon is relatively electron-rich because of the operation of anomeric effects. Thus, the electron density of the oxygen is transferred to the antibonding orbitals of the benzylic carbon.

For aliphatic brominations, the succinimyl radical appears to be involved as the hydrogen-abstracting radical. This mode of reaction pathway can be explained as follows: electron-withdrawing groups diminish greatly the suitability of adjacent positions for reactions with free halide radicals. Halide radicals are highly electrophilic and as a result react at positions of high electron density. On the other hand, the succinimyl radical is nucleophilic and reacts at positions of low electron density. Thus, bromine radicals will not react with the hydrogens of a sugar ring because of the many electron-withdrawing groups. However, the succinimyl radical is highly reactive with the relatively electron-poor sugar ring. Also, a radical adjacent to a heteroatom is stabilised by resonance stabilisation.

Finally, halogens can be introduced by electrophilic addition to a double bond (Scheme 3.4b). For example, the reaction of glucal with xenon difluoride in the presence of a catalytic amount of $BF_3 \cdot OEt_2$ gives a 2-deoxy-2-fluoro-glucosyl fluoride as a mixture of anomers.[19]

3.3 Unsaturated sugar derivatives

3.3.1 Introduction

Unsaturated sugar derivatives are important and versatile intermediates for organic synthesis.[20] Introduction of a double bond can result in the formation of alkenes, enol ethers or enediols. In this subsection, several general methods for the preparation of these compounds will be discussed. Furthermore, some important reactions using unsaturated sugar derivatives will be covered.

3.3.2 Glycals

Glycals are saccharide derivatives having a double bond between the anomeric carbon and the adjacent carbon atom. Formally, these derivatives are highly electron-rich enol ethers, which can undergo many reactions with high regioselectivity and stereoselectivity.

Traditionally, glycals are prepared by the reductive elimination of a peracetylated glycosyl bromide in the presence of activated zinc (Scheme 3.5a).[21] The proposed mechanistic picture is that of reductive removal of

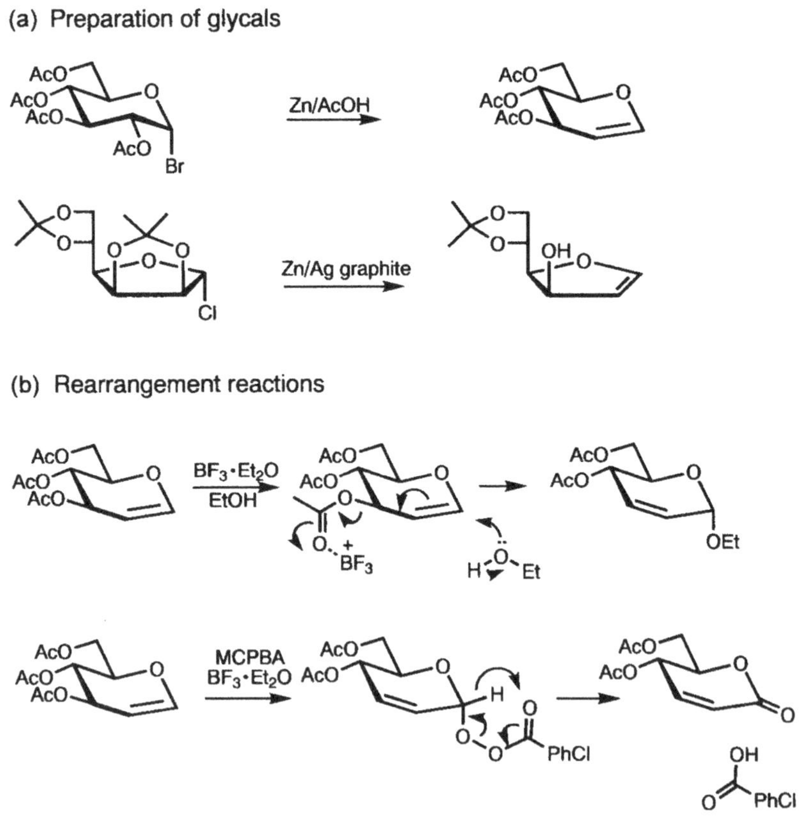

(a) Preparation of glycals

(b) Rearrangement reactions

Scheme 3.5 Preparation and reactions with glycals.

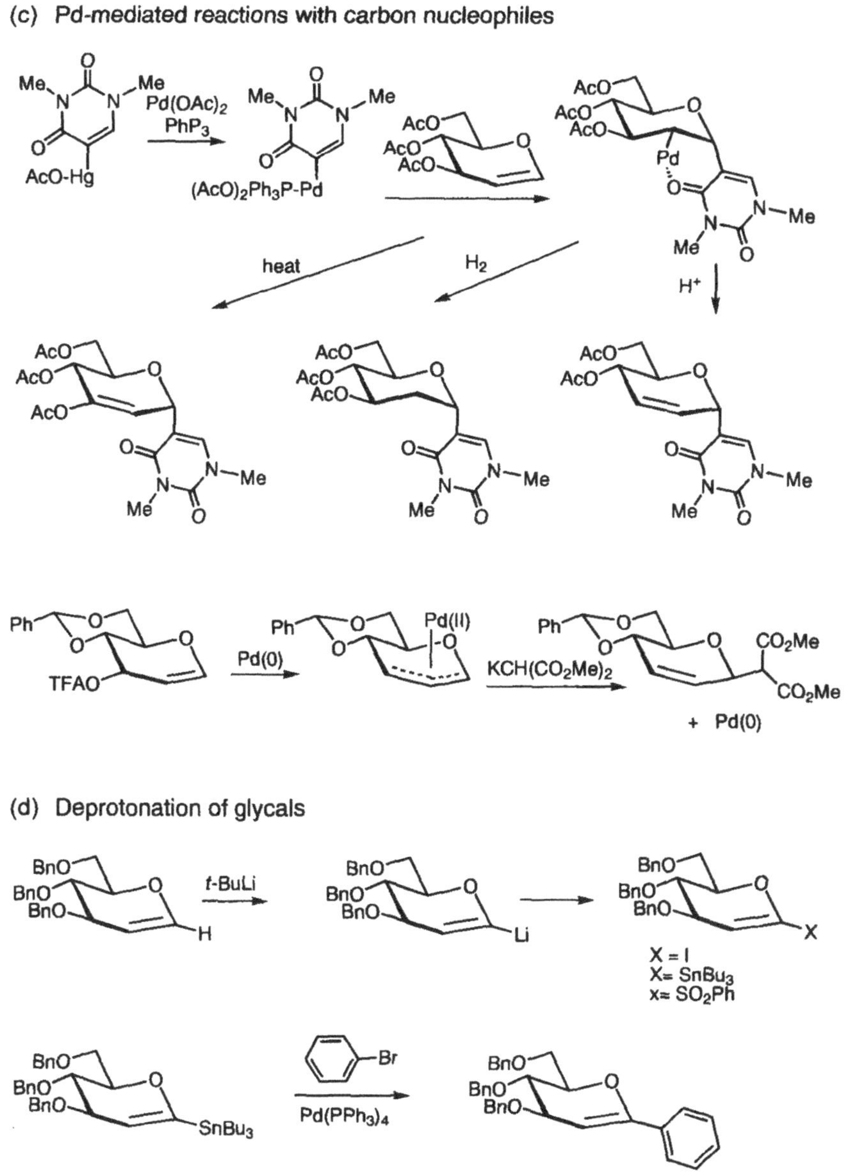

(c) Pd-mediated reactions with carbon nucleophiles

(d) Deprotonation of glycals

Scheme 3.5 (Continued).

the anomeric bromide by zinc followed by elimination of the C(2) O-acetate to give a glycal derivative. This relatively harsh procedure is incompatible with many functionalities and protecting groups and cannot be applied to the preparation of certain furanoid glycals. However, glycofuranosyl chlorides can be converted into glycals by the application of milder reducing conditions such as Zn/Ag graphite, lithium in liquid ammonia and sodium naphthalenide. Glycals can undergo a number of protecting-group manipulations but acidic conditions have to be avoided because of the acid sensitivity of the enol ether moiety. They have been used in electrophilic addition, rearrangement and cycloaddition reactions. Electrophiles add at the C(2) of glycals resulting in the formation of a well-stabilised oxocarbenium ion which can react with numerous nucleophiles. This type of reaction is of great importance for the preparation of 2-deoxy glycosides (see Chapter 4).

Reaction of Lewis acids with acetylated glycals in the presence of an alcohol results in allylic rearrangements to give 2,3-unsaturated compounds (Scheme 3.5b).[22] This reaction is named the Ferrier reaction and has found widespread use in natural product synthesis. The reaction is initiated by coordination of the Lewis acid with the acyl moiety at C(3), converting it into a good leaving group. Nucleophilic attack at the anomeric centre results in migration of the double bond and departure of the activated acyl moiety. A wide range of nucleophiles have been used in the Ferrier reaction. For example, 2,3-unsaturated lactones are formed when m-chloroperbenzoic acid is used as the nucleophile.[23] In this reaction, a peroxy ester is formed at the anomeric centre, which rearranges through a cyclic six-membered ring transition state to a lactone and m-chlorobenzoic acid. The Ferrier reaction makes it possible to add thiols, azides, cyanides and purines to an anomeric centre.

Transition-metal-catalysed reactions of glycals with carbon nucleophiles provide an important route to C-glycosides.[24] These reactions may proceed by two different pathways depending on the oxidation state of the transition metal used and the nature of the carbon nucleophile (Scheme 3.5c). For example,[25] transmetallation of $Pd(OAc)_2$ with pyrimidinyl mercuric acetate results in the formation of a pyrimidinyl palladium complex, which adds to the double bond of glycal derivatives. The resulting intermediate can undergo different transformations depending on the reaction conditions.

Alternatively, Pd(0) adds oxidatively to the double bond of a glycal derivative resulting in the formation of a π-allyl complex, which may react with carbon nucleophiles to give C-glycosides with a double bond between C(2) and C(3).[26] A π-allyl complex may also be formed starting from a Ferrier rearrangement product (2,3-unsaturated sugar derivative).[27]

Properly protected glycals can be selectively deprotonated at C(1) by very strong bases such as t-butyllithium (Scheme 3.5d). The resulting anions can be quenched with electrophiles such as Bu_3SnCl, I_2 and $PhSO_2Cl$.[28] The anomeric tin and iodo derivatives have been used in

Pd-catalysed cross couplings to give C-glycosides that have a double bond between C(1) and C(2).

3.3.3 Isolated double bonds

In the previous subsection, it was shown that the Ferrier reaction offers an opportunity to convert glycal derivatives into unsaturated sugar derivatives, which have an isolated double bond between C(2) and C(3). The Tipson–Cohen reaction is another important reaction for the introduction of isolated double bonds.[29] In this procedure, α *cis* or *trans* diols are converted into disulfonates (mesylates or tosylates) which are reductively eliminated with sodium iodide and zinc in refluxing DMF (Scheme 3.6a). In this reaction, the C(3) sulfonate is substituted by an iodide, which then is reductively removed by zinc with concomitant elimination of the second sulfonate moiety, introducing a double bond. Stereoelectronic effects make nucleophilic substitutions at C(3) more favourable than similar reactions at C(2) (see Section 3.2.3). Probably, the elimination proceeds through a boat conformation. In this case, the iodide and tosylate are in a *syn* relation. In most cases, E2 elimination proceeds via a transition state involving an *anti* orientation. Nevertheless, *syn* elimination becomes the dominant mode of reaction when structural features prohibit an *anti* orientation.

Deoxygenation of sugar α-diols has also successfully been performed using radical conditions (Scheme 3.6b). Thus, treatment of an α-dixanthate with the radical promoter azobisisobutyronitrile (AIBN) in the presence of diphenylsilane gives easy access to unsaturated sugar derivatives.[30] The first xanthate is removed by radical reduction to give a sugar radical. Next, this intermediate undergoes a radical elimination to give a double bond between C(2) and C(3).

Unsaturated sugar derivatives can undergo a wide range of transformations, in particular, when an enone is involved.[31] Examples include cyclopropanations, Diels-Alder cycloadditions and photohydroxymethylation. Many of these reactions proceed with high regio- and stereoselectivity (Scheme 3.6c).

3.3.4 6-Deoxy-hex-5-enopyranose derivatives

The preparation of 6-deoxy-hex-5-enopyranoses needs special attention since these sugar derivatives are important starting materials for the preparation of carbocyclic compounds (Scheme 3.7).[32] 6-Deoxy-hex-5-enopyranoses are enol ethers but, unlike glycals, the exocyclic double bond is located between C(5) and C(6). They can easily be prepared by

(a) Elimination of α-disulfonates

(b) Radical mediated elimination of dithiocarbonates

(c) Addition reactions to sugar double bonds

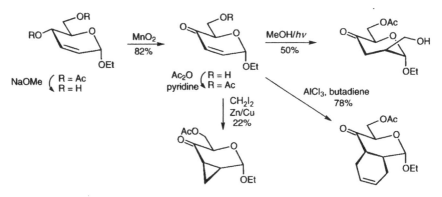

Scheme 3.6 Saccharides that have isolated double bonds.

elimination of a 6-deoxy-6-iodo-pyranoside. Treatment of the unsaturated sugar derivative with mercury(II) salts in refluxing aqueous acetone
results in the formation of β-hydroxycyclohexanones. Probably, the
reaction proceeds by hydroxymercuration of the double bond followed
by hemiacetal ring-opening to give a linear mercury-containing inter-

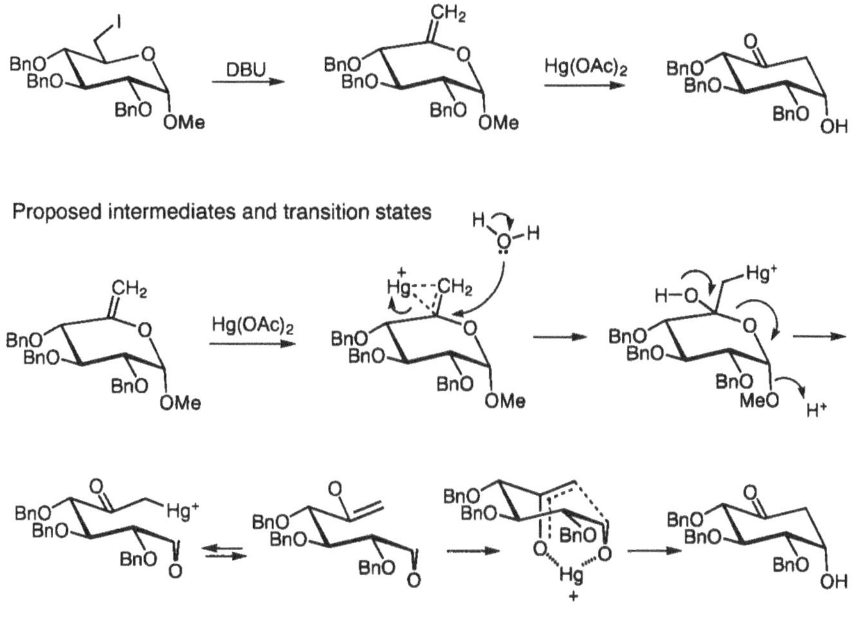

Scheme 3.7 The chemistry of 6-deoxy-hex-5-enopyranosides.

mediate which cyclises by intramolecular aldol-type reaction. The regioselectivity of the hydroxymercuration originates from assistance of an endocyclic oxygen lone pair in the departure of the cyclic mercury ion. A notable feature of this reaction is that it proceeds with high relative stereoselectivity, a *trans* relationship emerging between the newly formed asymmetric centre and the former C(3) position. It has been proposed that the stereochemical outcome of the reaction originates from a cyclic intermediate.

3.4 Deoxy sugars

3.4.1 Introduction

The deoxy sugars are a class of saccharides in which one or more hydroxyl group is replaced by a hydrogen atom. Deoxy aldoses are prevalent constituents of many naturally occurring oligosaccharides.[33] Some deoxy sugars, such as rhamnose (6-deoxy-L-mannose) and fucose (6-deoxy-L-galactose) are commercially available; others have to be prepared synthetically. Deoxygenation of sugar hydroxyls is most commonly performed by metal hydride reduction of halides, sulfonates or epoxides but radical mediated deoxygenations have also been reported. The preparation of 2-deoxy-glycosides is discussed in Chapter 4.

3.4.2   Reduction of halides, sulfonates and epoxides

The reduction of halides or sulfonates is one of the most conventional methods for the synthesis of deoxy sugars. This method is particularly useful for the deoxygenation of primary hydroxyls of carbohydrates. Reduction has been accomplished employing one of the following three methods.

- Halides and sulfonates can be reduced by catalytic hydrogenolysis using palladium on charcoal[34] or Raney nickel[35] (Scheme 3.8a). The reaction needs to be performed in the presence of a base because the reduction results in the formation of a mineral acid.
- Radical-induced reduction of halides with tributyltin hydride affords the corresponding deoxy sugars.[36] This method works equally well for primary and secondary halides but is not feasible with sulfonates (Scheme 3.8b). Interestingly, radical reduction of anomeric bromides containing an acyl or phosphotriester moiety at C(2) results in migration of the ester functionality to give the corresponding 2-deoxy-glycosyl ester or phosphate, respectively.[36d, 37]
- Treatment of halides or sulfonates with hydride donors such as tetrabutylammonium borohydride,[38] lithium aluminium hydride,[39] lithium triethylborohydride[40] or sodium borohydride generate deoxy sugar derivatives (Scheme 3.8c).[41] When sodium borohydride is employed, a transition metal catalyst ($PdCl_2$ or $NiCl_2$) may be added.

An alternative method[17c,d,g] for the preparation of 6-deoxy sugars is based on the elimination of 6-bromo-6-deoxy pyranosides to give 5,6-unsaturated derivatives followed by catalytic hydrogenation of the double bond to give methyl 2,6-di-deoxy-3-O-methyl-β-L-*lyxo*-hexopyranoside (Scheme 3.8d). In this manner, an inversion of configuration at C(5) is accomplished, converting a D into an L saccharide. Note that the starting material has a $^4C_1$ conformation and the product a $^1C_4$ conformation. The stereochemical outcome is based on complexation of the palladium catalyst with the hydroxyl at C(4) followed by delivery of the hydrogen to the bottom face of the molecule. Thus, it is crucial for a high diastereoselectivity that there is a free hydroxyl at C(4) of the elimination product.

As mentioned earlier, the introduction of deoxy moieties by reduction has mainly been applied at primary positions. This fact can be attributed

(a) Catalytic hydrogenation of halides and sulfonates

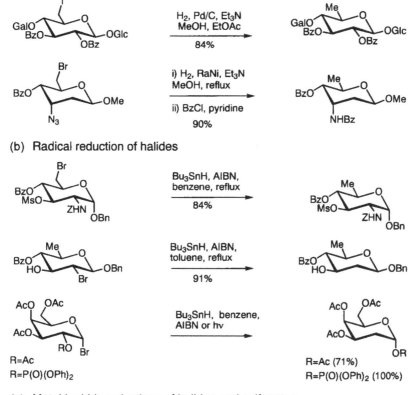

(b) Radical reduction of halides

(c) Metal hydride reductions of halides and sulfonates

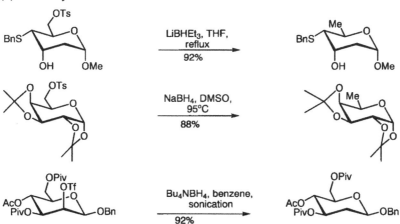

Scheme 3.8 Deoxygenation of saccharides.

(d) Pd-catalysed hydrogenation of double bonds

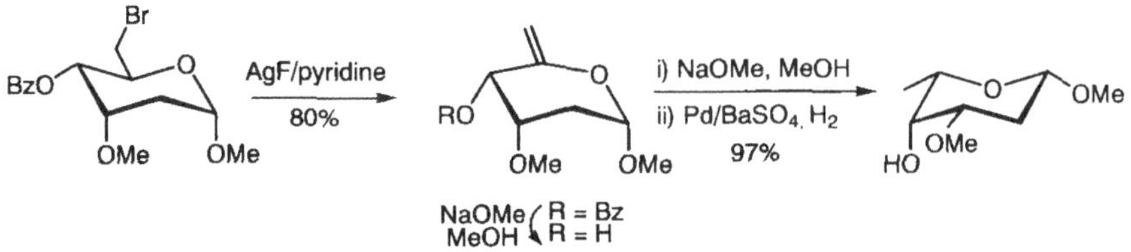

NaOMe ⎧ R = Bz
MeOH ⎩ R = H

(e) Metal hydride reduction of epoxides

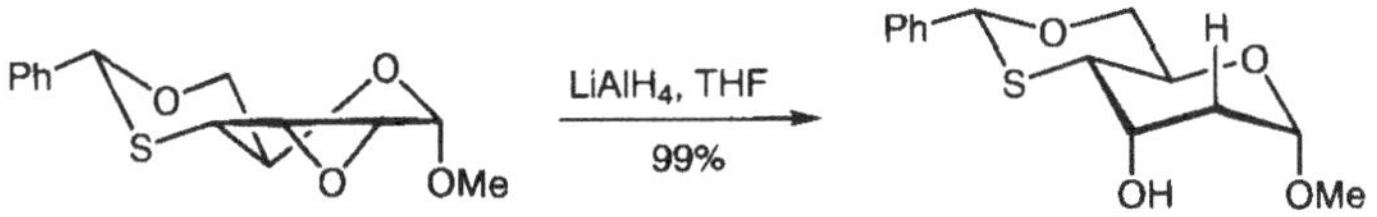

Scheme 3.8   (Continued).

to the relatively high reactivity of primary halides and sulfonates compared with that of secondary halides and sulfonates. Secondary deoxy functions can be obtained by reductive opening of epoxides with lithium aluminium hydride,[40b, 42] tetrabutylammonium borohydride[38a] or lithium triethylborohydride.[40a,c] In general, epoxide openings give *trans*-diaxial products (Scheme 3.8e).[40b] Interestingly, a 6-*O*-tosyl group can be reduced concomitantly with the opening of a 3,4-epoxide,[40a,c] whereas a 6-bromo functionality can be reduced selectively in the presence of a 2,3-anhydro functionality.[38a]

3.4.3   Radical deoxygenation of thiocarbonyl derivatives[43]

In 1975, Barton and McCombie described[44] the tributyltin hydride mediated reduction of thiocarbonyl derivatives of (sugar) hydroxyls (Scheme 3.9a) which proved especially suitable for the deoxygenation of secondary alcohols. The reaction starts by an attack of a tributyltin radical on the thiocarbonyl moiety with formation of an intermediate radical having a tin–sulfur bond. This intermediate fragments into the desired carbon radical and a thiocarbonyltin derivative. Finally, the radical is reduced by hydrogen atom transfer to give the desired product with reformation of a tributyltin radical. Thiocarbonyl derivatives, which have been widely employed, are *S*-methyldithiocarbonate,[45] imidazoylthiocarbonyl[46] and phenoxythiocarbonyl.[47] The latter group is more reactive towards reduction and has also been used for the deoxygenation

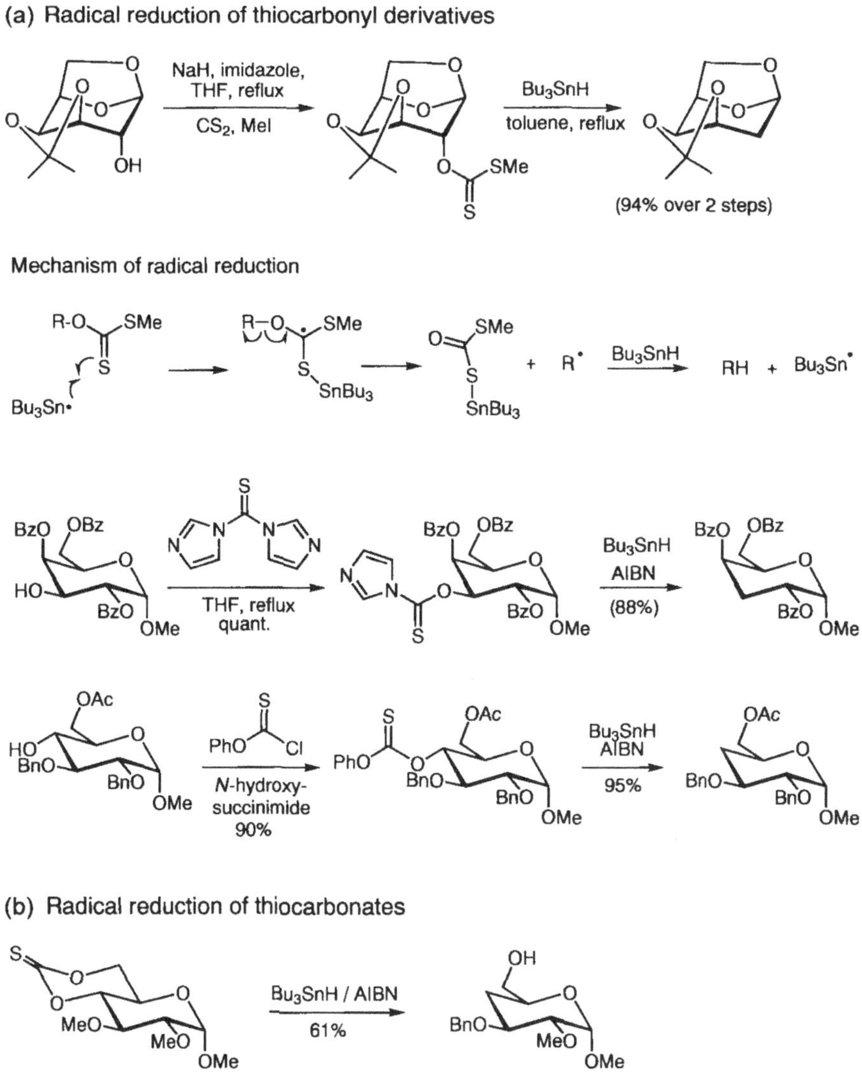

(a) Radical reduction of thiocarbonyl derivatives

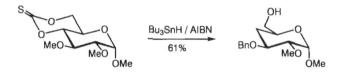

(94% over 2 steps)

Mechanism of radical reduction

(b) Radical reduction of thiocarbonates

Scheme 3.9 Radical deoxygenation of thiocarbonyl derivatives.

of primary sugar hydroxyls. Substituting the phenyl ring with fluorine can further enhance the reactivity of the phenoxythiocarbonyl group. Thus, pentafluorophenoxythionocarbonyl esters are reduced considerably faster than are the corresponding phenoxy derivatives.[48] Although a hydroxyl group vicinal to a thiocarbonyl group requires protection, other alcohol functions may be left unprotected. Radical deoxygenation is compatible with most commonly employed protective groups such as benzyl ethers, esters, acetals and ketals, and also with the p-toluenesulfonyl group. Thiocarbonyl esters can be introduced regioselectively[49] by the dibutyltin oxide method (Chapter 2). In addition, it has been shown that thioacylation of unreactive hydroxyl functions (e.g. 2-OH of galactopyranoside) can be accomplished via the corresponding tributyltin ether.

In the original procedure of Barton and McCombie the reduction was performed with tributyltin hydride,[50] but now the reaction is generally performed in the presence of a radical initiator such as 2,2'-azobisisobutyronitrile (AIBN), 2,2'-azobis(2-methylpropionitrile)[46b] or ultraviolet (UV) light. It has been reported[47c, 48c] that the addition of a radical initiator considerably decreases the reaction time, while the yields are increased. Less toxic reducing agents such as triethylsilane, phenylsilane[51] and tris(trimethylsilyl)silane[52] have also been employed.

Cyclic thiocarbonates offer another class of substrates for radical deoxygenation (Scheme 3.9b). In particular, thiocarbonates formed from a diol derived from a primary and secondary hydroxyl are of particular interest, since they can be deoxygenated regioselectively with tributyltin hydride and AIBN.[53] In these cases, the secondary position is deoxygenated owing to the higher stability of secondary over primary radicals. As expected, radical reduction of thiocarbonates derived from two secondary hydroxyls leads to a mixture of deoxygenated isomers.[52b,53]

3.5 Amino sugars

3.5.1 Introduction

Amino sugars are widely distributed in living organisms and occur as constituents of glycoproteins,[54] glycolipids and proteoglycans.[55] Furthermore, amino sugars are essential units of various antibiotics such as amino glycosides, macrolides, anthracyclines and cyclopeptide antibiotics.[56]

Several amino sugars are commercial available and the reactive amino functionality requires protecting if they are to be utilised synthetically. The N-phthalamido group is the most commonly applied amino protecting group, especially for amino groups at the C(2) position. Apart from protected amines and amides, azido groups are often used as an amino masking functionality.

Several efficient methods for the preparation of amino sugars have been developed[57] and the following general methods can be identified: nucleophilic displacement reactions, epoxide opening, additions to glycals, reduction of oximes and intramolecular substitutions.

Amino protecting groups were briefly discussed in Chapter 2, and the use of amino-containing glycosyl donors for oligosaccharide synthesis is covered in Chapter 4.

3.5.2 The preparation of amino sugars by nucleophilic displacement

One of the most widely employed methods for the introduction of an amino functionality is by nucleophilic substitution of an appropriate leaving group (i.e. sulfonates or halides) with nitrogen-containing nucleophiles, such as ammonia, hydrazine and phthalimide or azide ions. Generally, polar aprotic solvents are used and dimethylformamide (DMF) is often the solvent of choice. When the displacement is performed with tetrabutylammonium azide, the reaction can be conducted in toluene or benzene.[58] Other factors that determine the success of displacement reactions include the nature of the leaving group, the type of attacking species and the structure of the substrate.

Primary leaving groups are more reactive towards S_N2 displacements than are secondary leaving groups, although the reactivity can be strongly influenced by the configuration of the sugar. For example, displacements at the C(6) position of galactose can be cumbersome owing to the axially orientated substituent at C(4) (see Section 3.2.3).[59] When a primary amine is the desired product, chlorides or bromides often are sufficiently reactive leaving groups[60] (Scheme 3.10a).

When a secondary amine is required, mesylates,[61] tosylates[62] and triflates[63] are most commonly employed as leaving groups. These reactions proceed with inversion of configuration (Scheme 3.10b). Mesylates and tosylates can be obtained cheaply by treatment of an alcohol with sulfonyl chloride in pyridine. Triflation, however, is an expensive process that is performed by reaction of an alcohol with triflic anhydride in pyridine. In addition, unlike triflates, mesylates and tosylates are easier to handle, have better shelf-lives and can be purified by silica gel column chromatography. However, triflates can be up to 10^3 times more reactive towards substitution than tosylates.[64] In many cases, superior results are obtained when triflates are used instead of the corresponding mesylates or tosylates. In this respect, the sulfonate leaving group imidazolylsulfonate is worth discussing.[36c] Imidazylates are easily and cheaply prepared and are relatively stable and have an excellent shelf-life. Furthermore, they are reported to be more readily displaced by nucleophiles than the corresponding mesylates or tosylates and have reactivities comparable to that of triflates.

The reactivity of a leaving group also depends on the substitution and configuration of the saccharide.[59] For example, displacement at the C(2) position of mannosides is hindered by unfavourable dipolar interactions

(a) Nucleophilic substitution of primary bromide

(b) Nucleophilic substitution of triflates

(c) Nucleophilic ring opening of epoxides

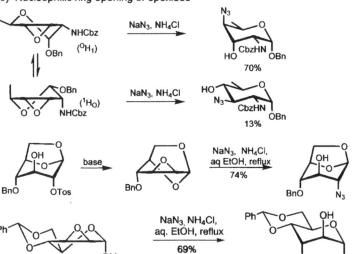

Scheme 3.10 Preparation of azido sugars by nucleophilic substitution.

(d) Nucleophilic ring opening of cyclic sulfates

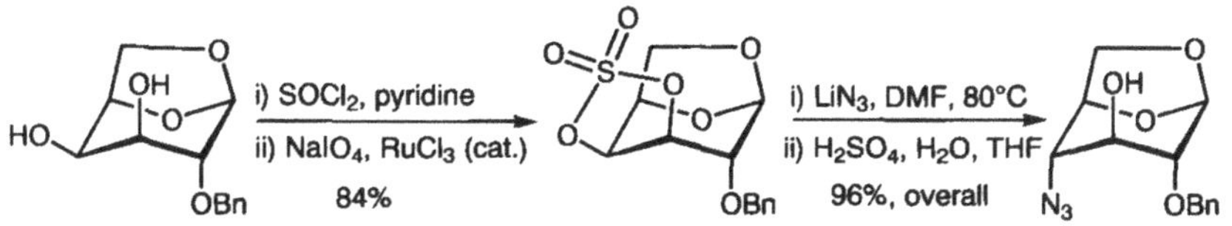

i) SOCl₂, pyridine
ii) NaIO₄, RuCl₃ (cat.)
84%

i) LiN₃, DMF, 80°C
ii) H₂SO₄, H₂O, THF
96%, overall

Scheme 3.10 (Continued).

developed in the transition state, which in turn are influenced by the nature and configuration of the anomeric substituent. For example (Scheme 3.10b), reaction of an α-mannoside with lithium azide gave the desired 2-azido-2-deoxy-D-mannoside in 39% yield together with an elimination product (44%),[58] whereas a similar reaction with a β-mannoside afforded the 2-azido-2-deoxy-glucose derivative in 80% yield.[61b] In both cases, the triflate and the proton at C(3) occupy axial positions and such an arrangement is favourable for a β elimination. In the case of an axial anomeric substituent (α-glycoside), the incoming nucleophile experiences unfavourable dipolar and steric interactions (see Section 3.2.3). These effects make the alternative β elimination the preferred mode of reaction. In the case of the β anomer, the incoming nucleophile does experience fewer unfavourable steric and electronic interactions and as a result the reaction proceeds smoothly through an S_N2 mechanism.[65] 1,6-Anhydro mannosides are also suitable derivatives for the synthesis of 2-azido-2-deoxy-glucose.[47c, 63c,e,f] The $^1C_4$ conformation of these sugar derivatives orientate C(2) leaving groups in an equatorial position disfavouring β eliminations. In addition, unfavourable steric and electronic interactions are precluded in the transition state and therefore displacement of the 2-O-triflate of 1,6-anhydro-β-mannoside gives substitution products in high yields.

Azide ions are by far the most common nucleophilic species employed in substitution reactions for the preparation of amino sugars. An azido moiety is stable under many reaction conditions but can be reduced to an amino group by a variety of reagents. The nucleophilicity of azido ions can be increased by the addition of a suitable crown ether to complex the counterion.[36c, 63b] In the past, ammonia and hydrazine were used as nucleophiles to overcome unfavourable dipolar interactions that arise when charged nucleophiles were used. However, a drawback of the use of these nucleophiles is that the product is still nucleophilic and can perform a second displacement. Phthalimide ions have successfully been applied in displacement reactions to yield a protected amino sugar derivative.[58, 63f]

Nucleophilic ring-opening of epoxides with nitrogen nucleophiles provides another route to amino sugars. Analogous to the above-discussed nucleophilic substitutions, epoxide openings are predominantly conduced with azide ions. The regioselectivity of the cleavage of an aldose epoxide is determined by the propensity of the reaction to proceed *trans*-diaxially.[16] Epoxides of monocyclic aldopyranoses can adopt two half-chair conformations. *Trans*-diaxial opening of each of these conformers will lead to different regioisomeric products. For example, treatment of benzyl 3,4-anhydro-2-benzyloxycarbonylamino-2,6-dideoxy-α-D-allopyr-anoside (Scheme 3.10c) with sodium azide gave the gulo (70%) and gluco epimers (13%).[17b] The product ratio indicates that the preferred conformation of reaction is the $^{O}H_1$ half-chair, in which the alkyl substituent at C(5) is positioned equatorially. Indeed, the general preference for a pyranose ring to adopt a conformation whereby the C(5) substituent is equatorially positioned substantiates the observed product ratio (see Chapter 1).

In general, the regioselectivity and stereoselectivity of sugar epoxide rings can be improved by the introduction of additional rigidity in the starting material forcing the pyranose ring in one particular conformation. For example, reaction of conformationally rigid 1,6-anhydro- or 4,6-*O*-benzylidene sugar epoxides furnish a single product in most cases.

During the opening of an epoxide, an alkoxide is liberated which may then be the cause of undesired side reactions. Ammonium chloride is often added to the reaction mixture to neutralise the alkoxide produced.

In an analogous fashion to epoxide ring-opening, nucleophilic ring-opening of cyclic sulfites,[66] sulfates[67] and sulfamidates[68] with azide ions proceeds in a highly regioselective manner affording mainly the expected *trans*-diaxial products (Scheme 3.10d). In general, cyclic sulfites react slower and cyclic sulfates faster than a corresponding epoxide. Furthermore, the cyclic sulfites or sulfates are prepared from corresponding *cis* diols, whereas epoxides are usually synthesised from *trans* diols. Cyclic sulfates can conveniently be prepared by the reaction of a diol with thionyl chloride to give a cyclic sulfite which can be oxidised to a cyclic sulfate using a catalytic amount of ruthenium trichloride and a stochiometric amount of sodium periodate as the cooxidant. Cyclic sulfates are much more reactive than their acyclic counterparts. For example, it has been shown that cyclic sulfates in an alkaline medium hydrolyse 10^7 times faster than the corresponding acyclic analogues. The high reactivity of cyclic sulfates has been attributed to ring strain, even though the origin of this ring stain is not well understood.

3.5.3 Addition to glycals

In 1979, Lemieux and Ratcliffe described[69] the azidonitration of peracetylated galactal leading to 2-azido-2-deoxy-galactosyl nitrates (Scheme 3.11a). The azidonitration is initiated by the oxidation of an azide ion with Ce(IV) to give an azido radical and Ce(III). Addition of the azido radical to the double bond of the glycal results in an anomeric radical and an azido substituent at C(2). The regioselectivity can be explained by resonance stabilisation of the radical by the endocyclic oxygen. Next, a second electron transfer gives an oxocarbenium ion, which in turn reacts with nitrate to give a 2-azido-glycosyl nitrate. The process is rather stereoselective for galactal derivatives, presumably because of the quasi-axial C(4) substituent, affording mainly products with the C(2) azido moiety in an equatorial position. In general, azidonitration of other glycal derivatives gives lower stereochemical outcomes. The ratio between equatorial and axial substitution depends

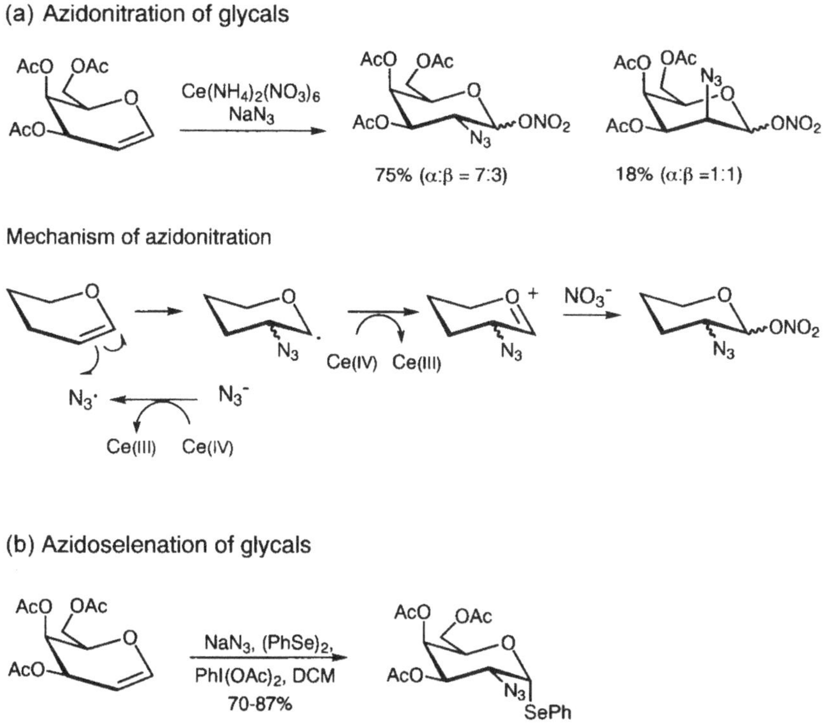

(a) Azidonitration of glycals

75% (α:β = 7:3)   18% (α:β =1:1)

Mechanism of azidonitration

(b) Azidoselenation of glycals

70-87%

Scheme 3.11 Preparation of azido sugars by azidonitration and selenation.

on the reaction temperature, solvent and ratio of reactants employed.[70] The anomeric nitrate can be readily replaced by a halide, acetyl or hydroxyl[69, 71, 72] functionality. Azidonitration is compatible with a wide range of protecting groups, including benzyl ethers, *tert*-butyldimethylsilyl ethers, methoxyethoxymethyl ethers, trityl ethers, benzylidene acetals, pyruvate acetals and isopropylidene ketals.

Several groups have reported the azidophenylselenylation of glycals via radical azido addition.[73] In this case, sodium azide and diphenyl diselenide in the presence of (diacetoxy)iodobenzene reacts with a peracetylated galactal to give stereoselectively the α-phenylselenyl glycoside in high yield (Scheme 3.11b). An epimeric mixture of products with gluco and manno-configuration was obtained when peracetylated glucal was used as the starting material. The procedure is low yielding when benzyl ethers or a benzylidene acetal are present probably because of oxidative cleavage of the protecting groups. However, azidophenylselenylation of perbenzylated glycals may successfully be achieved by using azidotrimethylsilane, tetra-*n*-butylammonium fluoride and *N*-phenylselenophthalimide.

3.5.4 Reduction of oximes

Oxime formation from alduloses and subsequent reduction represents yet another route to amino sugars.[61a, 74] The stereochemical outcome of the reduction depends *inter alia* on the reducing agent employed. For example, hydrogenation with Adams' catalyst (PtO_2) of the oxime in Scheme 3.12a furnished, after acetylation, only the L-*ribo* diastereoisomer with the amino group in an axial orientation. The stereochemical outcome of this reduction is believed to arise from complexation of the platinum with the sterically more accessible underside of the oximino moiety. Hydrogen delivery from this face of the molecule would lead to a product with an amino group in an axial orientation. The same product could be obtained when the oxime was treated with lithium aluminium hydride or sodium cyanoborohydride. In these reductions, equatorial attack of the hydride donor occurred. On the other hand, an L-*arabino* derivative, with an equatorial amino substituent, was obtained as the main product when the acetylated oxime was reduced with borane–tetrahydrofuran complex (Scheme 3.12b).[47] The latter stereoselectivity can be explained by coordination of the borane with the acetate group at O(4) followed by intramolecular hydride delivery from the sterically accessible top face (axial hydride transfer). Alternative, hydride delivery from the bottom face (equatorial attack) is hindered by the hydrogen at C(4). The resulting acetylated hydroxylamine was further reduced to an amino moiety.

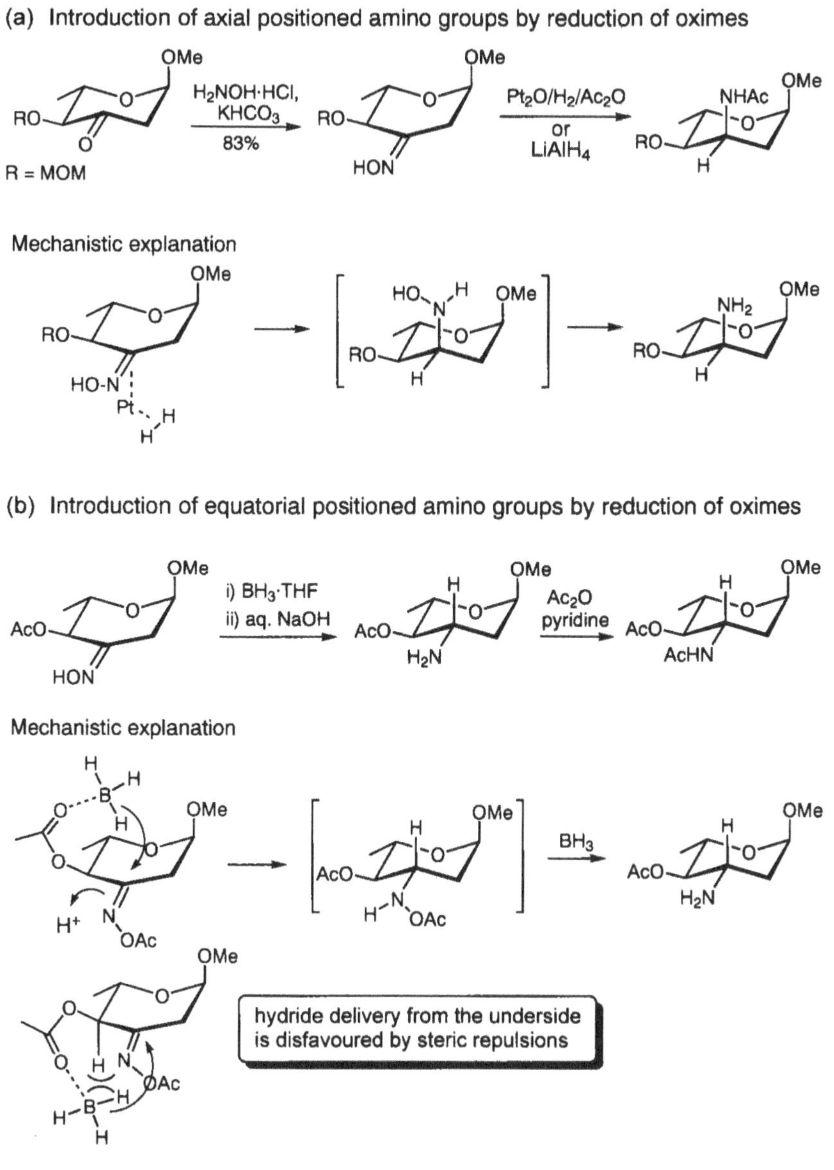

(a) Introduction of axial positioned amino groups by reduction of oximes

Mechanistic explanation

(b) Introduction of equatorial positioned amino groups by reduction of oximes

Mechanistic explanation

hydride delivery from the underside is disfavoured by steric repulsions

Scheme 3.12 Preparation of amino sugars by reduction of oximes.

It has been shown[75] that the stereochemical outcome of the reduction of 2-oxime sugars is influenced by the configuration of the anomeric centre. An interesting approach developed by Lichtenhaler et al.,[75a, 76] entails the use of 2-(benzoyloxyimino)glycosyl bromides as glycosyl donors. Stereoselective formation of β-glycosidic linkages follows stereo-selective reduction of the oxime to an amino functionality, and provides a route to 2-amino-2-deoxy-β-D-mannoside-containing oligosaccharides.

3.5.5 Intramolecular substitutions

Intramolecular substitutions offer a convenient and stereoselective method for the introduction of amino functionalities. A strategy for the preparation of vicinal cis-hydroxy amino moieties entails the halocyclisation of allylic trichloroacetimidates.[77] Conversion of the hydroxyl of unsaturated sugar derivatives into a trichloroacetimidate, followed by N-bromosuccinimide (NBS) or N-iodosuccinimide (NIS) mediated intramolecular cyclisation, gives bromo- and iodo-oxazoline derivatives (Scheme 3.13a). The oxazoline can be hydrolysed with mild acid to unmask the amino functionality, and the halogen can be removed by treatment with tributyltin hydride.

Sodium hydride mediated cyclisation of 3-O-benzoylcarbamate-2-O-triflate glucopyranoside (Scheme 3.13b) led to the formation of an N-benzoyloxazolidone, which was hydrolysed by base to the 2-amino-2-deoxy mannoside. The intramolecular displacement could be achieved with various substrates,[78] although in some cases O- instead of N-cyclisation was observed, presumably because of steric hindrance. An interesting approach for the preparation of 2-azido-2-deoxy-saccharides involves (diethylamino)sulfur trifluoride (DAST) mediated 1,2-migration of an anomeric azide (Scheme 3.13c).[79] Treatment with DAST converts the C(2) hydroxyl of a mannosyl azide into a good leaving group, which results in azido-group migration to the C(2) position, presumably via a cyclic intermediate. Attack by fluoride on C(1) gives the 2-azido-2-deoxyglycosyl fluoride. The usefulness of this approach was demonstrated by the preparation of a number of important 2-azido glycosides, and in all cases complete inversion at C(2) being observed.[79,80]

3.6   Epoxy sugars[81]

Nucleophilic ring opening of epoxy sugars is a valuable method for the synthesis of many modified sugar derivatives. The reaction is accompanied by Walden inversion, and a wide range of nucleophiles can be used. The cyclic nature of epoxides renders the competing elimination process stereoelectronically unfavourable. For asymmetric epoxides, in principle, two regio-isomeric products can be formed; however, in

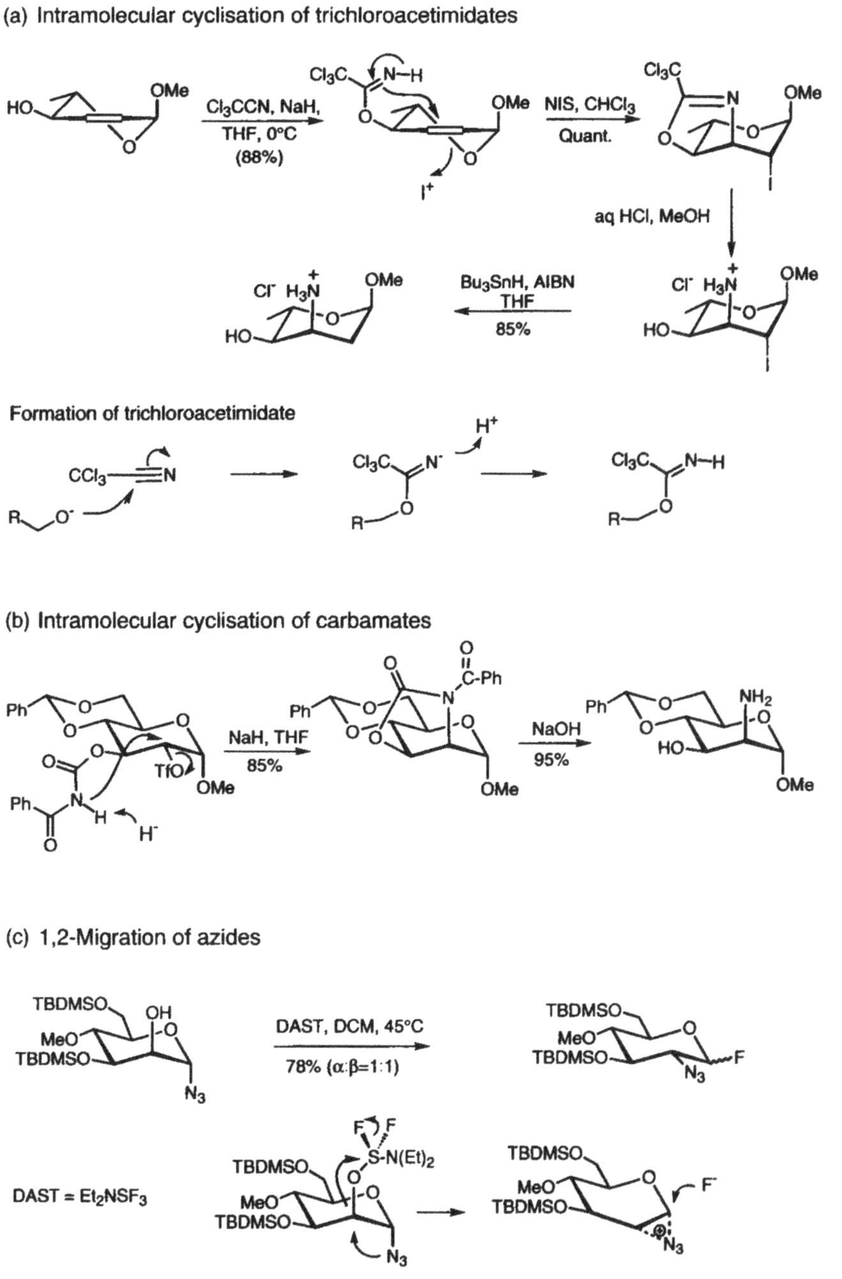

(a) Intramolecular cyclisation of trichloroacetimidates

Formation of trichloroacetimidate

(b) Intramolecular cyclisation of carbamates

(c) 1,2-Migration of azides

DAST = Et$_2$NSF$_3$

Scheme 3.13 Preparation of azido sugars by intramolecular substitution.

six-membered rings only the *trans*-diaxial products are usually seen. This feature needs further explanation. In the case of an unsymmetrical conformationally rigid epoxide, a diequatorial and diaxial product is theoretically possible (Scheme 3.14a). The transition state of the S_N2 reaction leading to the diequatorial products proceeds through a twist-boat conformation whereas the transition state leading to the diaxial product gives a chair. Often, the formation of a highly strained twist boat

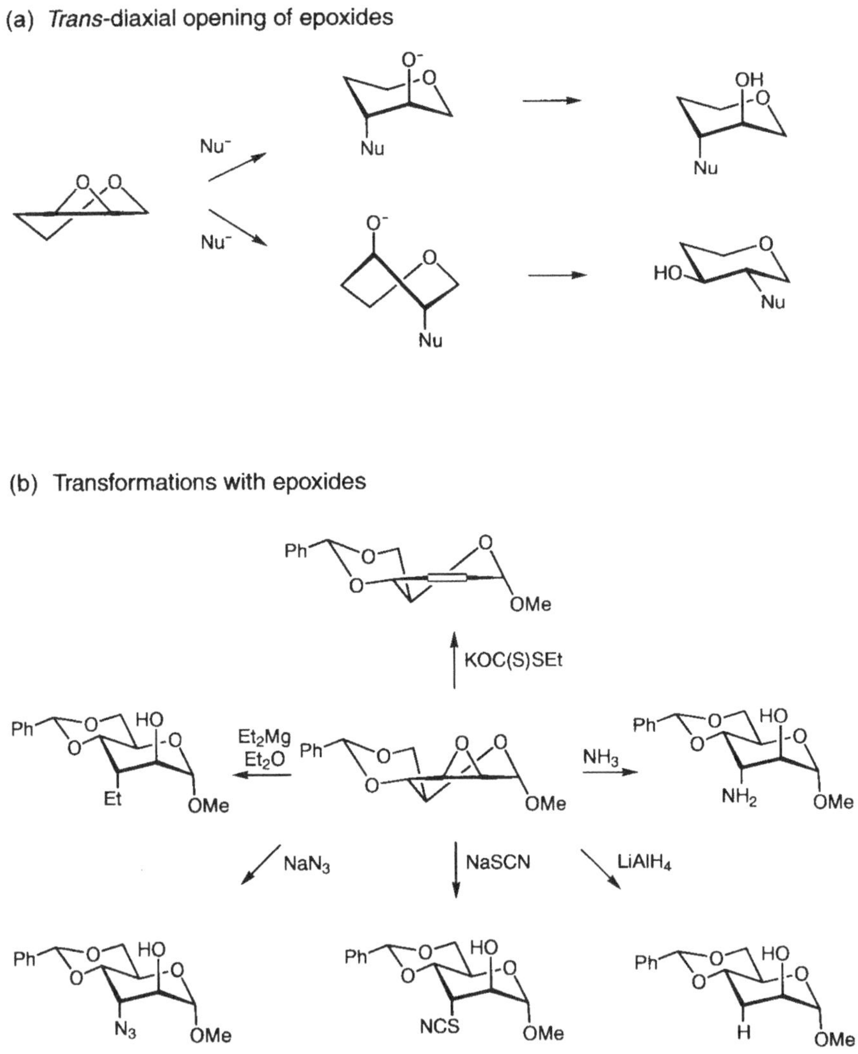

(a) *Trans*-diaxial opening of epoxides

(b) Transformations with epoxides

Scheme 3.14 The chemistry of epoxides.

is energetically unfavourable and hence the diequatorial product is usually not formed. Many examples of nucleophilic ring-opening of epoxides are covered in the previous sections of this chapter. Several other examples are depicted in Scheme 3.14b. In the following section, the most important synthetic procedures for the preparation of epoxy sugars will be covered.

Sugar epoxides are usually obtained by base treatment of α-hydroxysulfonates (Scheme 3.15a).[82] An important requirement of this reaction is that the hydroxyl and sulfonate ester are in a *trans* relationship. Such an arrangement can attain the coplanar S_N2 transition state necessary for epoxide formation. For the pyranoside ring, this means that both the groups should be in a *trans*-diaxial relationship. In some cases, it is necessary for the sugar ring to adopt a different conformation to satisfy this conformational arrangement of the two groups. For example, when the hydroxyl and sulfonyl group are in a *trans*-diequatorial arrangement, a conformational flip to a boat conformation allows an alkoxy and a sulfonate group to become *trans*-diaxial with respect to one other.

Sometimes, it is possible to prepare sugar epoxides from ditosylates (Scheme 3.15b).[83] For epoxide formation to occur, it is essential that one of the tosylates undergoes desulfonylation by S—O bond fission. Usually, this reaction occurs at the least hindered sulfonate. Then, the formed alkoxy will displace the remaining tosylate to provide the epoxide. Again, it is critical that the two reacting groups can attain a *trans*-diaxial relationship.

The Mitsunobu reaction has been successfully applied to the synthesis of carbohydrate epoxides directly from diols (Scheme 3.15c).[84] The more accessible and nucleophilic hydroxyl is converted into an alkoxyphosphonium ion, which in turn is intramolecularly displaced by a hydroxyl to give an epoxide. Again it is of critical importance that a coplanar S_N2 transition state be attained.

3.7 Sulfated saccharides

3.7.1 Introduction[85]

The glycosaminoglycans are a large family of highly glycosylated glycoproteins. They have an anionic and polydispersed polysaccharide moiety that is capable of binding to various proteins of biological interest, including antithrombin II and fibroblast growth factors. These polysaccharides are composed of disaccharide repeating units comprised of a 2-amino-2-deoxy sugar and a uronic acid moiety which are often sulfated. Many other sulfated oligosaccharides are known. Hormones

(a) Base treatment of α-hydroxysulfonates

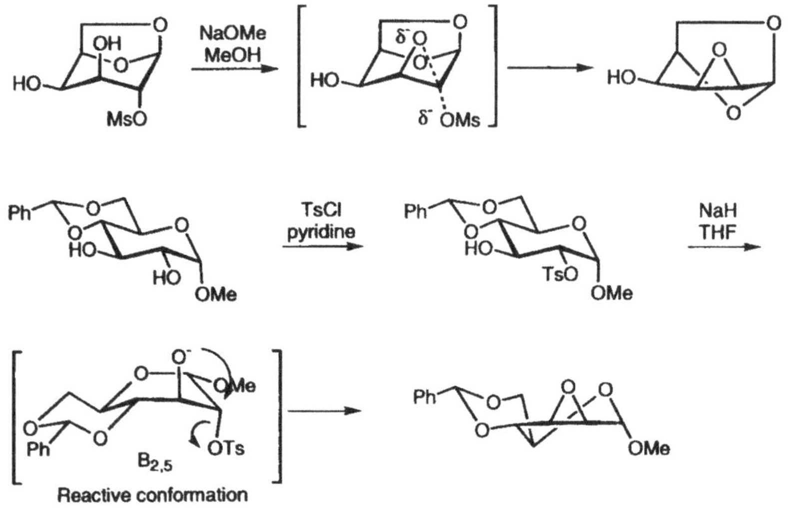

(b) Epoxides from ditosylates

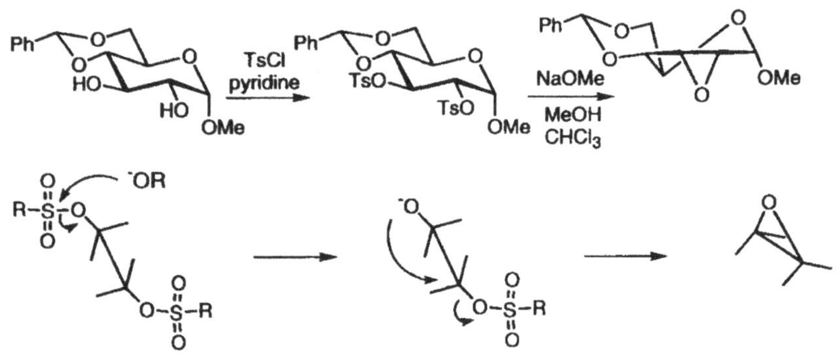

(c) Epoxide formation by a Mitsunobu reaction

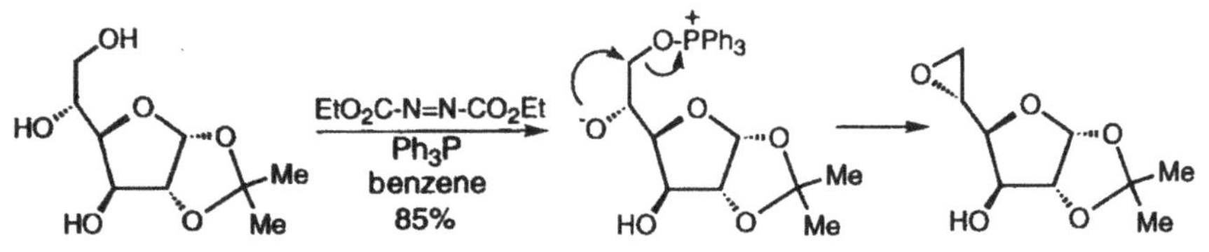

Scheme 3.15 Preparation of epoxy saccharides.

such as lutropin require sulfated saccharide moieties for expression of biological activity. Naturally occurring saccharides can be *O*- and *N*-sulfated.

In general, sulfates are introduced at the last stage of a synthetic scheme prior to deprotection. Such a strategy makes it unnecessary to introduce a sulfate in a protected form, although the use of phenyl sulfates has been described.[86] Although often the appropriately protected saccharide is treated with a sulfur trioxide containing reagent to yield the requisite sulfated sugar, some examples of regioselective sulfation have been reported which abate the necessity for extensive protecting group manipulations. Sulfate esters are stable under mild acid and basic conditions but are hydrolysed at low or high pH.[87]

3.7.2 *O and* N *sulfation*

Monosulfation of (oligo)saccharides is most commonly achieved by treatment of a suitably protected substrate containing one free hydroxyl function with a sulfur trioxide–pyridine[88] or sulfur trioxide–trialkylamine complex[89] in either pyridine or DMF (Scheme 3.16a). These reactions proceed smoothly and give the corresponding sulfates in good yields.

The simultaneous sulfation of several hydroxyls has been accomplished with the same reagents discussed above, that is, sulfur trioxide–pyridine[90] or sulfur trioxide–trialkylamine complexes.[91] Prolonged reaction times or repeated treatment with the sulfating agent may be needed to achieve complete sulfation.

The regioselective sulfation of a 6-hydroxyl of a partially protected saccharide can be achieved by employing the above-mentioned reagents, albeit at lower reaction temperatures.[89a,h,i, 92]

As shown in Scheme 3.16b, amino groups can be selectively sulfated in the presence of hydroxyl functions by employing the usual sulfur trioxide methodology but conducting the reaction in an alkaline (pH > 9), aqueous medium.[89c, 91b,d,e]

The acidity of sulfates may cause hydrolysis of glycosidic linkages and therefore the products are usually isolated as salts.

3.8 Phosphorylated saccharides

3.8.1 *Introduction*[93]

Sugar phosphates are widely found in nature and several classes can be identified. First, a distinction needs to be made between anomeric and non-anomeric sugar phosphates. Furthermore, natural sugar phosphates

(a) O-Sulfation

(b) N-Sulfation

occur as monoesters and diesters. In addition, phosphotriesters are often synthetic intermediates in the preparation of monoesters or diesters. Phosphotriesters are relatively labile intermediates that can be cleaved by base or acid and can easily migrate to other hydroxyls. The preparation of the different classes of phosphate esters requires different methodologies.

3.8.2 Non-anomeric sugar phosphates

Several chemical methods are available for the phosphorylation of suitably protected saccharides. The following approaches, classified according to the initial phosphorylation product, can be distinguished: phosphotriester, phosphite triester and H-phosphonate methods.

Phosphotriester methods

Diphenyl phosphorochloridate is one of the most widely employed phosphorylating reagents for the preparation of phosphomonoesters (Scheme 3.17a).[94] The chloride acts as a leaving group that can be displaced by hydroxyls. The phenyl groups act as protecting groups that

(a) Preparation of phosphomonoester

(b) Preparation of phosphodiester

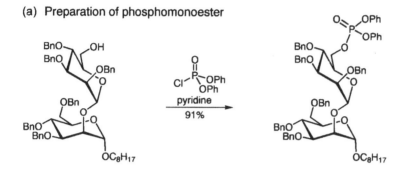

Scheme 3.17 Phosphorylation by phosphotriester method.

can be easily removed by benzaloxime and tetramethyl guanidine. This reagent also permits primary hydroxyl groups to be selectively converted into a phosphotriester in the presence of secondary hydroxyl groups.[95] Furthermore, the simultaneous phosphorylation of several hydroxyls can be achieved by using an excess amount of reagent. However, the formation of cyclic phosphotriesters may be a competitive side reaction when two adjacent hydroxyl groups in a saccharide have a *cis* configuration. In some cases, N,N-dimethylaminopyridine (DMAP) or imidazole is added as a promoter.

Properly protected phosphorochloridates have been employed, and examples include; bis(2,2,2-trichloroethyl) phosphorochloridate[96] and dibenzyl phosphorochloridate.[97] Phosphoroiodates prepared *in situ* by the oxidation of the corresponding trialkylphosphites with iodine can also be used.[98]

Bifunctional phosphorylating reagents are employed when unsymmetrical phosphodiesters are the target compounds. In this case, the reagent contains two leaving groups and one protecting group. A typical reaction sequence is depicted in Scheme 3.17b[99] and commences with the phosphorylation of a ribosyl sugar hydroxyl with bis[1-benzotriazolyl]-2-chlorophenylphosphate leading to an intermediate monophosphate. The remaining benzotriazolyl moiety of the monophosphate can be activated by the addition of N-methylimidazole, and reaction with the second ribosyl alcohol results in the formation of an unsymmetrical phosphotriester. Several other reagents have been reported that are based on a similar reaction sequence.[100]

Phosphite triester methods

Phosphites have a lower oxidation state (III) than corresponding phosphates (V) and are generally more reactive than phosphates. These derivatives can easily be oxidised to a phosphate. As illustrated in Scheme 3.18a,[101] phosphite triesters can be obtained by reaction of a sugar alcohol with dibenzyl diisopropylphosphoramidite[102] in the presence of $1H$-tetrazole. In this reaction, the mild acid $1H$-tetrazole protonates the the —N(i-Pr)$_2$ moiety converting it into a good leaving group, which is then displaced by an alcohol. Oxidation of the phosphite to afford the phosphotriester can be accomplished in the same pot with m-chloroperbenzoic acid,[102a] hydrogen peroxide,[102b] t-butylhydroperoxide[103] or iodine.[104] The benzyl protecting groups can be removed by catalytic hydrogenation.

The efficacy of bifunctional phosphitylating reagents has been widely demonstrated, *inter alia*, by the preparation of various phosphodiester-containing fragments of capsular polysaccharides,[103a,b,d,e,104,105] interglycosidic phosphodiesters[103d,e,106] and other biologically interesting

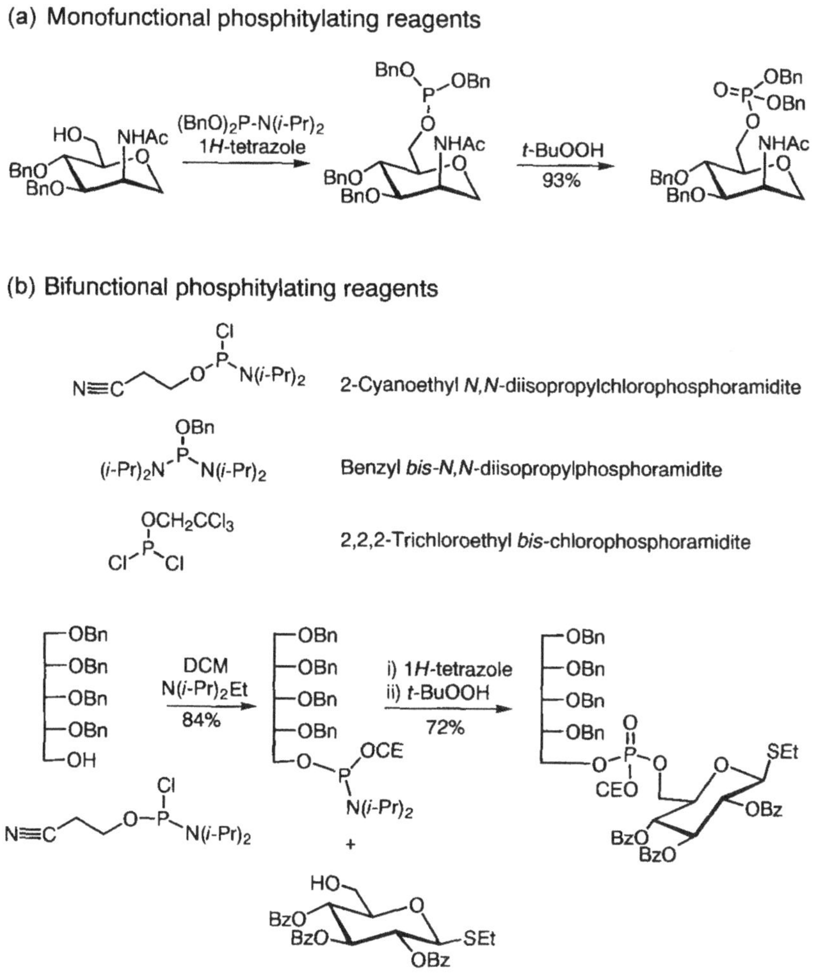

(a) Monofunctional phosphitylating reagents

(b) Bifunctional phosphitylating reagents

2-Cyanoethyl *N,N*-diisopropylchlorophosphoramidite

Benzyl *bis-N,N*-diisopropylphosphoramidite

2,2,2-Trichloroethyl *bis*-chlorophosphoramidite

Scheme 3.18 Phosphorylation by phosphite method.

saccharide phosphates.[101,103c,104c,107] In the first step of this type of methodology, an intermediate phosphoramidite (Scheme 3.18b)[103b] is formed by reaction of reagent 2-cyanoethyl *N,N*-diisopropylchlorophosphoramidite with a sugar alcohol in the presence of diisopropylethylamine (DIPEA). The resulting intermediate is sufficiently stable to be purified by silica gel column chromatography and stored for a considerable period of time. Next, the amidite is coupled with a second hydroxyl group in the presence of the mild acid 1*H*-tetrazole, and subsequent *in situ* oxidation of the obtained phosphite triester leads to the formation of the phosphotriester. The cyanoethyl protecting group can be removed by β elimination in the presence of a mild base.

H-*phosphonate methods*

An important advantage of the *H*-phosphonate method is that at no stage of the synthetic route, is a labile phosphotriester prepared.[103e,104b,108] The reaction of salicylchlorophosphite (Scheme 3.19) with a glucosamine derivative furnished a phosphite triester which was subsequently hydrolysed to the *H*-phosphonate monoester. The latter compound was coupled

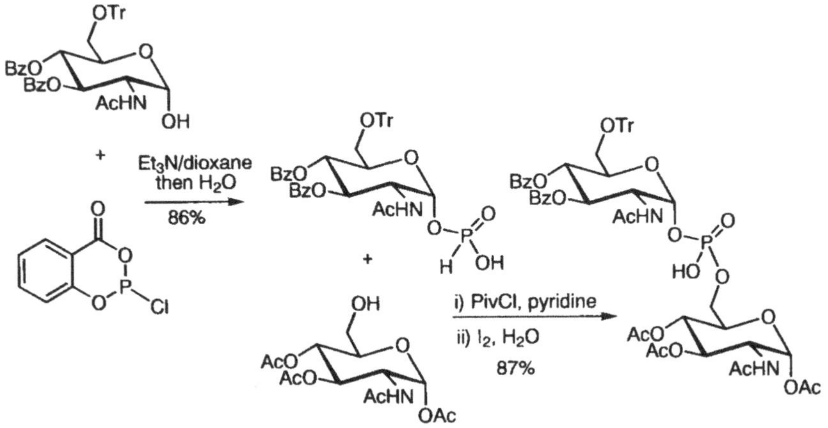

Scheme 3.19 Phosphorylation by the *H*-phosphonate method.

with another monosaccharide under the agency of pivaloyl chloride (PivCl) and the resulting *H*-phosphonate diester oxidised *in situ* to afford a phosphodiester.

Another reagent for the synthesis of *H*-phosphonates is tris(imidazolyl)phosphine,[109] prepared from trichlorophosphine and imidazole in the presence of triethylamine. Again, the intermediate phosphite derivative is hydrolysed and the obtained *H*-phosphonate is converted to a phosphodiester as described above.

3.8.3   Anomeric phosphates

Additional problems of anomeric phosphorylation are the control of anomeric selectivity and the increased lability of the products. The preparation of anomeric sugar phosphates can be accomplished via two distinct methods. The first commences with a free anomeric hydroxyl and can be considered as a standard phosphorylation. The second approach employs a substrate bearing a suitable leaving group at the anomeric centre and may be described as the glycosylation of a phosphate species.

Phosphorylation of anomeric hydroxyls
Not surprisingly the procedures described in Section 3.7.2 for the phosphorylation of nonanomeric hydroxyl groups, that is, the phospho-triester, phosphite triester, and *H*-phosphonate approaches, have been applied to the preparation of anomeric phosphates.[103d,e, 110]

An important aspect of the synthesis of anomeric phosphates is the control of the anomeric configuration. In the phosphotriester method the stereochemical outcome is mainly governed by the anomeric configuration of the starting material, although temperature and reaction time take on secondary effects.[111] In addition, the configuration of the saccharide (axial/equatorial substituents) determines the relative reactivity of the two anomers and hence effects the stereochemical course of the phosphorylation. For example, anomerisation of the β-D-gluco- or β-D-galactopyranosyl phosphotriesters to the thermodynamically more stable α anomers has been reported. This process can be avoided by immediate deprotection of the phosphotriesters to give the more stable phosphate monoester.

Anomeric phosphorylation by glycosylation methods
In the above-discussed phosphorylation methods, the sugar hydroxyl acts as the nucleophile and the phosphorylating agent as the electrophile. However, phosphorylation of the anomeric centre may also be performed by another approach where the saccharide contains a leaving group.

Bromides and chlorides have frequently been used as anomeric leaving groups, most commonly in combination with dibenzyl phosphate as the nucleophile to give anomeric phosphates.[110d,112] The phosphorylation may be activated by the addition of silver(I) carbonate or silver(I) oxide. Alternatively, a silver phosphate salt, prepared *in situ* from an appropriately protected phosphate and silver(I) nitrate, reacts readily with glycosyl halides. In addition, anomeric phosphorylation of glycosyl halides under phase-transfer catalysis has been described (Scheme 3.20). Most commonly, the sugar halides have been prepared from peracetylated substrates and hence the phosphorylations proceed by neighbouring group participation, leading to 1,2-*trans* substituted products. On the other hand, reaction of peracetylated α-L-fucopyranosyl bromide with dibenzyl phosphate furnished either an anomeric mixture of products or only the α anomeric product. Thus, the acetyl protecting group at C(2) does not perform efficiently neighbouring group participation or the initial formed β-phosphate anomerises to the thermodynamically more stable α-anomer. A more satisfactory outcome was obtained when

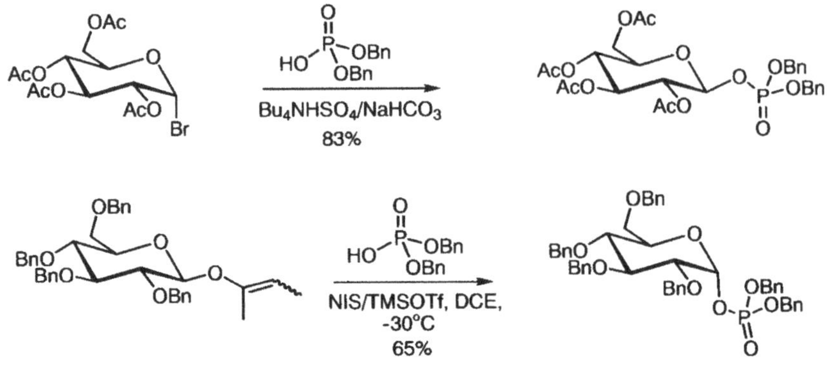

Scheme 3.20 Anomeric phosphorylation.

benzoates were used as protecting groups and in this case only the formation of the β anomer was observed.

Another interesting reaction is the anomeric phosphorylation of 2-acetamido-3,4,6-tri-O-acetyl-2-deoxy-α-D-glucopyranosyl chloride. Although it was shown[109] that treatment of the latter compound with dibenzyl phosphate, under phase-transfer conditions, gave only the corresponding oxazoline derivative, others found that oxazolines can be phosphorylated with dibenzyl phosphate[110,114,115] or 2,2,2-tribromoethyl-phosphoric acid[111] to afford anomeric phosphates. The β-phosphate is the initial product of this reaction, but prolonged reaction time afforded only the α anomer via anomerisation.

Among other leaving groups which have been employed for the synthesis of anomeric phosphates are the trichloroacetimidate,[116] 4-pentenyl,[117] ethylthio,[118] phosphorodithioate[112e] and 2-buten-2-yl.[119] Analogous to the above-described glycosylations, condensation of dibenzyl phosphate (DBP) with glycosyl donors bearing a participating group at C(2) furnish 1,2-*trans* products, and α-phosphates are mainly obtained when the C(2) position of the glycosyl donor is protected by a nonparticipating protecting group.

References

1. (a) A.A.E. Penglis, 1981, *Adv. Carbohydr. Chem. Biochem.*, 38, 195; (b) T. Tsuchiya, 1990, *Adv. Carbohydr. Chem. Biochem.*, 48, 91; (c) J.E.G. Barnett, 1967, *Adv. Carbohydr. Chem.*, 22, 177; (d) S. Hanessian, 1968, *Adv. Chem. Ser.*, 74, 159; (e) W.A. Szarek, 1973, *Adv. Carbohydr. Chem. Biochem.*, 28, 225.

2. N.F. Taylor, 1988, *Fluorinated Carbohydrates. Chemical and Biochemical Aspects, ACS Symposium Series*, 374.

3. T.J. Tewson, 1989, *Nucl. Med. Biol.*, 16, 533.

4. (a) A.K.M. Anisuzzaman and R.L. Whistler, 1978, *Carbohyd. Res.*, 61, 511; (b) C. R. Haylock, L.D. Melton, K.N. Slessor and A.S. Tracey, 1971, *Carbohyd. Res.*, 16, 375; (c) R.L. Whistler and A.K.M. Anisuzzaman, 1980, *Methods Carbohydr. Chem.*, VIII, 227; (d) J.P.H. Verheyden and J.G. Moffatt, 1972, *J. Org. Chem.*, 37, 2289.

5. S. Hanessian, M.M. Ponpidom and P. Lavallee, 1972, *Carbohydr. Res.*, 24, 45.

6. (a) J. Alais and S. David, 1992, *Carbohydr. Res.*, 230, 79; (b) P.J. Garegg and B. Samuelsson, 1980, *J. Chem. Soc. Perkin I*, 2866; (c) J. Kihlberg, T. Frejd, K. Jansson, A. Sundin and G. Magnusson, 1988, *Carbohydr. Res.*, 176, 271; (d) Z. Zhiyuang and G. Magnusson, 1995, *J. Org. Chem.*, 60, 7304.

7. H.H. von Brandstetter and E. Zbiral, 1980, *Helv. Chim. Acta*, 63, 327.

8. J.P.H. Verheyden and J.G. Moffatt, 1970, *J. Org. Chem.*, 35, 2319.

9. (a) T.G. Back, D.H.R. Barton and B.L. Rao, 1977, *J. Chem. Soc. Perkin I*, 1715; (b) S. Hanessian and N.R. Plessas, 1969, *J. Org. Chem.*, 34, 2163.

10. H.J. Jennings and J.K.N. Jones, 1965, *Can. J. Chem.*, 43, 3018.

11. (a) S.-H. An and M. Bobek, 1986, *Tetrahedron Lett.*, 27, 3219; (b) P.J. Card and G.S. Reddy, 1983, *J. Org. Chem.*, 48, 4734; (c) P.J. Card, 1983, *J. Org. Chem.*, 48, 393; (d) M.G. Straatman and M.J. Welch, 1977, *J. Labelled Compd. Radiopharm.*, 13, 210; (e) C.W. Somawardhana and E.G. Brunngraber, 1983, *Carbohydr. Res.*, 121, 51.

12. (a) R.S. Tipson, 1953, *Adv. Carbohydr. Chem.*, 8, 107; (b) D.H. Ball and F.W. Parrish, 1968, *Adv. Carbohydr. Chem.*, 23, 233; (c) D.H. Ball and F.W. Parrish, 1969, *Adv. Carbohydr. Chem.*, 24, 139; (d) A.C. Richardson, 1969, *Carbohydr. Res.*, 10, 395; (e) M. Miljkovic, M. Gligorijevic and D. Glisin, 1974, *J. Org. Chem.*, 39, 3223

13. (a) R.W. Binkley, M.G. Ambrose and D.G. Hehemann, 1980, *J. Org. Chem.*, 45, 4387; (b) R.W. Binkley and D.G. Hehemann, 1979, *Adv. Carbohydr. Chem.*, 24, 139.

14. (a) T. Tsuchiya, Y. Takahaski, M. Endo, S. Umezawa and H. Umezawa, 1985, *J. Carbohydr. Chem.*, 4, 587; (b) A. Hasegawa, M. Goto and M. Kiso, 1985, *J. Carbohydr. Chem.*, 4, 627; (c) A.B. Foster and R. Hems, 1967, *Carbohyd. Res.*, 5, 292.

15. (a) J. Pacak, Z. Tocik and M. Cerny, 1969, *Chem. Commun.*, 77; (b) A.D. Barford, A.B. Foster, J.H. Westwood, L.D. Hall and R.N. Johnson, 1971, *Carbohydr. Res.*, 19, 48; (c) P.A. Beeley, W.A. Szarek, G.W. Hay and M.M. Perlmutter, 1984, *Can. J. Chem.*, 62, 2709; (d) W.A. Szarek, G.W. Hay, B. Doboszewski and M.M. Perlmutter, 1986, *Carbohydr. Res.*, 155, 107.

16. N.R. Williams, 1970, in *Adv. Carbohydr. Chem. Biochem.*, (R.S. Tipson, ed.), Academic Press, New York, p. 109.

17. (a) S. Hanessian and N.R. Plessas, 1969, *J. Org. Chem.*, 1035; (b) A. Banaszek, Z. Pakulski and A. Zamojski, 1995, *Carbohydr. Res.*, 279, 173; (c) J. Boivin, C. Monneret and M. Pais, 1981, *Tetrahedron*, 37, 4219; (d) M. Du and O. Hindsgaul, 1996, *Carbohydr. Res.*, 286, 87; (e) T. Ekberg and G. Magnusson, 1993, *Carbohydr. Res.*, 146, 119; (f) D. Horton and W. Eckerle, 1975, *Carbohydr. Res.*, 44, 227; (g) C. Monneret, C. Conreur and Q. Khuonghuu, 1978, *Carbohydr. Res.*, 65, 35; (h) I. Pelyvás, A. Hasegawa and R.L. Whistler, 1986, *Carbohydr. Res.*, 146, 193; (i) J. Thiem and H. Karl, 1980, *Chem. Ber.*, 113, 3039; (j) T.D. Lowary and O. Hindsgaul, 1993, *Carbohydr. Res.*, 249, 163; (k) J. Thiem, M. Gerken and K. Bock, 1983, *Liebigs Ann. Chem.*, 462.

18. (a) R. Blattner and R.J. Ferrier, 1980, *J. Chem. Soc. Perkin I*, 1523; (b) R.J. Ferrier, S.R. Haines, G.J. Gainsford and E.J. Gabe, 1984, *J. Chem. Soc. Perkin I*, 1683.

19. W. Korytnyk, S. Valentekovic-Horvath and C.R. Petrie III, 1982, *Tetrahedron*, 38, 2547.

20. R.J. Ferrier, 1969, *Adv. Carb. Chem. Biochem.*, 24, 199.

21. W. Roth and W. Pigman, 1963, *Methods Carbohydr. Chem.*, 2, 405.

22. R.J. Ferrier and N. Prasad, 1969, *J. Chem. Soc. (C)*, 570.

23. F.W. Lichtenthaler, S. Ronninger and P. Jarglis, 1989, *Liebigs Ann. Chem.*, 1153.

24. D.E. Levy and C. Tang, 1995, *Tetrahedron Org. Chem. Ser.*, 13.

25. (a) J.C.-Y. Cheng and G.D.J. Daves, 1987, *J. Org. Chem.*, 52, 3083; (b) I. Ari, T.D. Lee, R. Hanna and G.D.J. Daves, 1982, *Organometallics*, 1, 742.

26. (a) S. Yougai and T. Miwa, 1982, *J. Chem. Soc. Chem. Comm.*, 68; (b) M. Brakta, F. Le Borgne and F. Sinou, 1987, *J. Carbohydr. Chem.*, 6, 307; (c) M. Brakta, P. Lhoste and D. Sinou, 1989, *J. Org. Chem.*, 54, 1890; (d) T.V. Rajanbabu, 1985, *J. Org. Chem.*, 50, 3642.

27. (a) L.V. Dunkerton, J.M. Euske and A.J. Serino, 1987, *Carbohydr. Res.*, 171, 89; (b) G. J. Engelbrecht and C.W. Holzapfel, 1991, *Heterocycles*, 32, 1267.

28. (a) E. Dubois and J.-M. Beau, 1990, *J. Chem. Soc. Chem. Comm.*, 1191; (b) E. Dubois and J.-M. Beau, 1992, *Carbohydr. Res.*, 228, 103; (c) R.W. Friesen and F.C. Sturino, 1990, *J. Org. Chem.*, 55, 2572; (d) M.A. Tius, G.J. Gomez, X.Q. Gu and J.H. Zaidi, 1991, *J. Am. Chem. Soc.*, 113, 5775; (e) R.W. Friesen and R.W. Loo, 1991, *J. Org. Chem.*, 56, 4821.

29. (a) L.H.B. Baptistella, A.Z. Neto, H. Onaga and E.A.M. Godoi, 1993, *Tetrahedron Lett.*, 34, 8407; (b) K. Takeo, T. Fukatsu and T. Yasato, 1982, *Carbohydr. Res.*, 107, 71.

30. D.H.R. Barton, D.O. Jang and J.C. Jaszberenyi, 1993, *Tetrahedron*, 49, 7193.

31. (a) B. Fraser-Reid, A. McLean, E.W. Usherwood and M. Yunker, 1970, *Can. J. Chem.*, 48, 2877; (b) B. Fraser-Reid and B.J. Carthy, 1972, *Can. J. Chem.*, 50, 2928; (c) B. Fraser-Reid, N.L. Holder and M.B. Yunker, 1972, *J. Chem. Soc. Chem. Commun.*, 1286; (d) J. L. Primeau, R.C. Anderson and B. Fraser-Reid, 1980, *J. Chem. Soc. Chem. Commun.*, 6.

32. (a) K. Sato, N. Kubo, R. Takada, A. Aqeel, H. Hashimoto and J. Yoshimura, 1988, *Chem. Lett.*, 1703; (b) R. Blattner, R.J. Ferrier and P. Prasit, 1980, *J. Chem. Soc., Chem. Commun.*, 944; (c) R. Blattner, R.J. Ferrier and S.R. Haines, 1985, *J. Chem. Soc. Perkin I*, 2413; (d) S. Adams, 1988, *Tetrahedron. Lett.*, 29, 6589.

33. (a) N.R. Williams and J.D. Wander, 1980, in *The Carbohydrates* (W. Pigman and D. Horton, eds.), Academic Press, New York, p. 761; (b) P. Collins and R. Ferrier, 1995, *Monosaccharides*, Wiley, Chichester.

34. (a) D.R. Bundle, M. Gerken and T. Peters, 1988, *Carbohydr. Res.*, 174, 239; (b) S. Hanessian and N.R. Plessas, 1969, *J. Org. Chem.*, 1045; (c) A. Malik, N. Afza and W. Voelter, 1983, *J. Chem. Soc. Perkin I*, 2103; (d) T.-H. Lin, P. Kovác and C.P.J. Glaudemans, 1989, *Carbohydr. Res.*, 188, 228.

35. D. Horton, G. Rodemeyer and H. Saeki, 1977, *Carbohydr. Res.*, 59, 607.

36. (a) F.-I. Auzanneau, H.R. Hanna and D.R. Bundle, 1993, *Carbohydr. Res.*, 240, 161; (b) Z. Györgydeák, 1991, *Liebigs Ann. Chem.*, 1291; (c) S. Hanessian and J.-M. Vatèle, 1981, *Tetrahedron Lett.*, 22, 3579; (d) K.C. Nicolaou, R.A. Daines, T.K. Chakraborty and Y. Ogawa, 1987, *J. Am. Chem. Soc.*, 109, 2821; (e) A. Koch and B. Giese, 1993, *Helv. Chim. Acta*, 76, 1687; (f) J. Thiem and B. Schöttmer, 1987, *Angew. Chem. Int. Ed. Engl.*, 26, 555.

37. (a) A. Koch, C. Lamberth, F. Wetterich and B. Giese, 1993, *J. Org. Chem.*, 58, 1083; (b) B. Giese, S. Gilges, K.S. Gröninger, C. Lamberth and T. Witzel, 1988, *Liebigs Ann. Chem.*, 615.

38. (a) K.-I. Sato, T. Hoshi and Y. Kajihara, 1992, *Chem. Lett.*, 1469; (b) K.-I. Sato and A. Yoshitomo, 1995, *Chem. Lett.*, 39.

39. (a) J. Thiem, V. Duckstein, A. Prahst and M. Matzke, 1987, *Liebigs Ann. Chem.*, 289; (b) M.J. Eis and B. Ganem, 1988, *Carbohydr. Res.*, 176, 316.

40. (a) H.H. Baer and D.J. Astles, 1984, *Carbohydr. Res.*, 126, 343; (b) A. Claßen and H.-D. Scharf, 1993, *Liebigs Ann. Chem.*, 183; (c) T.L. Lowary, E. Eichler and D.R. Bundle, 1995, *J. Org. Chem.*, 60, 7316.

41. (a) D.H.G. Crout, J.R. Hanrahan and D.W. Hutchinson, 1993, *Carbohydr. Res.*, 139, 305; (b) E.-P. Barrette and L. Goodman, 1984, *J. Org. Chem.*, 49, 176; (c) L.M. Lerner, 1993, *Carbohydr. Res.*, 241, 291; (d) T. Nishio, Y. Miyake, K. Kubota, M. Yamai, S. Miki, Y. Ito and T. Oku, 1996, *Carbohydr. Res.*, 180, 357; (e) J. Thiem and B. Meyer, 1980, *Chem. Ber.*, 113, 3067; (f) J. Thiem and A. Sievers, 1980, *Chem. Ber.*, 113, 3505.

42. K. Jones and W.W. Wood, 1988, *J. Chem. Soc. Perkin I*, 999.

43. W. Hartwig, 1983, *Tetrahedron*, 39, 2609.
44. D.H.R. Barton and S.W. McCombie, 1975, *J. Chem. Soc. Perkin I*, 1574.
45. (a) J. Kihlberg, T. Frejd, K. Jansson and G. Magnusson, 1986, *Carbohydr. Res.*, 152, 113; (b) M. Trumtel, P. Tavecchia, A. Veyrières and P. Sinaÿ, 1989, *Carbohydr. Res.*, 191, 29.
46. (a) J.R. Rasmussen, C.J. Slinger, R.J. Kordish and D.D. Newman-Evans, 1981, *J. Org. Chem.*, 46, 4843; (b) E. Petráková, P. Kovác and C.P.J. Glaudemans, 1992, *Carbohydr. Res.*, 233, 101; (c) L.A. Mulard, P. Kovác and C.P.J. Glaudemans, 1994, *Carbohydr. Res.*, 251, 213.
47. (a) Y. Ogawa, P.-S. Lei and P. Kovác, 1995, *Carbohydr. Res.*, 277, 327; (b) K. Okamoto, T. Kondo and T. Goto, 1986, *Tetrahedron Lett.*, 27, 5229; (c) M.J. Robins, J.S. Wilson and F. Hansske, 1983, *J. Am. Chem. Soc.*, 105, 4059; (d) E. Petráková and C.P.J. Glaudemans, 1995, *Carbohydr. Res.*, 279, 133; (e) J.P.G. Hermans, C.J.J. Elie, G.A. van der Marel and J.H. van Boom, 1987, *J. Carbohydr. Chem.*, 6, 451.
48. (a) O. Kanie, S.C. Crawley, M.M. Palcic and O. Hindsgaul, 1993, *Carbohydr. Res.*, 243, 139; (b) J. Gervay and S. Danishefsky, 1991, *J. Org. Chem.*, 56, 5448; (c) D.H.R. Barton and J.C. Jaszberenyi, 1989, *Tetrahedron Lett.*, 30, 2619.
49. M.E. Haque, T. Kikuchi, K. Kanemitsu and Y. Tsuda, 1987, *Chem. Pharm. Bull.*, 35, 1016.
50. W.P. Neumann, 1987, *Synthesis*, 665.
51. D.H.R. Barton, D.O. Jang and J.C. Jaszberenyi, 1993, *Tetrahedron*, 49, 2793.
52. (a) D.H.R. Barton, 1992, *Tetrahedron*, 48, 2529; (b) O. Kjolberg and K. Neumann, 1993, *Acta Chem. Scand.*, 47, 721.
53. D.H.R. Barton and R. Subramanian, 1977, *J. Chem. Soc. Perkin I*, 1718.
54. R.J. Sturgeon, 1988, in *Carbohydrate Chemistry* (J.F. Kennedy, ed.), Clarendon Press, Oxford, p. 263.
55. J.F.K. Kennedy and C.A. White, 1988, in *Carbohydrate Chemistry* (J.F. Kennedy, ed.), Clarendon Press, Oxford, p. 303.
56. (a) F.M. Hauser and S.R. Ellenberger, 1986, *Chem. Rev.*, 35; (b) A.K. Malans, 1988, in *Carbohydrate Chemistry* (J.F. Kennedy, ed.), Clarendon Press, Oxford, p. 73.
57. (a) J. Banoub, P. Boullanger and D. Lafont, 1992, *Chem. Rev.*, 92, 1167; (b) D. Horton and J.D. Wander, 1980, in *The Carbohydrates* (W. Pigman and D. Horton, eds.), Academic Press, New York, p. 644.
58. W. Karpeisiuk, A. Banaszek and A. Zamojski, 1989, *Carbohydr. Res.*, 186, 156.
59. D.H. Ball and F.W. Parrish, 1969, in *Adv. Carbohydr. Chem. Biochem.* (M.L. Wolfrom and R.S. Tipson, eds), Academic Press, New York, p. 139.
60. (a) C. Monneret, R. Gagnet and J.-C. Florent, 1993, *Carbohydr. Res.*, 240, 313; (b) A. M. Cimecioglu, D.H. Ball, D.L. Kaplan and S.H. Huang, 1994, *Macromolecules*, 27, 2917.
61. (a) N.A.L. Al-Masoudi and N.J. Tooma, 1993, *Carbohydr. Res.*, 239, 273; (b) M. Gotoh and P. Kovác, 1994, *J. Carbohydr. Chem.*, 13, 1193; (c) V. Pavliak and P. Kovác, 1991, *Carbohydr. Res.*, 210, 333.
62. (a) J. Ariza, M. Díaz, J. Font and R.M. Ortuño, 1993, *Tetrahedron*, 49, 1315; (b) T. Ercégovic and G. Magnusson, 1996, *J. Org. Chem.*, 61, 179; (c) G. Janairo, A. Malik and W. Voelter, 1985, *Liebigs Ann. Chem.*, 653; (d) C. Kolar, K. Dehmel and H. Moldenhauer, 1990, *Carbohydr. Res.*, 208, 67.
63. (a) R.W. Binkley and M.G. Ambrose, 1984, *J. Carbohydr. Chem.*, 3, 1; (b) H.H. Baer and Y. Gan, 1991, *Carbohydr. Res.*, 210; (c) F. Dasgupta and P.J. Garegg, 1988, *Synthesis*, 626; (d) G.W.J. Fleet, M.J. Gough and P.W. Smith, 1984, *Tetrahedron Lett.*, 25, 1853; (e) M. Kloosterman, M.P. de Nijs and J.H. van Boom, 1984, *Recl. Trav. Chim. Pays-Bas*, 103, 243; (f) M. Kloosterman, P. Westerduin and J.H. van Boom, 1986, *Recl. Trav. Chim. Pays-Bas*, 105, 136; (g) I. Pelyvás, F. Sztaricskai, L. Szilágyi and R. Bognár, 1979, *Carbohydr. Res.*, 68, 321; (h) W. Kowollik, G. Janaito and W. Voelter, 1988, *Liebigs Ann. Chem.*, 427; (i) K. Zegelaar-Jaarsveld, S. van der Plas, G.A. vander Marel and J.H. van Boom, 1996, *J. Carbohydr. Chem.*, 15, 591.

64. (a) L.D. Hall and D.C. Miller, 1976, *Carbohydr. Res.*, 47, 299; (b) R.D. Howells and J.D. McGown, 1977, *Chem. Rev.*, 77, 69.
65. J.N. Vos, J.H. van Boom, C.A.A. van Boeckel and T. Beetz, 1984, *J. Carbohydr. Chem.*, 3, 117.
66. A. Guiller, C.H. Gagnieu and H. Pacheco, 1985, *Tetrahedron Lett.*, 26, 6343.
67. (a) P.A.M. van der Klein, W. Filemon, G.H. Veeneman, G.A. van der Marel and J.H. van Boom, 1992, *J. Carbohydr. Chem.*, 11, 837; (b) K. Vanhessche, E. van der Eycken and M. Vandewalle, 1990, *Tetrahedron Lett.*, 31, 2337.
68. B. Aguilera and A. Fernández-Mayoralas, 1996, *J. Chem. Soc., Chem. Commun.*, 127.
69. R.U. Lemieux and R.M. Ratcliffe, 1979, *Can. J. Chem.*, 57, 1244.
70. (a) H. Paulsen and J.P. Lorentzen, 1984, *Carbohydr. Res.*, 133, C1; (b) H. Hashimoto, K. Araki, Y. Saito, M. Kawa and J. Yoshimura, 1986, *Bull. Chem. Soc. Jpn.*, 59, 3131.
71. M. Forsgren and T. Norberg, 1983, *Carbohydr. Res.*, 116, 39.
72. A. Marra, F. Gauffeny and O. Sinaÿ, 1991, *Tetrahedron*, 47, 5149.
73. (a) S. Czernecki, E. Ayadi and D. Randriamandimby, 1994, *J. Chem. Soc., Chem. Commun.*, 35; (b) S. Czernecki and D. Randriamandimby, 1993, *Tetrahedron Lett.*, 49, 7915; (c) E. Chelain and S. Czernecki, 1996, *J. Carbohydr. Chem.*, 15, 571; (d) F. Santoyo-González, F.G. Calvo-Flores, P. García-Mendoza, F. Hernández-Mateo, J. Isac-García and R. Robles-Dias, 1993, *J. Org. Chem.*, 58, 6122; (e) M. Tingoli, M. Tiecco, L. Testaferri and A. Temperini, 1994, *J. Chem. Soc., Chem. Commun.*, 1883.
74. (a) R.M. Giuliano, V.E. Manetta and G.R. Smith, 1995, *Carbohydr. Res.*, 278, 345; (b) J.S. Brimacombe, R. Hanne, M.S. Saeed and C.N. Tucker, 1982, *J. Chem. Soc. Perkin I*, 2583; (c) R. Andersson, I. Gouda, O. Larm, M.E. Riquelme and E. Scholander, 1985, *Carbohydr. Res.*, 142, 141; (d) D.M. Clode, D. Horton and W. Weckerle, 1976, *Carbohydr. Res.*, 49, 305; (e) P. Smid, W.P.A. Jörning, A.M.G. van Duuren, G.J.P.H. Boons, G.A. van der Marel and J.H. van Boom, 1992, *J. Carbohydr. Chem.*, 11, 849.
75. (a) F.W. Lichtenhaler, E. Kaji and S. Weprek, 1985, *J. Org. Chem.*, 50, 3505; (b) A. Banaszek and W. Karpeisiuk, 1994, *Carbohydr. Res.*, 251, 233.
76. (a) E. Kaji, Y. Osa, K. Takahashi, M. Hirooka, S. Zen and F.W. Lichtenhaler, 1994, *Bull. Chem. Soc. Jpn.*, 67, 1130; (b) E. Kaji, Y. Osa, K. Takahashi and S. Zen, 1996, *Chem. Pharm. Bull.*, 44, 15.
77. (a) H.W. Pauls and B. Fraser-Reid, 1983, *J. Org. Chem.*, 48, 1392; (b) H.W. Pauls and B. Fraser-Reid, 1986, *Carbohydr. Res.*, 150, 111; (c) A. Bongini, G. Cardillo, M. Orena, S. Sandri and C. Tomasini, 1983, *Tetrahedron*, 39, 3801.
78. (a) S. Knapp, P.J. Kukkola, S. Sharma, T.G.M. Dahr and A.B.J. Naugthon, 1990, *J. Org. Chem.*, 55, 5700; (b) K.-I. Iida, T. Ishii, M. Hirama, T. Otani, Y. Minami and K.-I. Yoshida, 1993, *Tetrahedron Lett.*, 34, 4079; (c) S. Knapp, P.J. Kutkola, S. Sharma and S. Pietranico, 1987, *Tetrahedron Lett.*, 28, 5399.
79. K.C. Nicolaou, T. Ladduwahetty, J.L. Randall and A. Chucholowski, 1986, *J. Am. Chem. Soc.*, 108, 2466.
80. H.M. Zuurmond, P.A.M. van der Klein, J. de Wilt, G.A. van der Marel and J.H. van Boom, 1994, *J. Carbohydr. Chem.*, 13, 323.
81. (a) N.R. Williams, 1970, *Adv. Carbohydr. Chem. Biochem.*, 109; (b) R.E. Parker and N.S. Isaacs, 1959, *Chem. Rev.*, 59, 737; (c) J.A. Mills, 1955, *Adv. Carbohydr. Chem.*, 10, 51.
82. (a) R.J. Robertson and C.E. Griffiths, 1935, *J. Chem. Soc.*, 1193; (b) S.P. James, 1946, *J. Chem. Soc.*, 625.
83. N.K. Richtmeyer and C.S. Hudson, 1941, *J. Am. Chem. Soc.*, 63, 1727.
84. S. Achab and B.C. Das, 1982, *Syn. Comm.*, 12, 931.
85. R.J. Ferrier, 1988, in *Carbohydrate Chemistry* (J.F. Kennedy, ed.), Clarendon Press, Oxford, p. 443.
86. C.L. Penney and A.S. Perlin, 1981, *Carbohydr. Res.*, 93, 241.
87. (a) E. Percival, 1980, *Methods Carbohydr. Chem.*, VIII, 281; (b) K.R. King, J.M. Williams, J.R. Clamp and A.P. Corfield, 1992, *Carbohydr. Res.*, 235, C9.

88. (a) G.V. Reddy, R.K. Jain, R.D. Locke and K.L. Matta, 1996, *Carbohydr. Res.*, 280, 261; (b) L.-X. Wang, C. Li, Q.-W. Wang and Y.-Z. Hui, 1994, *J. Chem. Soc. Perkin I*, 621; (c) J.M. Coteron, K. Singh, J.L. Asensio, M. Dominguez-Dalda, A. Fernandez-Mayoralas, J. Jimenez-Barbero, M. Martin-Lomas, J. Abad-Rodriquez and M. Nietos-Ampedro, 1995, *J. Org. Chem.*, 60, 1502; (d) R.K. Jain, X.-G. Liu and K.L. Matta, 1995, *Carbohydr. Res.*, 268, 279; (e) H. Maeda, K. Ito, M. Kiso and A. Hasegawa, 1995, *J. Carbohydr. Chem.*, 14, 387; (f) V. Srivastava and O. Hindsgaul, 1989, *Carbohydr. Res.*, 185, 163; (g) O. Vandana, O. Hindsgaul and J.U. Baenzinger, 1987, *Can. J. Chem.*, 65, 1645.

89. (a) R.A. Field, A. Otter, W. Fu and O. Hindsgaul, 1995, *Carbohydr. Res.*, 276, 347; (b) A. Endo, M. Iida, S. Fujita, M. Numata, M. Sugimoto and S. Nunomura, 1995, *Carbohydr. Res.*, 270, C9; (c) T. Chiba, J.-C. Jacquinet, P. Sinaÿ, M. Petitou and J. Choay, 1988, *Carbohydr. Res.*, 174, 253; (d) J.-C. Jacquinet, 1990, *Carbohydr. Res.*, 199, 153; (e) A. Marra, X. Dong, M. Petitou and P. Sinaÿ, 1989, *Carbohydr. Res.*, 195, 39; (f) K.C. Nicolaou, N.J. Bokovich and D.R. Carcanague, 1993, *J. Am. Chem. Soc.*, 115, 8843l; (g) T. Nakano, Y. Ito and T. Ogwawa, 1991, *Tetrahedron Lett.*, 32, 1569; (h) S. Rio, J.-M. Beau and J.-C. Jacquinet, 1994, *Carbohydr. Res.*, 255, 103; (i) R. Vig, R.K. Jain, C.F. Piskorz and K.L. Matta, 1995, *J. Chem. Soc., Chem. Commun.*, 2073; (j) M. Zsiska and B. Meyer, 1991, *Carbohydr. Res.*, 215, 261.

90. T. Böcker, T.K. Lindhorst and V. Vill, 1992, *Carbohydr. Res.*, 230, 245.

91. (a) N.A. Kraaijeveld and C.A.A. van Boeckel, 1989, *Recl. Trav. Chim. Pays-Bas*, 108, 39; (b) Y. Ichikawa, R. Monden and H. Kuzuhara, 1988, *Carbohydr. Res.*, 1988, 37; (c) M. Kobayashi, F. Yamazaki, Y. Ito and T. Ogawa, 1990, *Carbohydr. Res.*, 201, 51; (d) H. Paulsen, A. Huffziger and C.A.A. van Boeckel, 1988, *Liebigs Ann. Chem.*, 419; (e) M. Petitou, P. Duchaussoy, I. Lederman, J. Choay, J.-C. Jacquinet, P. Sinaÿ and G. Torri, 1987, *Carbohydr. Res.*, 167, 67.

92. R.K. Jain, R. Vig, R.D. Locke, A. Mohammad and K.L. Matta, 1996, *J. Chem. Soc., Chem. Commun.*, 65.

93. J. Thiem and M. Franzkowiak, 1989, *J. Carbohydr. Chem.*, 8, 1.

94. (a) F.-I. Auzanneau, D. Charon and L. Szabó, 1991, *J. Chem. Soc. Perkin I*, 509; (b) C. Hällgren and O. Hindsgaul, 1994, *Carbohydr. Res.*, 260, 63; (c) T. Kusama, T. Soga, Y. Ono, E. Kumazawa, E. Shioya, Y. Osada, S. Kusumoto and T. Shiba, 1991, *Chem. Pharm. Bull.*, 39, 1994; (d) H. Regeling, B. Zwanenburg, G.J.F. Chittenden and N. Rehnberg, 1993, *Carbohydr. Res.*, 244, 187; (e) P. Szabó, S.R. Sarfati, C. Diolez and L. Szabó, 1983, *Carbohydr. Res.*, 111; (f) H. Rembold and R.R. Schmidt, 1993, *Carbohydr. Res.*, 246, 137.

95. O.P. Srivastava and O. Hindsgaul, 1986, *Can. J. Chem.*, 64, 2324.

96. (a) F.-I. Auzanneau, D. Charon, L. Szilágyi and L. Szabó, 1991, *J. Chem. Soc. Perkin I*, 803; (b) M.K. Christensen, M. Meldal and K. Bock, 1993, *J. Chem. Soc. Perkin I*, 1453.

97. D.D. Manning, C.R. Bertozzi, S.D. Rosen and L.L. Kiessling, 1996, *Tetrahedron Lett.*, 37, 1953.

98. J.K. Stowell and T.S. Widlanski, 1995, *Tetrahedron Lett.*, 36, 1825.

99. P. Hoogerhout, D. Evenberg, C.A.A. van Boeckel, J.T. Poolman, E.C. Beuvery, G.A. van der Marel and J.H. van Boom, 1987, *Tetrahedron Lett.*, 28, 1553.

100. (a) J.P.G. Hermans, D. Noort, G.A. van der Marel and J.H. van Boom, 1988, *Recl. Trav. Chim. Pays-Bas*, 107, 635; (b) F. Ramirez, S.B. Mandal and J.F. Marecek, 1983, *J. Org. Chem.*, 48, 2008.

101. H.J.G. Broxterman, G.A. van der Marel and J.H. van Boom, 1991, *J. Carbohydr. Chem.*, 10, 215.

102. (a) K. Fukase, T. Kamikawa, Y. Iwai, T. Shiba, E.T. Rietschel and S. Kusumoto, 1991, *Bull. Chem. Soc. Jpn.*, 64, 3267; (b) O. Eyrisch, G. Sinerius and W.-D. Fessner, 1993, *Carbohydr. Res.*, 238, 287.

103. (a) C.J.J. Elie, H.J. Muntendam, H. van den Elst, G.A. van der Marel, P. Hoogerhout and J.H. van Boom, 1989, *Recl. Trav. Chim. Pays-Bas*, 108, 219; (b) P. Smid, M. de Zwart,

W.P.A. Jörning, G.A. van der Marel and J.H. van Boom, 1993, *J. Carbohydr. Chem.*, 12, 1073; (c) R.R. Verduyn, C.M. Dreef-Tromp, G.A. van der Marel and J.H. van Boom, 1991, *Tetrahedron Lett.*, 32, 6637; (d) P. Westerduin, G.H. Veeneman, J.E. Marugg, G.A. van der Marel and J.H. van Boom, 1986, *Tetrahedron Lett.*, 27, 1211; (e) P. Westerduin, G.H. Veeneman, J.E. Marugg, G.A. van der Marel and J.H. van Boom, 1986, *Tetrahedron Lett.*, 27, 6271.

104. (a) G.H. Veeneman, H.F. Brugghe, H. van den Elst and J.H. van Boom, 1990, *Carbohydr. Res.*, 195, C1; (b) P. Westerduin, G.H. Veeneman and J.H. van Boom, 1987, *Recl. Trav. Chim. Pays-Bas*, 106, 601; (c) C. Le Bec and T. Huynh-Dinh, 1991, *Tetrahedron Lett.*, 32, 6553; (d) L. Chan and G. Just, 1990, *Tetrahedron*, 46, 151.

105. G.H. Veeneman, H.F. Brugghe, P. Hoogerhout, G.H. van der Marel and J.H. van Boom, 1988, *Recl. Trav. Chim. Pays-Bas*, 107, 610.

106. T. Ogawa and A. Seta, 1982, *Carbohydr. Res.*, 110, C1.

107. (a) A.S. Campbell and B. Fraser-Reid, 1995, *J. Am. Chem. Soc.*, 117, 10387; (b) K. Morisaki and S. Ozaki, 1996, *Carbohydr. Res.*, 286, 123; (c) A. De Nino, A. Liguori, A. procopio, E. Roberti and G. Sindona, 1996, *Carbohydr. Res.*, 186, 77.

108. (a) X. Pannecoucke, G. Schmitt and B. Luu, 1994, *Tetrahedron*, 50, 6569; (b) A.M.P. van Steijn, J.P. Kamerling and J.F.G. Vliegenthart, 1991, *Carbohydr. Res.*, 211, 261.

109. (a) A.V. Nikolaev, I.A. Ivanova and V. Shibaev, 1993, *Carbohydr. Res.*, 242, 91; (b) A.V. Nikolaev, T.J. Rutherford, M.A.J. Ferguson and J.S. Brimacombe, 1995, *J. Chem. Soc. Perkin I*, 1977; (c) M. Nilsson, J. Westman and C.-M. Svahn, 1993, *J. Carbohydr. Chem.*, 12, 23; (d) A.V. Nikolaev, J.A. Chudek and M.A.J. Ferguson, 1995, *Carbohydr. Res.*, 272, 179; (e) F. Guillod, J. Greiner and J.G. Riess, 1994, *Carbohydr. Res.*, 261, 37.

110. (a) S. Liemann and W. Klaffke, 1995, *Liebigs Ann. Chem.*, 1779; (b) S.J. Freese and W.F. Vann, 1996, *Carbohydr. Res.*, 281, 313; (c) S.J. Hecker, M.L. Minich and K. Lackey, 1990, *J. Org. Chem.*, 55, 4904; (d) M.M. Sim, H. Kondo and C.-H. Wong, 1993, *J. Am. Chem. Soc.*, 115, 2260; (e) Y. Ichikawa, M.M. Sim and C.-H. Wong, 1992, *J. Org. Chem.*, 57, 2943.

111. S. Sabesan and S. Neira, 1992, *Carbohydr. Res.*, 169.

112. (a) K. Adelhorst and G.M. Whitesides, 1993, *Carbohydr. Res.*, 242, 69; (b) T. Furuta, H. Torigai, T. Osawa and M. Iwamura, 1993, *J. Chem. Soc. Perkin I*, 3139; (c) U.B. Gokhale, O. Hindsgaul and M.M. Palcic, 1990, *Can. J. Chem.*, 68, 1063; (d) J. Niggemann and J. Thiem, 1992, *Liebigs Ann. Chem.*, 535; (e) J. Niggemann, T.K. Lindhorst, M. Waltfort, L. Laupichler, H. Sajus and J. Thiem, 1993, *Carbohydr. Res.*, 246, 173; (f) A. Kohen, V. Belakhov and T. Baasov, 1994, *Tetrahedron Lett.*, 35, 3179.

113. R. Roy, F.D. Tropper and C. Grand-Maître, 1991, *Can. J. Chem.*, 69, 1462.

114. (a) J.E. Heidlas, W.J. Lees, P. Pale and G.M. Whitesides, 1992, *J. Org. Chem.*, 57, 146; (b) M. Inage, H. Chaki, S. Kusumoto and T. Shiba, 1981, *Tetrahedron Lett.*, 22, 2281; (c) C. D. Warren, M.A.E. Shaban and R.W. Jeanloz, 1977, *Carbohydr. Res.*, 59, 427.

115. C.A.A. van Boeckel, J.P.G. Hermans, P. Westerduin, J.J. Oltvoort, G.A. van der Marel and J.H. van Boom, 1983, *Recl. Trav. Chim. Pays-Bas*, 102, 438.

116. (a) R.R. Schmidt and M. Stumpp, 1984, *Liebigs Ann. Chem.*; (b) R.R. Schmidt, B. Wegmann and K.-H. Jung, 1991, *Liebigs Ann. Chem.*, 111; (c) J.E. Pallanca and N.J. Turner, 1993, *J. Chem. Soc. Perkin I*, 3017.

117. P. Pale and G.M. Whitesides, 1991, *J. Org. Chem.*, 56, 4547.

118. (a) B.M. Heskamp, H.J.G. Broxterman, G.A. van der Marel and J.H. van Boom, 1996, *J. Carbohydr. Chem.*, 15, 611; (b) G.H. Veeneman, H.J.G. Broxterman, G.A. van der Marel and J.H. van Boom, 1991, *Tetrahedron Lett.*, 32, 6175.

119. (a) G.-J. Boons, A. Burton and P. Wyatt, 1996, *Synlett*, 310; (b) A. Burton, P. Wyatt and G.-J. Boons, 1997, *J. Chem. Soc. Perkin Trans I*, 2375.

4   Oligosaccharide synthesis

G.-J. Boons

4.1 Introduction

The chemical synthesis of oligosaccharides is much more complicated than the synthesis of other biopolymers such as peptides and nucleic acids. The difficulties in the preparation of complex oligosaccharides are a result of a greater number of possibilities for the combination of monomeric units to form oligosaccharides. In addition, the glycosidic linkages have to be introduced stereospecifically (α/β selectivity). To date, there are no general applicable methods or strategies for oligosaccharide synthesis and consequently the preparation of these compounds is very time consuming. Nevertheless, contemporary carbohydrate chemistry makes it possible to execute complex multistep synthetic sequences that give oligosaccharides consisting of as many as 20 monosaccharide units. The preparation of oligosaccharides of this size is only possible when each synthetic step in the assembly of the oligosaccharide is high yielding and the formation of each glycosidic linkage is highly stereoselective. Apart from this, the assembly of the monomeric units should be highly convergent. In order to solve problems associated with chemical oligosaccharide synthesis, enzymatic procedures have been developed. However, the number of enzymes available for glycosidic bond synthesis is still very limited.

4.2 Chemical glycosidic bond synthesis

Interglycosidic bond formation is generally achieved by condensing a fully protected glycosyl donor, which bears a potential leaving group at its anomeric centre, with a suitably protected glycosyl acceptor that contains often only one free hydroxyl group (Scheme 4.1).[1] Traditionally, the most widely used glycosylation methods have exploited anomeric halide derivatives of carbohydrates as glycosyl donors. However, these compounds often suffer from instability and require relatively drastic conditions for their preparation. The introduction of the orthoester[2] and imidate[3] procedures were the first attempts to find alternatives to the glycosyl halide methodologies. Since these original disclosures, many other leaving groups for the anomeric centre have been reported (Figure 4.1).[1k] However, from these glycosyl donors, the anomeric fluorides,

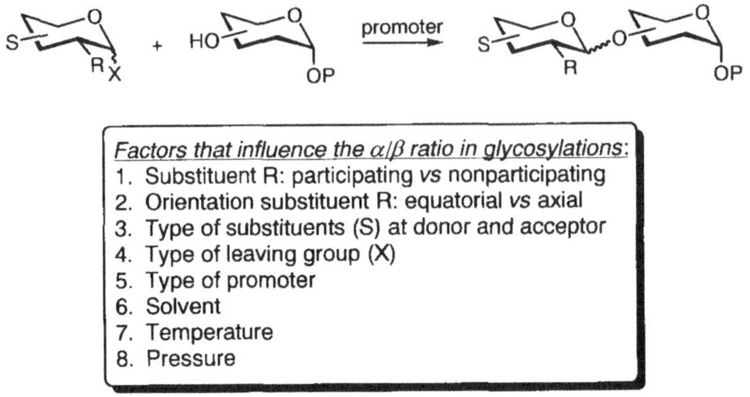

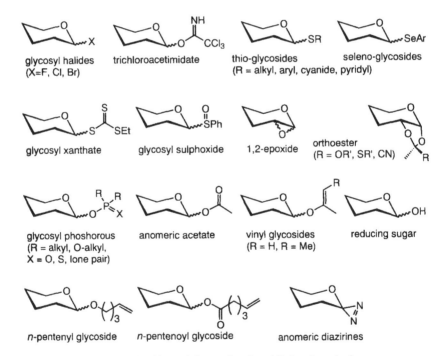

Scheme 4.1   General approach for chemical glycosylations.

glycosyl halides
(X=F, Cl, Br)

trichloroacetimidate

thio-glycosides
(R = alkyl, aryl, cyanide, pyridyl)

seleno-glycosides

glycosyl xanthate

glycosyl sulphoxide

1,2-epoxide

orthoester
(R = OR', SR', CN)

glycosyl phoshorous
(R = alkyl, O-alkyl,
X = O, S, lone pair)

anomeric acetate

vinyl glycosides
(R = H, R = Me)

reducing sugar

n-pentenyl glycoside

n-pentenoyl glycoside

anomeric diazirines

Figure 4.1   Glycosyl donors for glycosidic bond synthesis.

trichloroacetimidates and thioglycosides have been applied most widely. These compounds can be prepared under mild conditions, are sufficiently stable to be purified and stored for a considerable period of time and undergo glycosylations under mild conditions. By selecting the appropriate reaction conditions, high yields and good α/β ratios can be obtained.

The next section describes briefly the procedures for the preparation of anomeric halides, trichloroacetimidates and thioglycosides and their modes of activation.

4.2.1 Glycosyl halides

Koenigs and Knörr introduced the use of glycosyl bromides and chlorides as glycosyl donors in 1901.[4] This classical approach uses heavy metal salts (mainly silver and mercury salts) or alkyl ammonium halides as activators.[1a,b] Complexation of the anomeric bromide or chloride with a silver or mercury salt greatly improves their leaving group ability. Departure of the anomeric leaving group will give an oxocarbenium ion, which in turn will react with an alcohol to give a glycoside (Scheme 4.2a). Alternatively, an activated anomeric halide may be substituted by an alcohol by an S_N2 mechanism leading to a glycoside. Detailed reaction mechanisms of glycosylations are discussed in Section 4.3.

The reactivity of glycosyl halides is determined by the protecting group pattern and in general ether protected derivatives are more reactive than analogous ester protected glycosyl donors. Furthermore, the protecting group at C(2) has the greatest effect on the reactivity of a glycosyl donor. These observations can easily be rationalised. Departure of the anomeric leaving group results in a partial positive charge at the anomeric centre. Esters are strongly electron-withdrawing and will destabilise the resulting positively charged intermediate and hence displacement of such halides is energetically less favourable. Protecting groups also determine the stability of glycosyl halides. For example, 2,3,4,6-tetra-O-acetyl-α-D-glycosyl bromide is a reasonably stable compound, which can be stored for a considerable period of time. However, the analogous O-benzylated derivative will decompose within several hours after preparation. Glycosyl bromides are more reactive than glycosyl chlorides but are also more labile. In general, glycosyl iodides are too labile to be used in glycosylations.

For a long time, it was believed that glycosyl fluorides were too stable to be used as glycosyl donors. However, in 1981 Mukaiyama and co-workers demonstrated that these compounds can be activated with $AgClO_4/SnCl_2$.[5] Subsequent reports[1k,o] expanded the repertoire of activators for glycosyl fluorides, and commonly used promoters are $BF_3 \cdot OEt_2$, $Cp_2MCl_2 - AgClO_4$ (M = Hf, Zr) and $Cp_2HfCl_2 - AgOTf$. Glycosyl fluorides can be purified by silica gel column chromatography and have a good shelf-life. They can even undergo a limited number of protecting group manipulations.

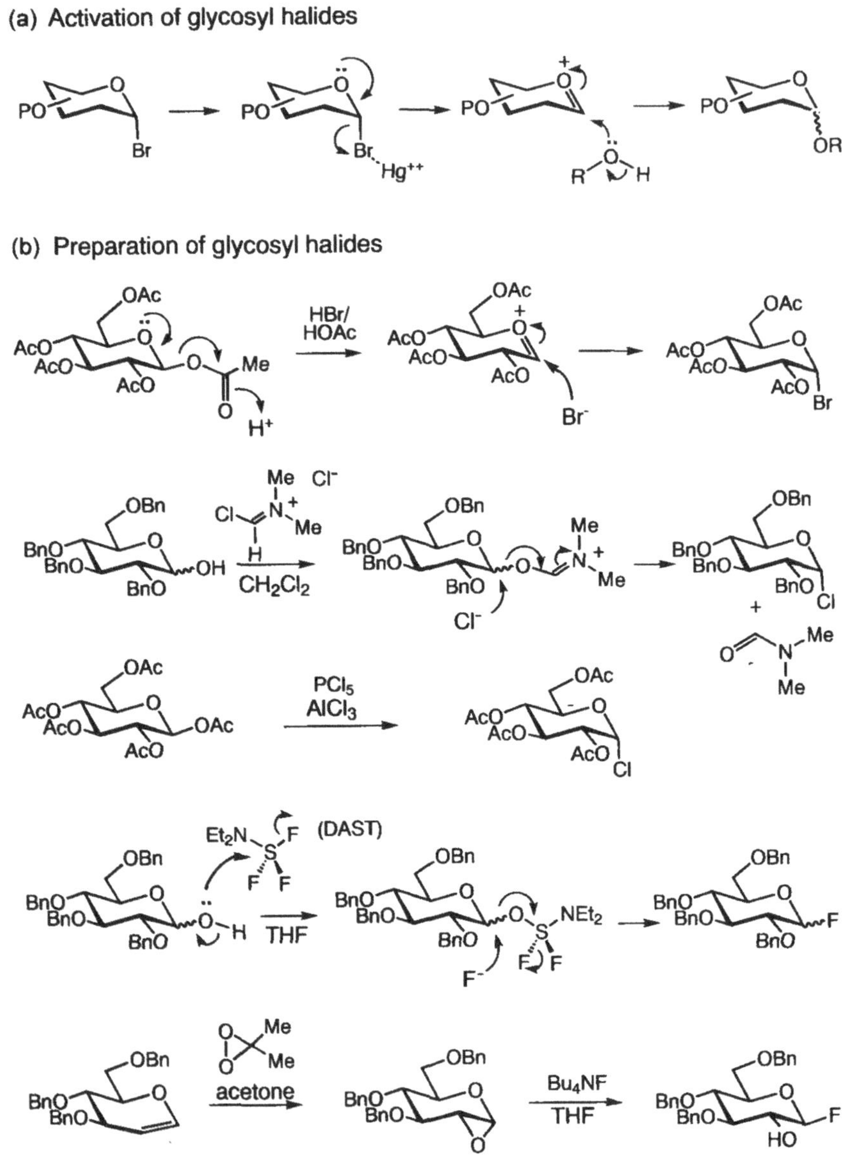

(a) Activation of glycosyl halides

(b) Preparation of glycosyl halides

Scheme 4.2 Preparation and glycosylation of glycosyl halides.

Glycosyl bromides are most commonly prepared by treatment of a per-*O*-acylated sugar derivative with a solution of HBr in acetic acid (Scheme 4.2b). The more stable α-anomer is usually obtained in high yield. In this reaction, the anomeric acetyl moiety is converted into a good leaving group by protonation. Next, an oxocarbenium ion is formed by departure of acetic acid, which is substituted by bromide. In the first instance, a mixture of anomeric bromides may be formed, which quickly equilibrates to the thermodynamically more stable α anomer. The latter compound is stabilised by a strong endoanomeric effect. Other acetyl moieties may also be protonated; however, departure of these groups will result in the formation of a highly unstable carbonium ion. Thus, normally this type of reaction will not occur.

Glycosyl chlorides can be obtained by treatment of aldosyl acetates with aluminum chloride or phosphorus pentachloride. This procedure is relatively harsh and many functionalities will not survive these conditions. Several milder methods have been described with the Vilmeier–Haack reagent[6] being one of the most useful reagents for the preparation of labile glycosyl chlorides or bromides. The Vilmeier–Haack reagent $(Me_2N^+{=}CHCl\,Cl^-)$ is formed by reaction of dimethylformamide (DMF) with oxalyl chloride [ClC(O)C(O)Cl]. The anomeric hydroxyl performs a 1,2-nucleophilic addition with concomitant elimination of a chloride ion. This reaction results in the introduction of a very good leaving group, which is displaced by chloride. An anomeric bromide will be formed when oxalyl bromide is used in the formation of the Vilsmaier–Haack reagent $(Me_2N^+{=}CHBr\,Br^-)$.

Several methods have been reported for the preparation of glycosyl fluorides[7] but the most common procedure is treatment of a thioglycoside with NBS (N-bromosuccinimide) and (diethylamino)sulfur trifluoride (DAST)[8] or the reaction of a lactol with DAST or 2-fluoro-1-methylpyridinium p-toluenesulfonate.[7,9] An alternative and interesting method is the treatment of a 1,2-anhydro-pyranoside with tetra-n-butylammonium fluoride (TBAF).

4.2.2   Trichloroacetimidates[1d,e]

In recent times, anomeric trichloroacetimidates have become the most widely used glycosyl donors. They can easily be prepared by a base-catalysed reaction of a lactol with trichloroacetonitrile. When the reaction is performed in the presence of the mild base potassium carbonate, the kinetic β-trichloroacetimidate is formed (Scheme 4.3). Under these conditions, the more reactive β-alkoxide forms preferentially. It then attacks the trichloroacetonitrile irreversibly. However, when a strong base such as NaH or 1,8-diazabicyclo[5.4.0]undec-7-ene (DBU) is employed, alkoxide equilibration occurs, with the more stable α-alkoxide predominating. It then goes on to react with the trichloroacetonitrile to give the α-trichloroacetimidate.

The higher nucleophilicity of a β alkoxide can be attributed to unfavourable dipole–dipole interactions resulting from repulsion of the

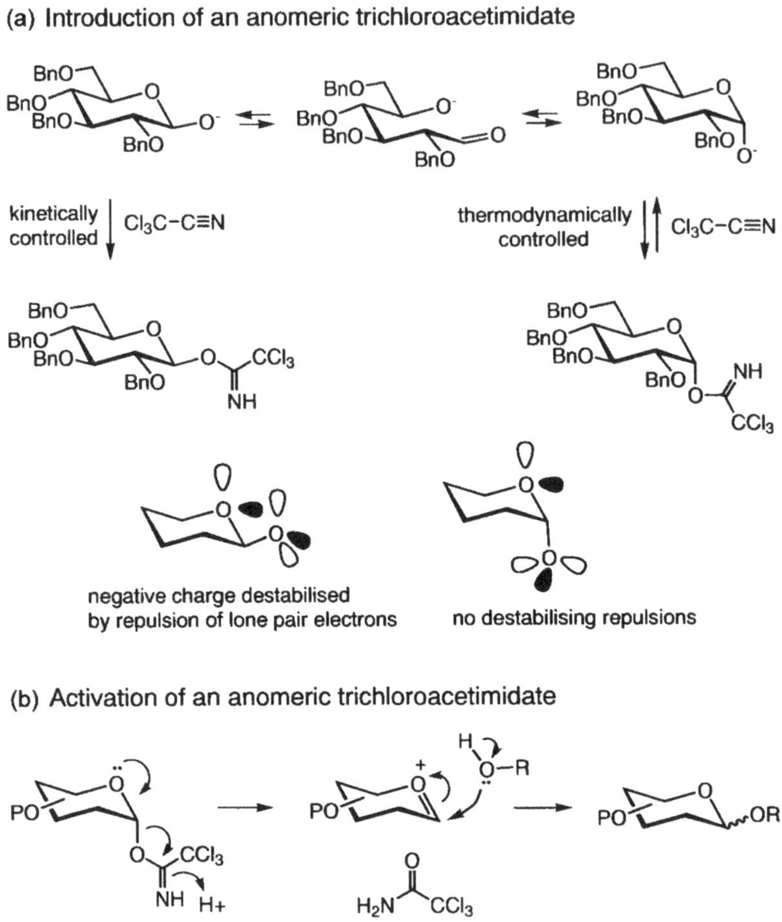

(a) Introduction of an anomeric trichloroacetimidate

kinetically controlled | $Cl_3C-C\equiv N$

thermodynamically controlled | $Cl_3C-C\equiv N$

negative charge destabilised by repulsion of lone pair electrons

no destabilising repulsions

(b) Activation of an anomeric trichloroacetimidate

Scheme 4.3 Preparation and glycosylation of trichloroacetimidates.

lone electron pairs of the exocyclic and endocyclic ring oxygen atoms. Thermodynamically, the α alkoxide is preferred because of additional stabilisation by the endoanomeric effect.

Anomeric trichloroacetimidates are usually activated by catalytic amounts of Lewis acid, with trimethylsilyl triflate (TMSOTf) and $BF_3 \cdot Et_2O$ being the reagents most commonly used. O-Trichloroacetimidate glycosylations can be performed at relatively low temperatures and generally give high yields. However, when the acceptor is very unreactive, substantial rearrangement to the corresponding trichloroacetamide may occur leading to a low recovery of the O-glycoside product. Anomeric trichloroacetimidates have a reasonable shelf-life but cannot usually undergo protecting group manipulations.

4.2.3 Thioglycosides[1,j,k,n]

Alkyl(aryl) thioglycosides have emerged as versatile building blocks for oligosaccharide synthesis. They can conveniently be prepared by a Lewis-acid-catalysed reaction of an anomeric acetate with a mercaptan (Scheme 4.4a). Instead of using mercaptans, more reactive alkylthiostannanes or alkylthiosilanes may be used. Alternatively, nucleophilic substitution of a glycosyl halide with mercaptides gives thioglycosides in high yield. Owing to their excellent chemical stability, anomeric alkyl(aryl) thio groups offer

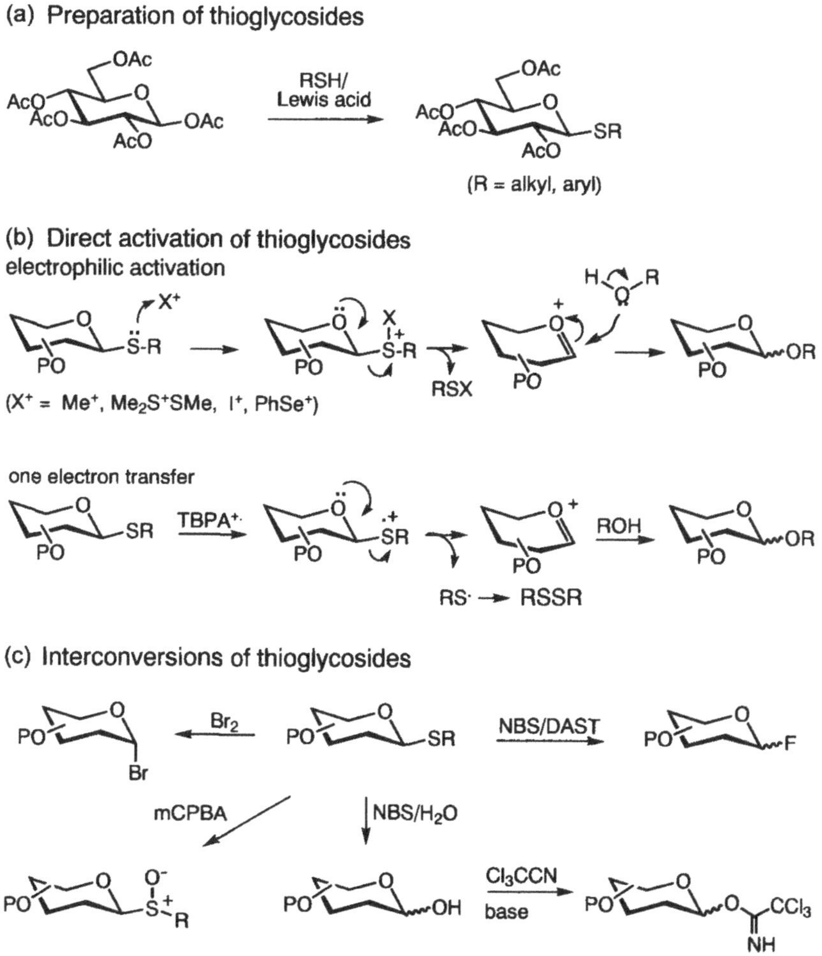

(a) Preparation of thioglycosides

(R = alkyl, aryl)

(b) Direct activation of thioglycosides
electrophilic activation

(X^+ = Me^+, Me_2S^+SMe, I^+, $PhSe^+$)

one electron transfer

$RS^{\cdot} \rightarrow RSSR$

(c) Interconversions of thioglycosides

Scheme 4.4 Preparation, activation and interconversion of thioglycosides.

efficient protection of anomeric centres and are compatible with many reaction conditions often employed in carbohydrate chemistry. However, in the presence of soft electrophiles (X^+), thioglycosides can be activated and used directly in glycosylations (Scheme 4.4b). The most commonly used activators include methyl triflate (MeOTf), dimethyl(methylthio) sulfonium triflate (DMTST), N-iodosuccinimide – triflic acid (NIS – TfOH), iodonium dicollidine perchlorate (IDCP)) and phenyl selenyl triflate (PhSeOTf).[1k] In these reactions, the electrophilic activator reacts with the lone pair on sulfur resulting in the formation of a sulfonium intermediate. The latter is an excellent leaving group and can be displaced by a sugar hydroxyl. Thioglycosides can also be activated by a one-electron transfer reaction from sulfur to the activating reagent tris-(4-bromophenyl)ammonium hexachloroantimonate (TBPA$^+$).[10]

Another attractive feature of thioglycosides is that they can be transformed into a range of other glycosyl donors (Scheme 4.4c). For example, treatment of a thioglycoside with bromine gives a glycosyl bromide which, after work-up, can be used in a Hg(II) or Ag(I) promoted glycosylation.[7] A glycosyl bromide can also be prepared and glycosylated *in situ* under the influence of (Bu$_4$N)$_2$CuBr$_4$ and AgOTf.[11] A thioglycoside can be converted into a glycosyl fluoride by treatment with N-bromosuccinimide/(diethylamino)sulfur trifluoride (NBS/DAST) and it can be hydrolysed to the 1-OH derivative, which is a suitable substrate for the preparation of an anomeric trichloroacetimidate. Finally, thioglycosides can be oxidised to the corresponding sulfoxides, and these then activated with triflic anhydride at low temperature.[12]

4.3 Stereoselective control in glycosidic bond synthesis

Anomeric linkages are classified according to the relative and absolute configuration at C(1) and C(2) (Figure 4.2), there being: 1,2-*cis* and 1,2-*trans* types. Miscellaneous other glycosidic linkages can also be identified, including 2-deoxy-glycosides and 3-deoxy-2-keto-ulo(pyranosylic) acids.

The stereoselective formation of a glycosidic linkage is one of the most challenging aspects of oligosaccharide synthesis. The nature of the protecting group at C(2) in the glycosyl donor plays a dominant role in controlling anomeric selectivity. A protecting group at C(2), which can perform neighbouring group participation during glycosylation, will give 1,2-*trans* glycosidic linkages. However, reaction conditions such as solvent, temperature and promoter can also determine anomeric selectivity when a nonassisting functionality is present at C(2). The constitution of a glycosyl donor and acceptor (e.g. type of saccharide,

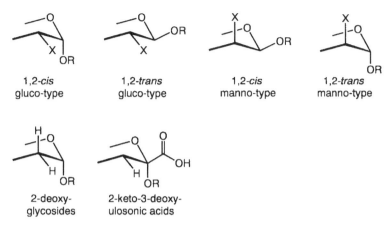

1,2-*cis*
gluco-type

1,2-*trans*
gluco-type

1,2-*cis*
manno-type

1,2-*trans*
manno-type

2-deoxy-
glycosides

2-keto-3-deoxy-
ulosonic acids

Figure 4.2   Different types of glycosidic linkages.

leaving group at the anomeric centre, protection and substitution pattern) can also have a major effect on the α/β selectivity.

4.3.1   *Neighbouring-group-assisted procedures*

The most reliable method for the construction of 1,2-*trans*-glycosidic linkages utilises neighbouring-group participation from a 2-*O*-acyl functionality. The principle of this approach is schematically illustrated in Scheme 4.5. Thus, a promoter (A) activates an anomeric leaving group, to assist in its departure. This results in the formation of an oxocarbenium ion. Subsequent neighbouring-group participation of a 2-*O*-acyl protecting group leads to the formation of a more stable acyloxonium ion. In the latter intermediate, additional resonance stabilisation of the positive charge is provided by two oxygen atoms. In the case of the oxocarbenium ion, only the ring oxygen atom gives resonance stabilisation and hence this is less stable. Attack of an alcohol at the anomeric centre of the acyloxonium ion results in the formation of a 1,2-*trans*-glycoside. Thus, in the case of glucosyl-type donors, β-linked products will be obtained whereas manno-type donors will give α-glycosides. The neighbouring-group-assisted glycosylation procedures are compatible with many different glycosylation protocols and most leaving groups depicted in Figure 4.1 can be used.

In some glycosylations, the alcohol will attack at the C(2) position of the dioxolane ring of the acyloxonium ion, resulting in the formation of an undesired orthoester. In some cases, the orthoester can be isolated as a

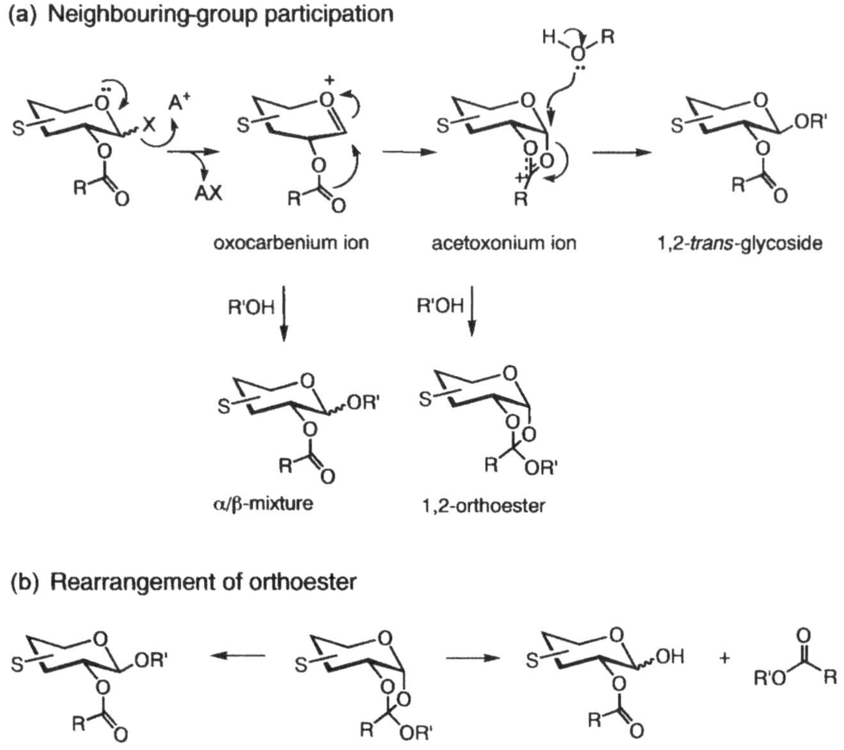

(a) Neighbouring-group participation

| oxocarbenium ion | acetoxonium ion | 1,2-*trans*-glycoside |

R'OH ↓   R'OH ↓

| α/β-mixture | 1,2-orthoester |

(b) Rearrangement of orthoester

Scheme 4.5 Synthesis of 1,2-*trans* glycosides by neighbouring-group participation.

moderately stable product but in other reactions it may rearrange to the desired glycoside or to an aldose and an acylated sugar alcohol. Orthoester formation may be prevented by the use of a C(2) benzoyl or pivaloyl group. In these cases, orthoester formation is disfavoured by the presence of the bulky phenyl or *tert*-butyl group attached to the dioxolane ring. In some cases, the glycosylation may also proceed via the oxocarbenium ion to give mixtures of anomers.

4.3.2 In situ *anomerisation*

The introduction of 1,2-*cis* linkages requires glycosyl donors with a non-participatory protecting group at C(2). An interesting approach for synthesis of α-gluco-type glycosides involves the direct nucleophilic substitution of a β halide by a sugar hydroxyl. Such a reaction will give inversion of configuration at the anomeric centre resulting in the formation of an α glycoside. However, most β halides are very labile and difficult to prepare. In addition, these derivatives equilibrate rapidly to the corresponding α isomer.

A major breakthrough in α-glycosidic bond synthesis came with the introduction of the *in situ* anomerisation procedure.[13] Lemieux and co-workers observed that a rapid equilibrium can be established between α and β halides by the addition of tetra-*n*-butyl ammonium bromide (Scheme 4.6). The anomerisation is believed to proceed through several intermediates. At equilibrium, there is a shift towards the α bromide since this compound is stabilised by an endoanomeric effect. Because, the β bromide is much more reactive towards nucleophilic attack by an alcohol,

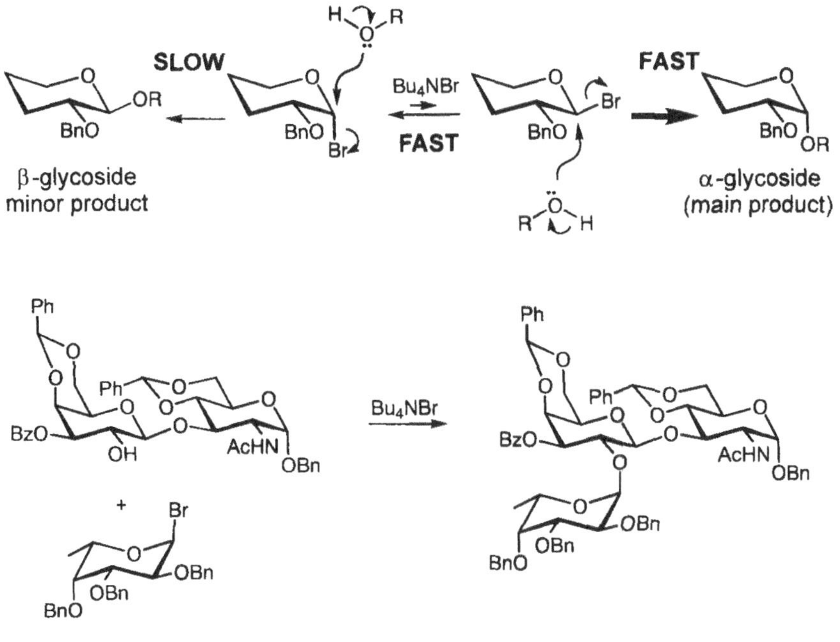

Scheme 4.6   α-Glycosidic bond synthesis by *in situ* anomerisation.

than the more stable α-bromide, glycosylation takes place preferentially on this intermediate in an S_N2 fashion to give mainly α glycosides. An important requirement for this reaction is that the rate of equilibration is much faster than that of glycosylation.

The anomeric outcome of these glycosylations can also be discussed in more general mechanistic terms. First, the product ratio is governed by competing rates of formation of the α and β glycoside and therefore the glycosylation is kinetically controlled. Second, the Curtin–Hammett principle describes that when two reactants are in fast equilibrium, the position of this equilibrium and therefore the reactant ratios will not determine the product ratio. The product ratio, however, will depend on the relative activation energies of the two reactants (α and β halide). In the case of the *in situ* anomerisation procedure, the activation energy for glycosylation of the β anomer is significantly lower than for the α anomer and therefore the reaction proceeds mainly through the β anomer. The origin of the higher reactivity of β halides is disputed but follows similar arguments as the explanation of the kinetic anomeric effect (see Chapter 1). Probably, the α anomer is less reactive because of ground-state stabilisation by an endoanomeric effect.

It is essential that the *in situ* anomerisation is performed in a solvent of low polarity. In polar solvents, the reaction proceeds via an oxocarbenium ion and the anomeric selectivity is reduced.

The efficacy of the *in situ* anomerisation procedure was demonstrated by the condensation of a fucosyl bromide with a glycosyl acceptor in the presence of tetra-*n*-butyl ammonium bromide to give a trisaccharide mainly as the α anomer (Scheme 4.6).

It should be noted that tetra-*n*-alkyl ammonium halides react only with very reactive glycosyl halides. More reactive activators are required for more demanding glycosylations and nowadays a range of activators with different reactivities are available, including $Hg(CN)_2$, $HgBr_2$, $AgClO_4$ and AgOTf.[1a,b]

High α-anomeric selectivities have been obtained with other anomeric leaving groups (Scheme 4.7). For example, trimethylsilyl-triflate-mediated couplings of benzylated trichloroacetimidates at low temperature give in many cases excellent α selectivities. Many examples have been reported in which thioglycosides and glycosyl fluorides also give high α selectivities. It has to be noted that the reaction mechanisms of these glycosylations have been less well studied. However, it is reasonable to assume that they proceed via an *in situ* anomerisation process and probably α and β ion pairs are formed as intermediates.

As mentioned above, it is very important that the equilibration between the two ion pairs be faster than the glycosylation, and many different parameters affect this requirement. Often many different reaction conditions have to be examined in order to obtain satisfactory results. Also, small changes in the constitution of the glycosyl donor or acceptor may have a dramatic effect on the stereochemical outcome of a glycosylation.

4.3.3 Glycosylation with inversion of configuration

The *in situ* anomerisation procedure requires a fast equilibrium being set up between an α and β halide or ion pairs. However, some glycosylation

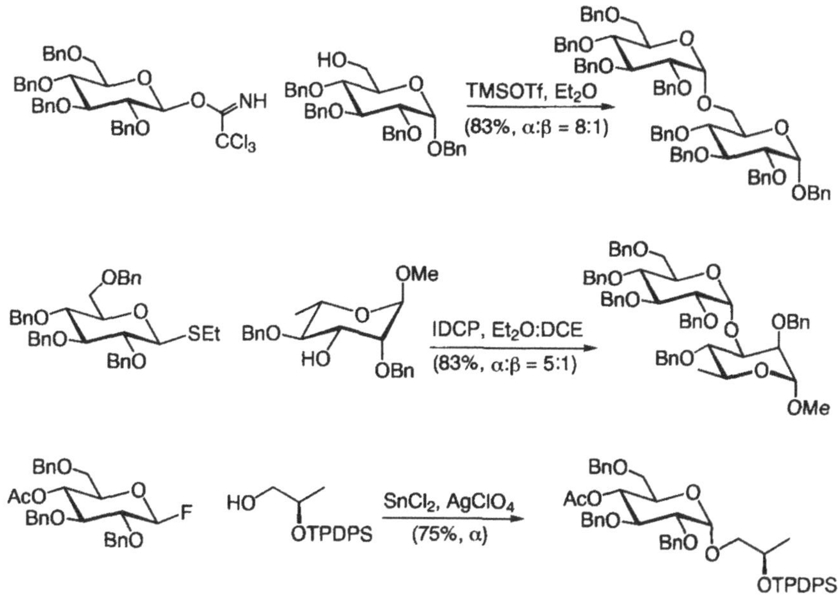

Scheme 4.7 *In situ* anomerisation procedure using different donors.

procedures are based on preventing this pre-equilibration. These protocols rely on glycosylation proceeding with inversion of configuration. For example, glycosylation of α halides in the presence of an insoluble silver salt results mainly in β glycoside formation (Scheme 4.8a).[14] In this case, anomerisation of the halide is prevented because halide nucleophiles are sequestered from the reaction mixture. As a consequence the reaction proceeds with inversion of configuration. Silver silicate and silver silicate-aluminate have often been applied in this capacity. These catalysts have proven to be valuable in the preparation of β-linked mannosides which cannot be prepared by neighbouring-group participation or *in situ* anomerisation. The presence of a nonparticipating substituent at C(2) is an important requirement for glycosylations using a heterogeneous catalyst, however, the nature of the substituents at C(3), C(4) and C(6) can also influence the anomeric ratios of the coupling products.[15]

Glycosylations may also proceed via inversion of configuration when performed in apolar solvents, especially if activation is accomplished with mild promoters. Toepfer and Schmidt showed[16] that BF$_3$ • Et$_2$O-mediated glycosylation of reactive α-glucosyl and α-galactosyl trichloroacetimidate donors in dichloromethane or mixtures of dichloromethane/hexane give mainly β glycosides (Scheme 4.8b).

β Mannosides have also been prepared by direct substitution of an α-manno-glycosyl tosylate donor.[17] However, this method has not been applied widely in oligosaccharide synthesis because of the difficulties involved in the generation of such intermediates. However, α-glycosyl triflates can conveniently be prepared *in situ* by the treatment of an

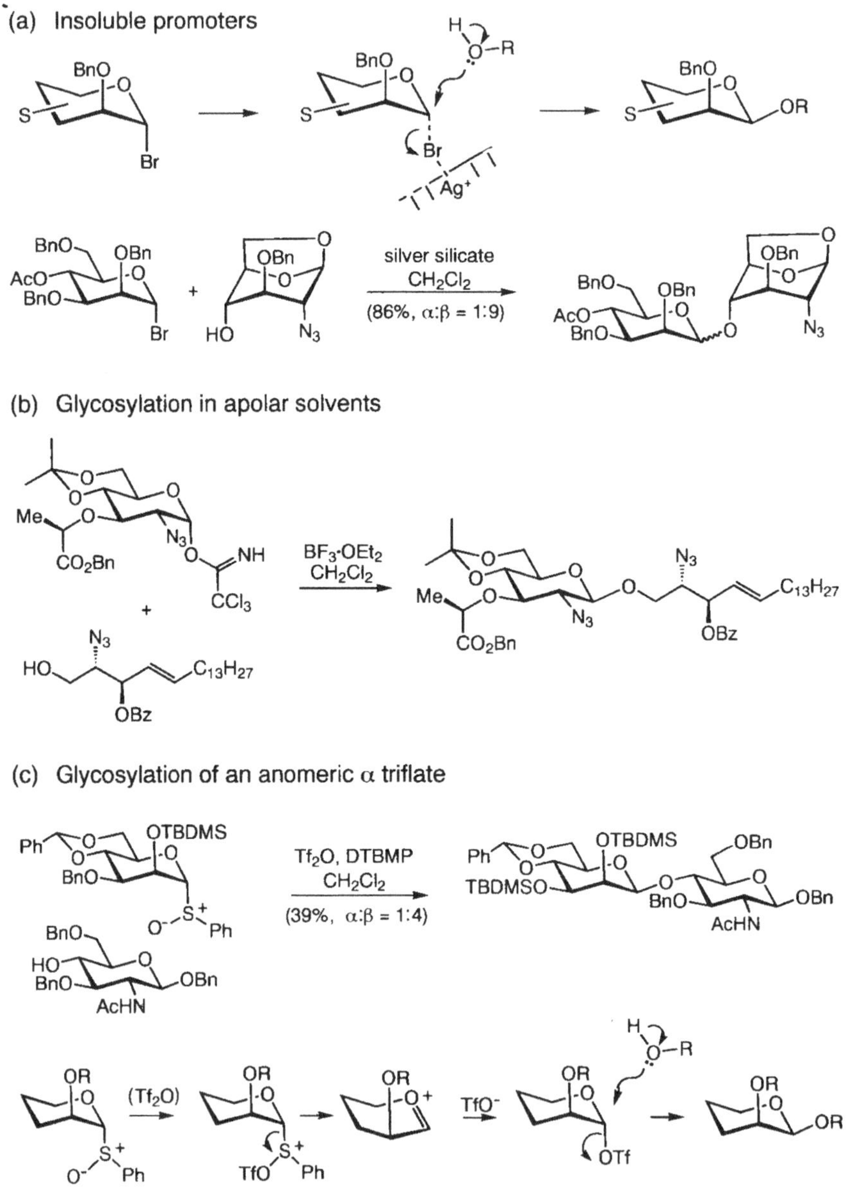

(a) Insoluble promoters

(b) Glycosylation in apolar solvents

(c) Glycosylation of an anomeric α triflate

Scheme 4.8 Glycosylation with inversion of configuration.

(d) Glycosylation of thiocyanates

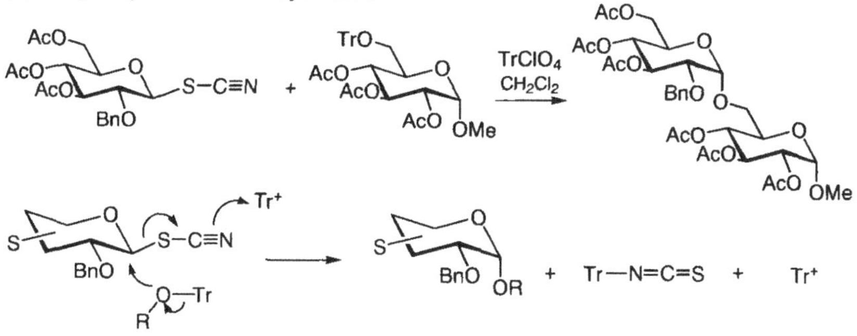

(e) Glycosylation of 1,2-epoxides

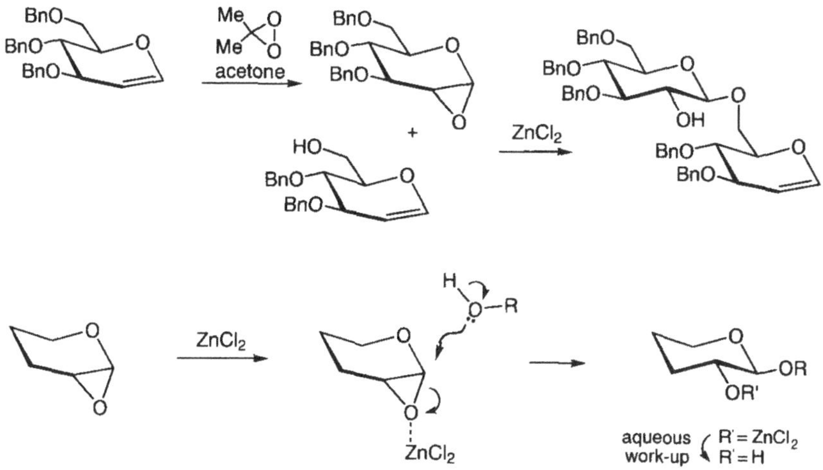

Scheme 4.8 (Continued).

anomeric sulfoxide with triflic anhydride (Tf$_2$O) (Scheme 4.8c). An α triflate is preferentially formed because this anomer is stabilised by an endoanomeric effect. On addition of an alcohol, the triflate is displaced in an S$_N$2 fashion resulting in the formation of a β mannoside.[18] Significantly a mixture of anomers is obtained when triflic anhydride is added to a mixture of a sulfoxide and alcohol. In this case, it is very likely that the glycosylation proceeds through an oxocarbenium ion since triflate formation is less likely because of the greater nucleophilicity of the alcohol. Another prerequisite for β mannoside formation is that the mannosyl donor be protected as a 4,6-O-benzylidene acetal. This observation is difficult to explain but it has been suggested that oxocarbenium ion formation is disfavoured because of torsional strain engendered on going to the half-chair conformation of this intermediate. It should be pointed out that β mannosides are difficult to produce. As already pointed out, methods based on neighbouring-group participation readily lead to α-mannosides. Furthermore, the axial C(2) substituent of a mannosyl donor sterically blocks incoming nucleophiles from the β-face.

Kochetkov et al. have reported[19] an efficient approach for the synthesis of 1,2-cis-pyranosides employing 1,2-trans-glycosyl thiocyanates as glycosyl donors and tritylated sugar derivatives as glycosyl acceptors (Scheme 4.8d). This coupling reaction is initiated by complexation of the nitrogen atom of the thiocyanate with a trityl cation, with simultaneous nucleophilic attack by the oxygen atom of the trityl protected sugar alcohol at the anomeric centre. It appears that this reaction proceeds by clean S_N2 inversion at the anomeric centre.

The substitution pattern of a glycosyl donor can also prevent in situ anomerisation and, under appropriate conditions, glycosylation will take place via S_N2 substitution. For example, Halcomb and Danishefsky have shown[20] that the reaction of 1,2-cis epoxides, obtained by epoxidation of glucals with dimethyldioxirane, with sugar alcohols in the presence of $ZnCl_2$ stereoselectively affords 1,2-trans glycosides (Scheme 4.8e). In this glycosylation, the epoxide is activated by chelation with a Lewis acid ($ZnCl_2$). Nucleophilic attack takes place at the anomeric centre for two reasons. First, trans-diaxial epoxide opening is stereoelectronically favoured over trans-equatorial ring opening. Second, the lone pair on the ring oxygen assists in the departure of an anomeric leaving group making this mode of attack more favourable (Chapter 1). $ZnCl_2$-mediated glycosylations of 1,2-epoxides give in some cases mixtures of anomers[21] if the 1,2-epoxides are derived from galactals. In these cases, the reactions proceed via an S_N1 mechanism. 1,2-Cyclic sulfites have been proposed[22] as alternatives for 1,2-epoxides as these compounds are more readily prepared and are less labile than the latter.

4.3.4 Solvent participation

The anomeric outcome of glycosylations with glycosyl donors having a nonparticipating group at C(2) is markedly influenced by the nature of the solvent. In general, solvents of low polarity are thought to increase the α-selectivity. In these cases, in situ anomerisation is facilitated and the formation of oxocarbenium ions suppressed. Solvents of moderate polarity, such as mixtures of toluene and nitromethane, are highly beneficial when glycosyl donors are used with neighboring group active C(2) substituents. Presumably, these solvents stabilise the positively charged intermediates.

Some solvents may form complexes with the oxocarbenium ion intermediates, thereby affecting the anomeric outcome of a glycosylation. For example, diethyl ether is known to increase the α anomeric selectivity. Probably, diethyl ether participates by the formation of diethyl oxonium ion (Scheme 4.9a). The β configuration of this intermediate is favoured because of the operation of the reverse anomeric effect (see Chapter 1). Nucleophilic displacement with inversion of configuration will give an α glycoside. Recently, it was shown that a mixture of toluene and dioxane provides a more efficient participating solvent mixture.

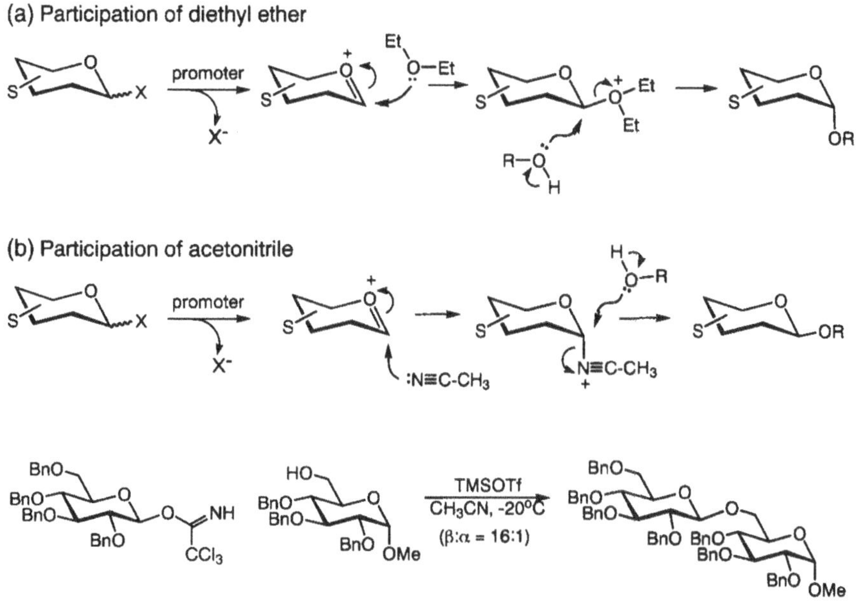

(a) Participation of diethyl ether

(b) Participation of acetonitrile

Scheme 4.9 Glycosylation by participating of solvents.

Acetonitrile is another participating solvent, that in many cases leads to the formation of an equatorially-linked glycoside.[23] It has been proposed that this reaction proceeds via an α-nitrilium ion which is generated under S_N1 conditions (Scheme 4.9b). It is not well understood why the nitrilium ion adopts an axial orientation; however, spectroscopic studies support the proposed anomeric configuration. Nucleophilic substitution of the α-nitrilium ion by an alcohol leads to a β-glycosidic bond.

It has been shown that different types of glycosyl donors (e.g. trichloroacetimidates, fluorides, phosphates and pentenyl-, vinyl- and thio-glycosides) feature the ability to form highly reactive nitrilium intermediates in the presence of acetonitrile. The best β selectivities are obtained with reactive alcohols at low reaction temperatures. Unfortunately, mannosides give poor anomeric selectivities under these conditions.

Finally, an important requirement for control of the anomeric centre by a solvent is the absence of a neighbouring participating functionality at C(2).

4.3.5 Intramolecular aglycon delivery

Recently, Stork and Kim[24] and Barresi and Hindsgaul[25] reported independently the preparation of β mannosides in a highly stereoselective manner by intramolecular aglycon delivery. In this approach, the sugar alcohol (ROH) is first linked via an acetal or silicon tether (Y = CH$_2$ or SiMe$_2$, respectively) to the C(2) position of a mannosyl donor, and subsequent activation of the anomeric centre of this adduct forces the aglycon to be delivered from the β face of the glycosyl donor. The remnant of the tether hydrolyses during the work-up procedure (Scheme 4.10).

A silicon tether can easily be introduced by conversion of a glycosyl donor into a corresponding chlorodimethyl silyl ether and subsequent reaction with the C(6) hydroxyl of an acceptor then gives the silicon connected compound. Oxidation of the phenylthio group yields a phenylsulfoxide which on activation with Tf$_2$O results in the selective formation of a β mannoside in good overall yield. Alternatively, a thioglycoside can be activated directly, also resulting in the formation of a β mannoside.

An acetal tethered compound can easily be prepared by treatment of equimolar amounts of a 2-propenyl ether derivative of a saccharide with a sugar hydroxyl in the presence of a catalytic amount of acid. Activation of the anomeric thio moiety of the tethered compound with N-iodosuccinimide (NIS) in dichloromethane results in the formation of the β-linked disaccharide. In this reaction, no α-linked disaccharide is usually detected. It is of interest to note that when this reaction was performed in the presence of methanol, no methyl glycosides are obtained. This experiment indicates that the glycosylation proceeds via a concerted reaction and not a free anomeric oxocarbenium ion.

The introduction of a methylene acetal tether needs some further discussion. The 2-propenyl ether is prepared by reaction of a C(2) acetyl group with Tebbe reagent, (C$_5$H$_5$)$_2$TiMe$_2$. Treatment of the resulting enol ether with p-toluenesulfonic acid results in the formation of an oxocarbenium ion, which upon reaction with an alcohol provides an acetal. As can be seen in the reaction scheme, the acid is regenerated, thus only a catalytic amount is required.

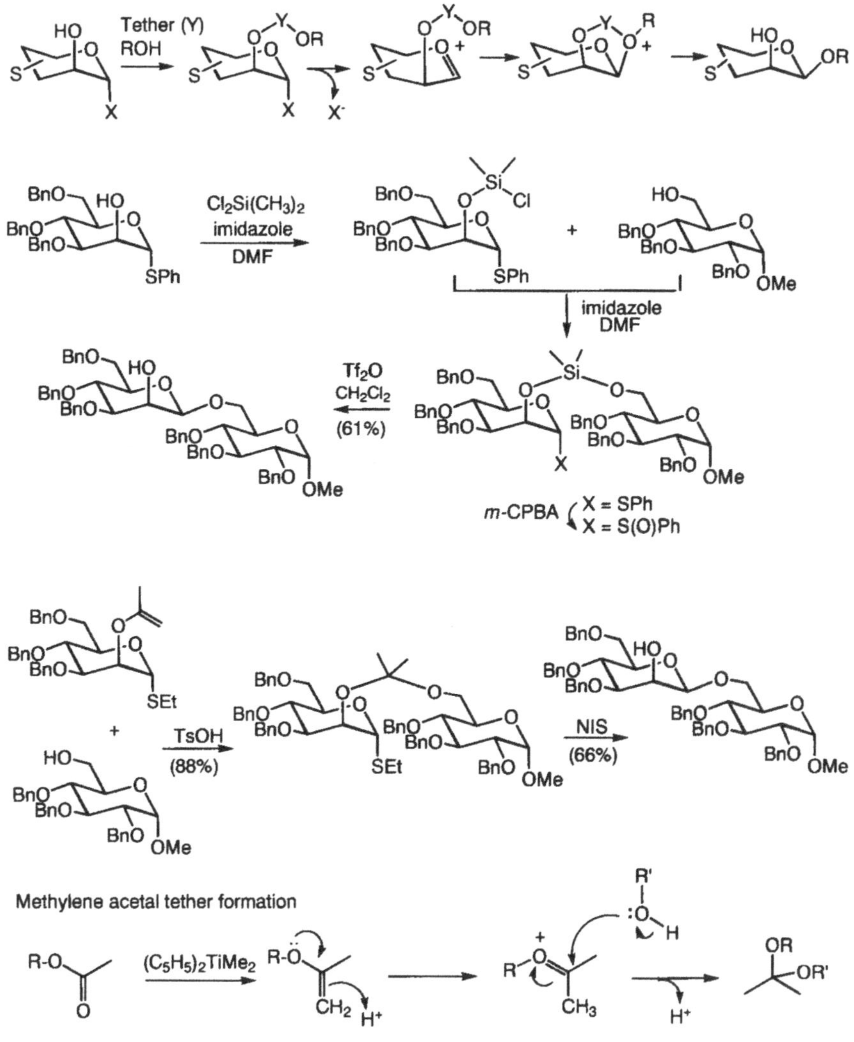

Scheme 4.10 Intramolecular aglycon delivery.

An intermolecular acetal can also be formed by treatment of a mixture of an alcohol and a mannoside, having a methoxybenzyl protecting group at C(2) with 2,3-dichloro-5,6-dicyano-1,4-benzoquinone (DDQ).[26] Intramolecular aglycon delivery has also been used for the preparation of 1,2-*cis*-glucosides.[27] Furthermore, glycosyl acceptors have also been tethered via the hydroxyls at C(4) and C(6).[28] However, in these cases the anomeric ratios in the glycosylations are usually rather disappointing.

Thus, it appears that a rigid five-membered ring transition state is important for high anomeric selectivities.

4.4   Preparation of 2-amino-2-deoxy-glycosides[29, 30]

Amino sugars are widely distributed in living organisms and occur as constituents of glycoproteins, glycolipids, bacterial lipopolysaccharides, proteoglycans and nodulation factors associated with leguminous plants. Glucosamine is the most common amino sugar and is generally found as an *N*-acetylated and β-linked glycoside. Among the bioactive amino sugars, the *N*-function can also be derivatised with fatty acids and sulfates, and several polyamino oligosaccharides possess variously differentiated *N*-acyl residues.

The chemical synthesis of complex oligosaccharides containing amino sugars is the focus of extensive research and requires amino protecting groups that are compatible with common protecting group manipulations and glycosylations. Such groups must be capable of being removed or exchanged readily and chemoselectively under mild conditions. Furthermore, as for the preparation of glycosides with a 2-hydroxyl, a participating protecting group is required when a 1,2-*trans* glycoside is desired, whereas a nonparticipating group is essential for the synthesis of a 1,2-*cis* 2-amino-glycoside.

Traditionally, 2-acetamido-2-deoxy-glycosyl donors were used for the synthesis of 1,2-*trans* glycosides. Activation of anomeric acetates or chlorides leads to the formation of relatively stable 1,2-oxazolines (Scheme 4.11a). These compounds can be isolated and used in a subsequent acid-catalysed glycosylation to give β-linked disaccharides. Yields are generally high for primary alcohols but modest for secondary alcohols. For the glycosylation of secondary alcohols, more reactive donors have to be used that do not react through 1,2-oxazolines.

To prevent oxazoline formation, 2-deoxy-phthalimido protected glycosyl donors have been introduced. These derivatives are often the compounds of choice for the preparation of 1,2-*trans*-glycosides of 2-amino-2-deoxy-glycosides (Scheme 4.11b). The phthalimido group is compatible with many glycosylation protocols and gives high yields when primary as well as secondary glycosyl acceptors are used. The introduction and cleavage of this group is discussed in Chapter 2. In some cases, the relatively harsh conditions required for the cleavage of the phthalimido group is problematic. Furthermore this protecting group can be damaged by strong basic conditions. Recently, the tetrachlorophthalimido, dichlorophthalimido, *n*-pentenoyl, dithiosuccinoyl, and *N*,*N*-diacetyl groups have been proposed as alternatives to the *N*-phthalimido group. These groups can be removed under milder reaction

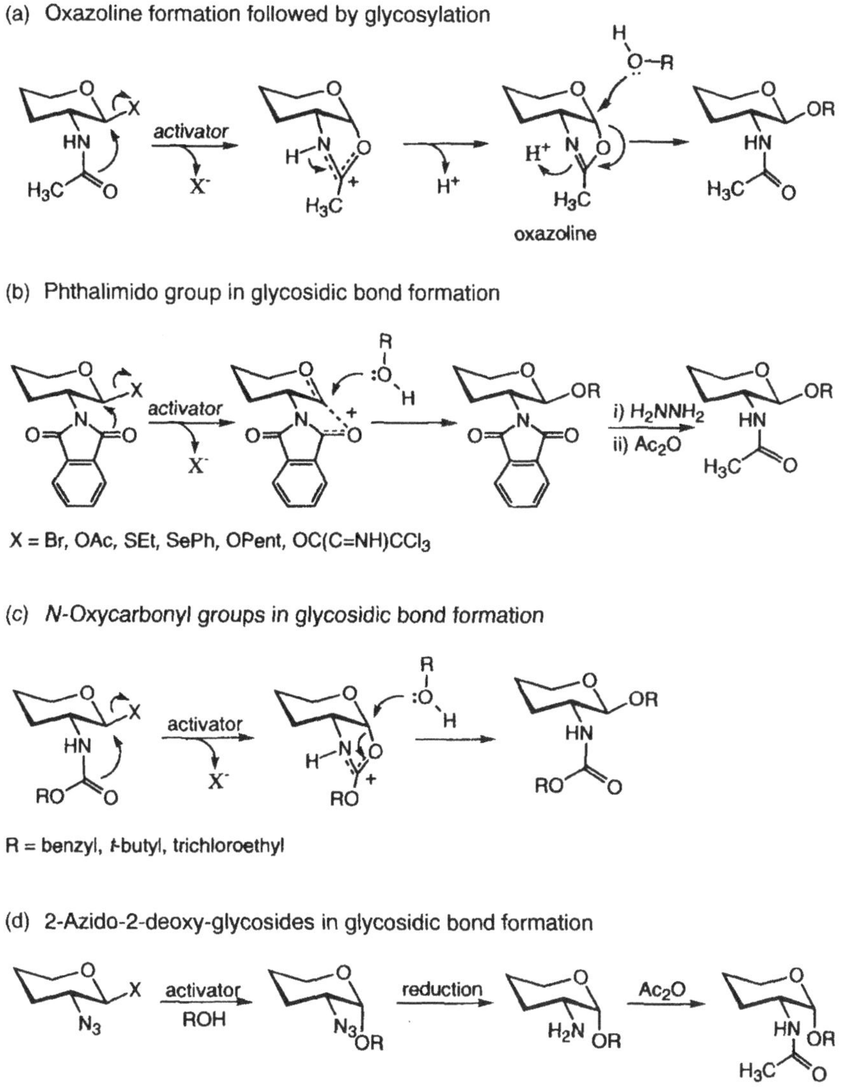

(a) Oxazoline formation followed by glycosylation

oxazoline

(b) Phthalimido group in glycosidic bond formation

X = Br, OAc, SEt, SePh, OPent, OC(C=NH)CCl₃

(c) *N*-Oxycarbonyl groups in glycosidic bond formation

R = benzyl, *t*-butyl, trichloroethyl

(d) 2-Azido-2-deoxy-glycosides in glycosidic bond formation

Scheme 4.11 Preparation of 2-amino-2-deoxy-glycosides.

conditions. The problem of these protecting groups, however, is that they are less stable to basic reaction conditions and will tolerate only a limited range of reaction conditions.

N-alkoxy carbonyl groups have also been applied for the synthesis of 1,2-*trans* glycosides (Scheme 4.11c). These groups are neighbouring-group active and effectively guide the glycosylation via a cyclic oxocarbenium ion to give 1,2-*trans* glycosides. In these cases, oxazoline formation is less likely because of the presence of the alkoxy substituent. As anomeric groups, bromides, trichloroacetimidates and thioglycosides have been applied. In some cases the formation of cyclic carbamates has been observed.

The synthesis of α linked *O*-glycosides of glucosamines and galactosamines requires a nonparticipating functionality at C(2) and traditionally the azido group has been selected for this purpose (Scheme 4.11d). To obtain high α selectivities, the reaction should proceed through an *in situ* anomerisation process. As with other nonparticipating functionalities at C(2), β-linked glycosides can be prepared by using acetonitrile as the solvent. An azido group can be introduced using several different methods, which are discussed in Chapter 3, and can be reduced using a wide range of reducing reagents to give an amino group. An azido group is compatible with many different glycosylation protocols. Furthermore, the group has the additional advantage of being stable to rather acidic and basic reaction conditions, making it compatible with many chemical manipulations.

4.5 Formation of glycosides of *N*-acetyl-neuraminic acid[31]

The aforementioned methods for stereoselective glycosylation are mainly applicable to aldoses with a substituent at C(2). However, there are other types of glycoside, the preparation of which requires special consideration. *N*-Acetyl-α-neuraminic acid (Neu5Ac) frequently terminates oligosaccharide chains of glycoproteins and glycolipids of cell membranes and plays vital roles in their biological activities. The use of derivatives of Neu5Ac as glycosyl donors is complicated by the fact that no C(3) functionality is present to direct the stereochemical outcome of glycosylation (Scheme 4.12a). In addition, the electron-withdrawing

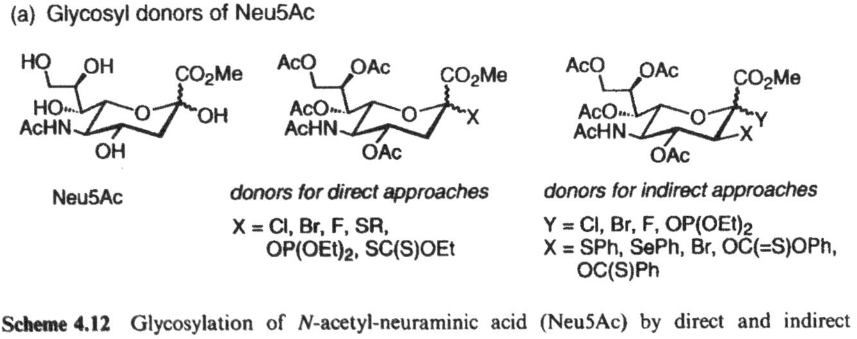

Scheme 4.12 Glycosylation of *N*-acetyl-neuraminic acid (Neu5Ac) by direct and indirect approaches.

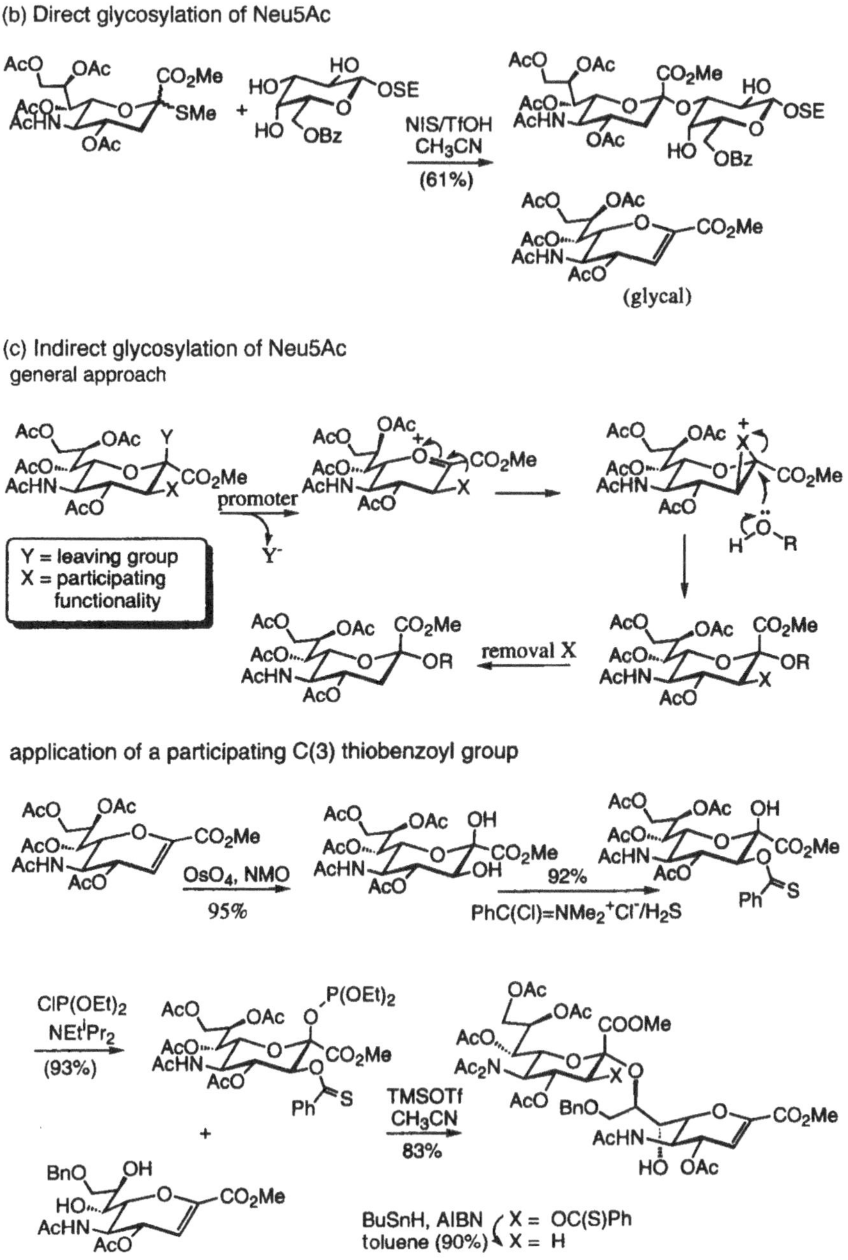

(b) Direct glycosylation of Neu5Ac

(c) Indirect glycosylation of Neu5Ac
general approach

Y = leaving group
X = participating
functionality

application of a participating C(3) thiobenzoyl group

Scheme 4.12   (Continued).

carboxyl group at the anomeric centre makes these derivatives prone to elimination. Finally, the glycosylation of Neu5Ac has to be performed at a tertiary oxocarbenium ion. Before discussing the approaches for glycosylation of Neu5Ac donors, a closer look needs to be taken at the sugar ring conformation and the nomenclature of the anomeric configuration of this monosaccharide derivative. As can be seen in Scheme 4.12a, the sugar ring of Neu5Ac adopts a $^1C_4$ conformation. In this case, the bulky side chain and C(5) acetamido derivative are in an equatorial orientation. In this particular case, the equatorial glycoside is the α anomer and the axial one the β glycoside. In natural oligosaccharides, glycosides of Neu5Ac are in an α configuration.

Glycosides of N-acetyl neuraminic acid can be introduced by direct or indirect approaches. In a direct approach, a fully protected Neu5Ac derivative having a leaving group at C(2) is coupled with a sugar alcohol (Scheme 4.12a,b). Silver or mercury salt promoted activation of bromides and chlorides of N,O-acylated neuraminic acid esters give, particularly with secondary sugar hydroxyl groups as acceptors, only modest yields of the desired α-linked coupling products. Thioglycosides of neuraminic acid derivatives have been used successfully as sialyl donors.[32] These compounds are readily available, stable under many different chemical conditions and undergo glycosylations in the presence of activators such as N-iodosuccinimide–triflic acid (NIS-TfOH), dimethyl(methylthio)sulfonium trifluoromethanesulfonate (DMTST) and phenyl selenyl triflate (PhSeOTf). For example, NIS–TfOH mediated coupling of a galactosyl acceptor having free hydroxyls at C(2), C(3) and C(4) with a thioglycoside of Neu5Ac in acetonitrile gave only the α-(2,3)-linked disaccharide in a yield of 61%. Apart from the coupling product, an elimination product (glycal) was also isolated. The regioselectivity of this reaction is due to the greater reactivity of the equatorial alcohol compared with the axial C(3) hydroxyl. Furthermore, the C(2) hydroxyl has a lower nucleophilicity as a result of the electron-withdrawing effect of the anomeric centre. When a similar glycosylation was performed with a galactosyl acceptor having only a free C(3) hydroxyl group, the yield and anomeric stereoselectivity was significantly reduced. It should be realised that the preparations of glycosyl acceptors with several free hydroxyls are often easier. Hence, such approaches may offer shorter routes to oligosaccharides. Sialylations have also been successfully performed using phosphites[33] or xanthates[34] as the anomeric leaving group. Furthermore, the reactivity of these donors can be further improved by conversion of the N-acetyl group into an N-acetyl acetamido functionality (NAc$_2$).[35]

Several indirect glycosylation methods have been described which take advantage of a temporary functionality at C(3) in Neu5Ac (Scheme

4.12a, c).[36] These C(3) functionalities perform neighbouring-group participation during a glycosylation and suppress possible elimination reactions. An efficient C(3) auxiliary is the thiobenzoyl moiety, which can be introduced via a five-step procedure from Neu5Ac. Thus, 2,3-dehydroneuraminic acid is stereoselectively dihydroxylated with a catalytic amount of OsO_4 and N-methylmorpholine N-oxide (NMO). A C(3)-thiobenzoyl group is then regioselectively introduced with N,N-dimethyl-α-chlorobenzimidium chloride and hydrogen sulfide, and the anomeric hydroxyl is converted into a phosphite by treatment with diethyl chlorophosphite. The resulting glycosyl donor has proven to be a suitable substrate for the glycosylation of the 8-OH of a Neu5Ac acceptor. This hydroxyl has a very low reactivity in glycosylations probably because of hydrogen bonding with the carboxyl group at the anomeric centre. Glycal derivatives of Neu5Ac can be used as glycosyl acceptors to minimise this unfavourable hydrogen bonding. The thiobenzoyl group is usually removed by reaction with tributyltin hydride and azobisisobutyronitrile (AIBN). Despite the fact that the indirect methods provide reliable approaches for the preparation of α-sialic acid derivatives, they are all hampered by the rather laborious nature of the synthetic sequence that is required. It should be noted that particular glycosides of Neu5Ac can efficiently be prepared by an enzymatic approach (see Section 4.10.1).

4.6 The introduction of 2-deoxy glycosidic linkages[37]

The macrolides, anthracyclines, cardiac glycosides and aureolic acids are important classes of glycosylated compounds that share the same feature; that is, they contain 2,6-di-deoxy-glycosides. The introduction of 2-deoxy-α/β-glycosidic linkages requires special consideration. The absence of a functionality at C(2) excludes neighbouring-group-assisted glycosylation procedures, and therefore control of anomeric selectivity is problematic. Additionally the glycosidic linkages of 2-deoxy derivatives are much more acid sensitive. This feature can easily be explained by the fact that cleavage of the anomeric moiety proceeds through an intermediate oxocarbenium ion, which is destabilised by an electron-withdrawing oxygen at C(2). Thus, replacement of a 2-hydroxyl by a hydrogen will result in less destabilisation of this intermediate. Glycosyl donors of 2-deoxy glycosides are much more reactive than their 2-hydroxy counterparts for the same reason that the oxocarbenium ions of these derivatives are of considerable stability and therefore easier to form.

While 2-deoxy-glycosyl halides have been employed in glycosidic bond synthesis, the yields and stereochemical outcomes can often be rather disappointing. The increased stability, ease of preparation and excellent reactivity of 2-deoxy-thioalkyl(phenyl) glycosides and glycosyl fluorides make these derivatives a better choice for use as glycosyl donors.

Reliable approaches for obtaining 2-deoxy glycosides are based on the introduction of a temporary directional functionality at C(2) and, in this respect, glycals are often employed (Scheme 4.13a).[38] Several electrophilic reagents add stereoselectively to the electron-rich enol-ether double bond of a glycal, leading to a three-membered cyclic cationic intermediate which can regioselectively be opened by alcohols leading to the formation of 1,2-*trans* glycosides. Removal of the anchimeric group (E) furnishes the 2-deoxy glycoside (Scheme 4.13a). The stereochemical outcome of this reaction needs further discussion. The electrophile can add to either the α or the β face of the glycal derivative, leading to intermediates **A** or **B**, respectively. Intermediate **B**, in which the electrophile adopts a pseudo-equatorial orientation at the anomeric centre, is stabilised by the reverse anomeric effect. This favours a charged substituent at the anomeric centre tends to adopt an equatorial orientation (see Chapter 1). Nucleophilic

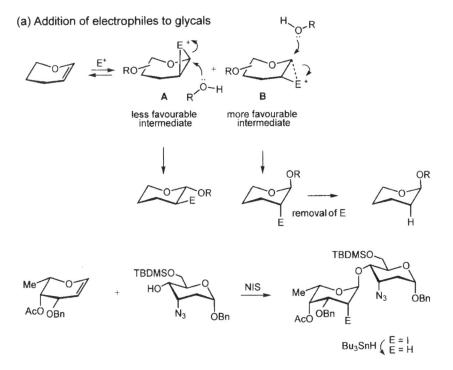

Scheme 4.13 Preparation of 2-deoxy glycosides.

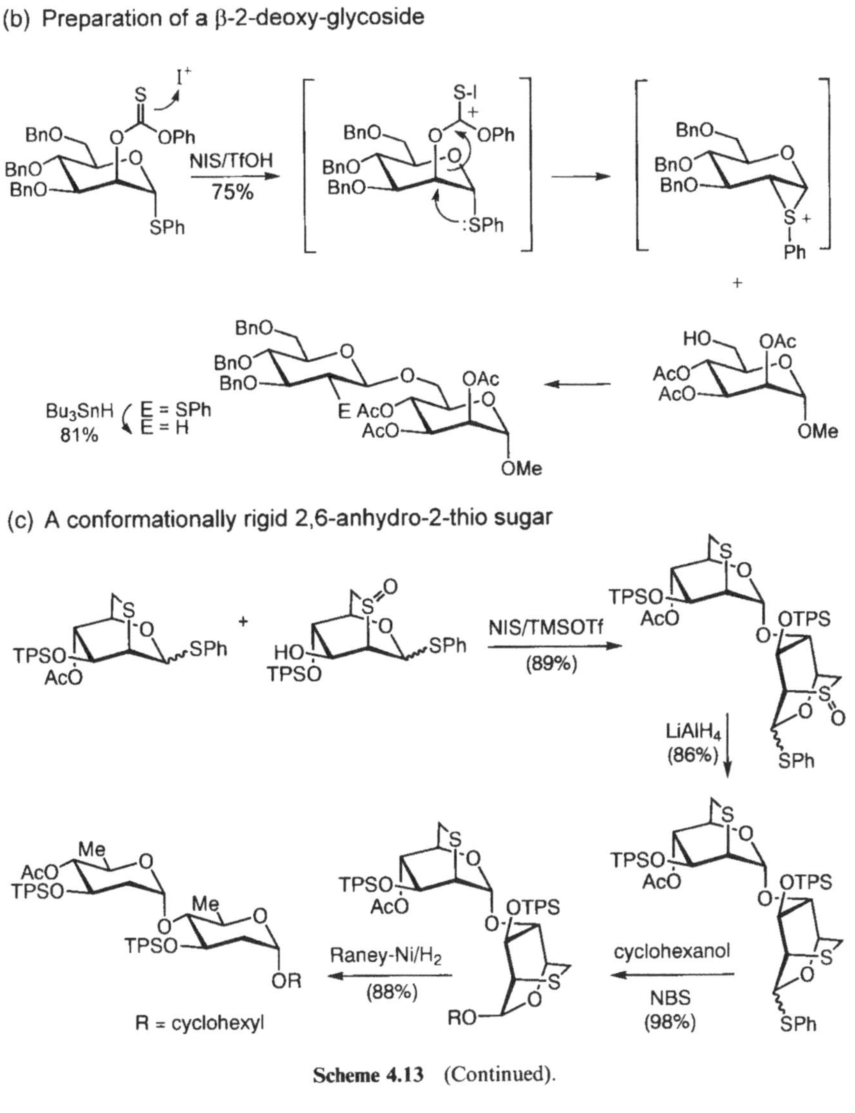

(b) Preparation of a β-2-deoxy-glycoside

(c) A conformationally rigid 2,6-anhydro-2-thio sugar

Scheme 4.13 (Continued).

ring opening of **B** by an alcohol would also be preferred over the ring opening of **A**. First, nucleophile attack at the anomeric centre is electronically much more favourable than attack at C(2). This observation can be explained by assistance of the ring oxygen in the departure of an anomeric leaving group making this the preferred mode of attack. Second, nucleophilic ring opening of intermediates **A** and **B** follows the same rules as epoxide opening. Thus, a 1,2-*trans* diaxial opening is the stereoelectronically preferred reaction mode. Nucleophilic ring opening at the anomeric centre of **A** would give a *trans*-diequatorial product and would be less favourable stereoelectronically. On the other hand, opening of **B** would give the preferred *trans*-diaxial product. Based on this discussion, it is clear that intermediate **B** will be the one that will lead to product formation. *N*-iodosuccinimide (NIS) is often used as the electrophilic reagent. In this case, the iodonium ion (I^+) adds to the double bond of the glycal derivative in a reversible manner. After glycosylation, the iodide at C(2) of the glycosylation product is usually removed by radical dehalogenation using Bu_3SnH.

The stereoselective synthesis of 2-deoxy-β-glycosides is still a major challenge, and the most reliable methods are based on an equatorially installed heteroatom substituent that guides the course of the glycosylation by anchimeric participation and is removed reductively to give the β linked 2-deoxy-glycosides.[39] The following C(2) substituents have been successfully applied: bromides, selenides, sulfenides and thiobenzoates. Alternatively, a thioglycoside having an axial phenoxythiocarbonyl ester on C(2) will, upon activation with iodonium ions, give a 1,2-cyclic sulfonium intermediate which can react with a glycosyl acceptor to afford a 1,2-*trans* glycoside having a β anomeric configuration (Scheme 4.13b). In this reaction, the phenoxythiocarbonyl ester is activated by an iodonium ion rather than the anomeric thio functionality. Raney-nickel-mediated desulfurisation of the SPh moiety gives the required product in high yield. α-Linked 2-deoxy-glycosides can be obtained when a gluco-type glycosyl donor is used in this reaction. The regioselective opening of the 1,2-cyclic sulfonium intermediate merits some further discussion. Normally, 1,2-cyclic sulfonium ions are opened in a *trans* diaxial fashion. In the present case, this would require attack to take place at C(2). However, electronically attack is preferred at the anomeric centre. The latter factor is the overriding determinant of the regioselective outcome of the reaction.

Toshiba *et al.* have designed conformationally rigid glycosyl donors which possess a thio ether bridge between the C(2) and C(6) position, and these compounds have been used for the stereoselective synthesis of 2,6-di-deoxy glycosides (Scheme 4.13c).[40] Chemoselective activation of the anomeric thiophenyl moiety with NIS/TMSOTf and reaction with the glycosyl acceptor gave mainly α-linked disaccharide in high yield. The chemoselectivity of this reaction is based on the greater reactivity of the 2,6-anhydro-2-thio-glycosyl donor compared with that of the 2,6-anhydrosulfenyl sugar. This observation can easily be explained: the fomation of an oxocarbenium ion is probably the rate-determining step in the glycosylation. The sulfenyl moiety is strongly electron-withdrawing and will disfavour positive charge development at the anomeric centre. As a result, such derivatives are of low reactivity in glycosylations (see Section 4.8 for other chemoselective glycosylations). The sulfoxide moiety was reduced with lithium aluminium hydride and the product was glycosylated with cyclohexanol. Finally, reductive cleavage of the thioethers gave 2,6-dideoxy glycosides.

4.7 Convergent block synthesis of complex oligosaccharides

Oligosaccharides can be prepared by a linear glycosylation strategy or by block synthesis. In a linear glycosylation strategy, monomeric glycosyl donors are added to a growing saccharide chain. Such an approach is less efficient than when oligosaccharide building blocks are used as glycosyl donors and acceptors (convergent approach). Glycosyl bromides have been used in block synthesis; however, results were often rather disappointing, especially with labile bromides.

Nowadays a variety of glycosyl donors are available which can be prepared under mild conditions, are sufficiently stable to be purified and stored for a considerable period of time. Many donors also undergo glycosylation under mild conditions, and by selecting appropriate reaction conditions they give high yields and good α/β ratios. These features allow the preparation of oligosaccharides by efficient block synthesis.

The favourable properties of the trichloroacetimidate methodology have been exploited in the block synthesis of the prominent tumour-associated dimeric antigen Lewis X (Le^x).[41] The retrosynthetic strategy is depicted in Figure 4.3. In order to make efficient use of common building blocks, it was decided to disconnect the octasaccharide into two trimeric units and a lactoside residue. The trisaccharide was further disconnected into a fucose and a lactosamine moiety. The lactoside was readily available from lactose. Thus, the strategy was designed in such a manner that optimal use could be made of the cheaply available disaccharide lactose. In such an approach, the number of glycosylation steps would be considerably reduced. The key building blocks for the preparation of the target compound I were 1, 2 and 3.

The azido-lactosyl building block 2 was prepared by azidonitration of lactal (see Chapter 3), followed by selective protection. The selectively protected lactoside 3 was readily available from lactose via a sophisticated protecting group interconversion strategy. α-Fucosylation of acceptor 2 with the very reactive fucosyl donor 1 gave trisaccharide 4 in an 89% yield (Scheme 4.14). The trisaccharide 4 was converted into the required glycosyl donor 5 and acceptor 6. Thus, removal of the anomeric tert-butyldimethyl silyl (TBDMS) protecting group of 4 with tetra-n-butylammonium fluoride (TBAF) and treatment of the resulting lactol

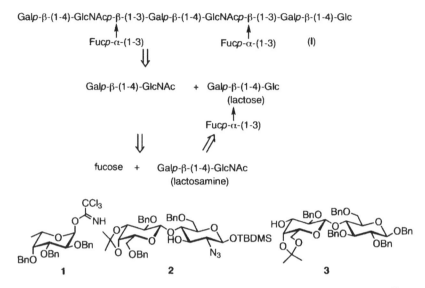

Figure 4.3 Retrosynthetic analysis and building blocks for the assembly of SLe^x.

with trichloroacetonitrile in the presence of DBU afforded trichloroacetimidate **5** in a good overall yield. However, cleavage of the isopropylidene moiety of **4** under mild acidic conditions furnished **6**. Coupling of glycosyl donor **5** with acceptor **6** in the presence of $BF_3 \cdot Et_2O$ as catalyst gave the hexasaccharide **7** in a 78% yield. In the latter reaction, the higher acceptor reactivity of the equatorial 3-OH group with respect to the axial 4-OH was exploited. The synthesis of octasaccharide **10** required a repetition of the above-described strategy, that is, conversion of the anomeric TBDMS group into a trichloroacetimidate functionality and coupling of the trichloroacetimidate **9** with lactoside unit **3** (64%). Finally, target molecule **I** was obtained by reduction of the azido group of **10**, followed by acetylation of the amino group and hydrogenation under acidic conditions. Using a similar approach, a spacer containing dimeric and trimeric Lewis X antigen was synthesised.[4fe]

The described glycosylation strategy was highly convergent and made optimal use of the common trisaccharide **4**. It also exploited the commercially available disaccharide lactose. Another noteworthy aspect of this synthesis was that the trichloroacetimidates were prepared in high yield, and these donors performed very well in the glycosylation reactions (high yields and high anomeric selectivities). The latter point deserves further comment. It should be realised that some types of glycosidic linkages can be constructed rather easily whereas others impose great difficulties. In planning a synthetic scheme, the disconnections have to be

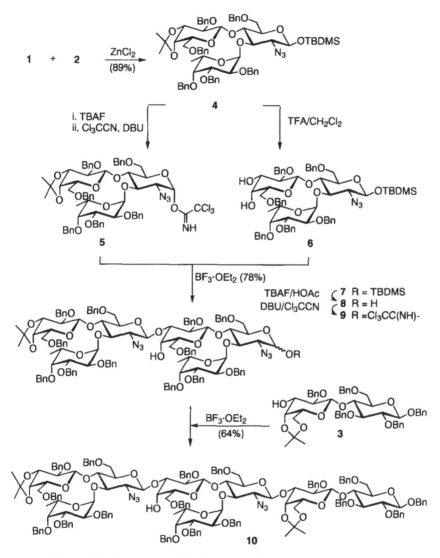

Scheme 4.14 Preparation of SLe[x] by the trichloroacetimidate methodology.

chosen in such a way that the block assembly will not impose problems. Furthermore, difficult glycosylations should be performed in an early stage in the synthesis.

For some oligosaccharide target molecules it may be advantageous to use a range of different anomeric leaving groups in the glycosylation strategy. For example, a fragment of the capsular polysaccharide of

Group B type III *Streptococcus* was prepared by employing thioglycosyl, cyanoethylidene, and pentenyl glycosides (Scheme 4.15).[35,42] The heptasaccharide **17** was assembled from the building blocks **11**, **12**, **14** and **16** and several important strategic aspects featured in the overall synthetic plan. For the preparation of building blocks **14** and **16**, readily available lactose was selected as the starting material as this reduced the

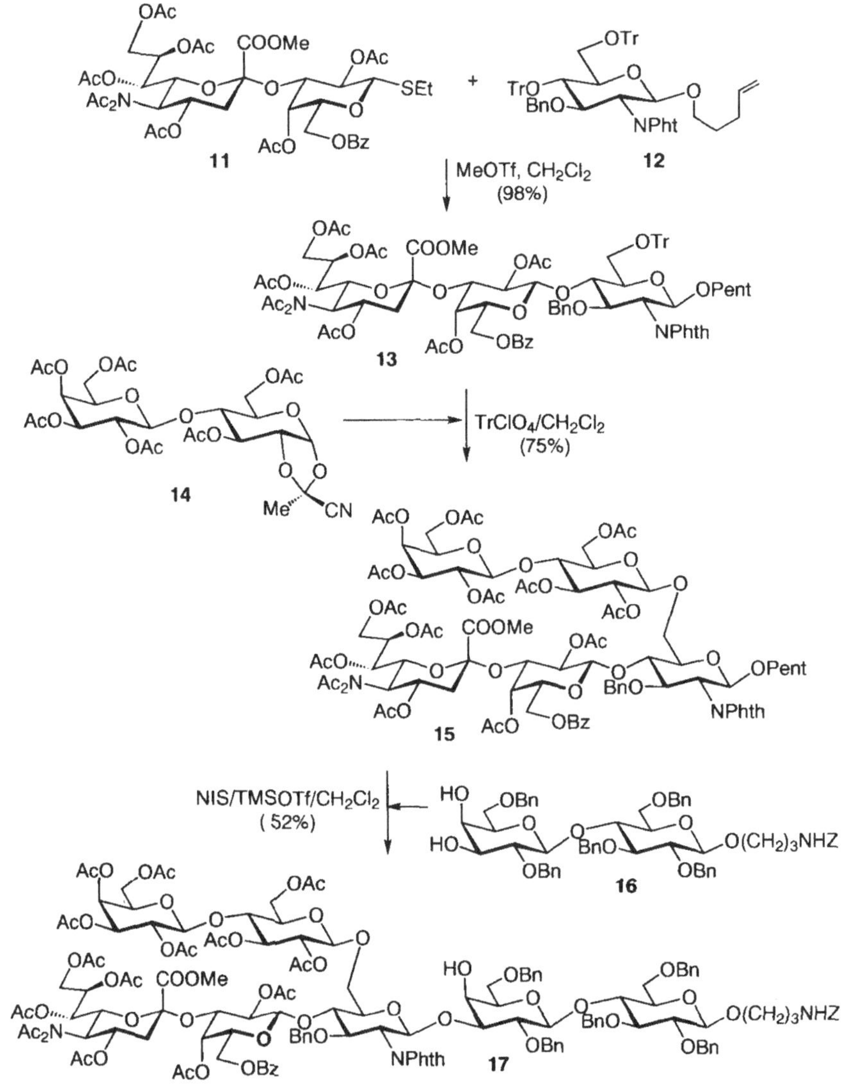

Scheme 4.15 Convergent synthesis of the oligosaccharide of Group B type III *Streptococcus.*

number of glycosylation steps that needed to be performed. Trityl ethers are convenient glycosyl acceptors and it has been shown that secondary trityl ethers are in general much more reactive than primary trityl ethers. These features permit first galactosylation of the trityl ether at C(4) in **12** followed by lactosylation of the trityl at C(6) with **14**. The different reactivities of *n*-pentenyl and thioglycosides allow thioglycosides to be activated in the presence of an *N*-pentenyl glycoside. These combinations of chemical feature facilitated the assembly of **10** in a convergent manner from the building blocks **11**, **12**, **14** and **16** without a single protecting group manipulation. Compound **11** was easily prepared by the strategy outlined in Section 4.5. Coupling of thioglycoside **11** with di-*O*-trityl thioglycosyl acceptor **12** in the presence of MeOTf proceeded with absolute regio- and stereoselectivity, and disaccharide **13** was obtained in a quantitative yield. To prevent cleavage of the trityl ethers, the reaction was performed in the presence of a relatively large amount of activated molecular sieves (3 Å). The preclusion of hydrolysis was very important as it was observed that an analogous saccharide having a free 4-hydroxyl group could not be glycosylated, confirming that tritylation activates a hydroxyl for glycosylation. It is also important to note that the C(5) acetamido group of the Neu5Ac derivative was protected as an *N*-acetyl-acetamido to prevent *N*-methylation by methyl triflate. The additional *N*-acetyl group also improved the preparation of disaccharide **11**. Next, the less reactive primary trityl ether of **13** was glycosylated with the cyanoethylidine derivative **14** in the presence of a catalytic amount of trityl perchlorate to furnish pentasaccharide **15** in a good yield. The pentenyl moiety of pentasaccharide **15** remained intact throughout the synthetic steps leaving it available to serve as an efficient anomeric leaving group. Thus, coupling of **15** with **16** in the presence of NIS/TMSOTf gave the requisite heptasaccharide **17** in an acceptable yield.

4.8 Chemoselective glycosylations and one-pot multistep glycosylations

An important requirement for convergent oligosaccharide synthesis is ease of accessibility of oligosaccharide building blocks. Fraser-Reid and co-workers introduced[43] a chemoselective glycosylation (armed – disarmed glycosylation strategy) which allows the preparation of this type of unit with a minimum number of protecting group manipulations. It was shown that pentenyl glycosides having a C(2) ether protection could be coupled chemoselectively to C(2) benzoylated pentenyl glycosides. The chemoselectivity relies on the fact that an electron-withdrawing C(2) ester deactivates (disarms) and an electron-donating C(2) ether activates (arms), the anomeric centre. Thus, coupling of an armed donor with a disarmed acceptor, in the presence of the mild activator iodonium dicollidine perchlorate (IDCP), gave a disaccharide as an anomeric mixture in a yield of 62% (Scheme 4.16). Next, the disarmed compound was further glycosylated with an acceptor, using the more powerful

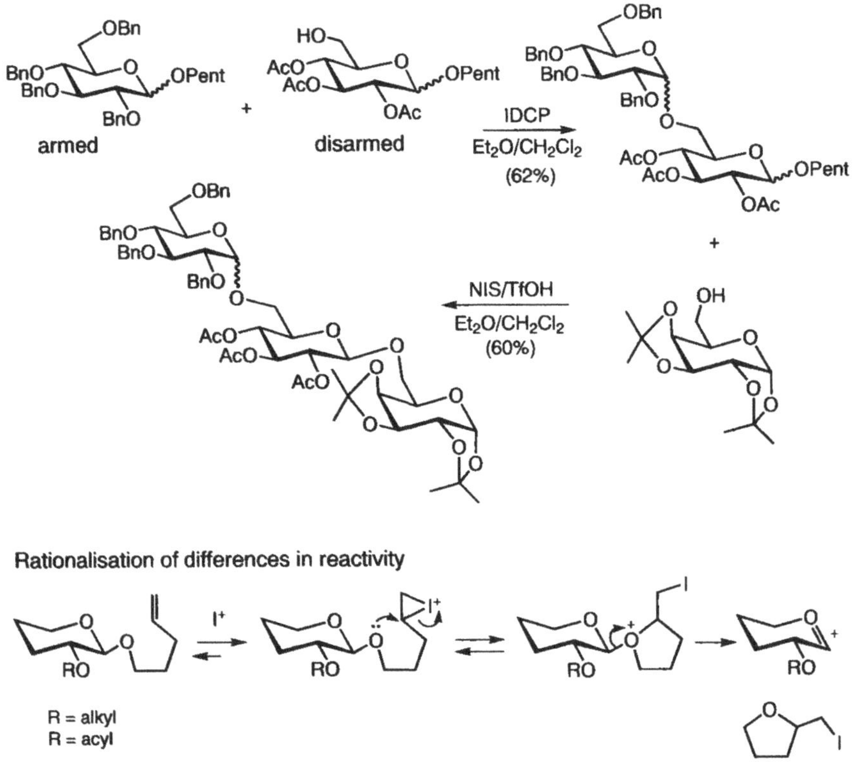

Rationalisation of differences in reactivity

R = alkyl
R = acyl

Scheme 4.16 Chemoselective glycosylations with pentenyl glycosides.

activating system N-iodosuccinimide/catalytic triflic acid (NIS/TfOH) to yield a trisaccharide (60%). Thus, this chemoselective glycosylation approach allows the preparation of a trisaccharide without a single protecting group manipulation during glycosylation.

The difference in reactivity between alkylated and acylated pentenyl glycosides can be rationalised as follows: the elctrophilic iodonium ion will add to the double bond of the pentenyl moiety to give a cyclic iodonium ion. Nucleophilic attack by the oxygen will lead to an oxonium ion intermediate which then forms an oxocarbenium ion and an iodo-tetrahydrofuran derivative. The aglycone oxygen will be of low nucleophilicity when the pentenyl derivative has an electron-withdrawing ester at C(2). On the other hand, protection of C(2) with an electron-donating ether substituent will increase the nucleophilicity and, hence, this substrate will react considerably faster. The effect of protecting groups upon anomeric reactivity has been known for many years; however, Fraser-Reid et al. were the first to exploit this effect in chemoselective glycosylation.

Chemoselective glycosylations have also been developed for thioglyco-sides[44], glycals[45], phosphorus-containing leaving groups[46] and fluor-ides.[47] In the case of thioglycosides, a relatively wide range of glycosyl donors and acceptors with differential reactivities has been developed which allows the preparation of larger oligosaccharide building blocks.[48]

Several methods have been reported to perform chemoselective glyco-sylations as a one-pot procedure. Kahne and co-workers described a glycosylation method that is based on activation of anomeric phenylsulf-oxides with triflic anhydride (Tf_2O) or triflic acid (TfOH).[12] Mechanistic studies revealed that the rate limiting step in this reaction is triflation of the sulfoxide; therefore the reactivity of the glycosyl donor could be influenced by the substituent in the *para* position of the phenyl ring and the following reactivity order was established OMe > H > NO_2. The reactivity difference between a *p*-methoxyphenyl sulfonyl donor and an unsubstituted phenylsulfonyl glycosyl acceptor is large enough to permit selective activation. In addition, silyl ethers are good glycosyl acceptors when catalytic triflic acid is the activating agent but react more slowly than the corresponding alcohol. These features opened the way for a one-pot synthesis of a trisaccharide from a mixture of three monosaccharides (Scheme 4.17).[49] Thus, treatment of the mixture with triflic anhydride resulted in the formation of the expected trisaccharide in a 25% yield. No other trisaccharides were isolated and the only other coupling product was a disaccharide. The products of the reaction indicate that the glycosylation takes place in a sequential manner. First, the most reactive *p*-methoxyphenylsulfenyl glycoside is activated to react with the alcohol but not with the silyl ether. In the second stage of the reaction, the less reactive silyl ether of the disaccharide reacts with the less reactive sulfoxide to give the trisaccharide. The phenylthio group of the trisaccharide can be oxidised to a sulfoxide, which is used in a subsequent glycosylation. The obtained product is part of the natural product Ciclumycin 0 and despite the relatively low yield of the coupling reactions this methodology provides a very efficient route to this compound.

Several variations of this concept have been reported.[50–52] For example, Ley and Priepke[50] prepared the trisaccharide unit, which is derived from the common polysaccharide antigen of a group B *Streptococci* by a facile one-pot two-step synthesis (Scheme 4.18). In this

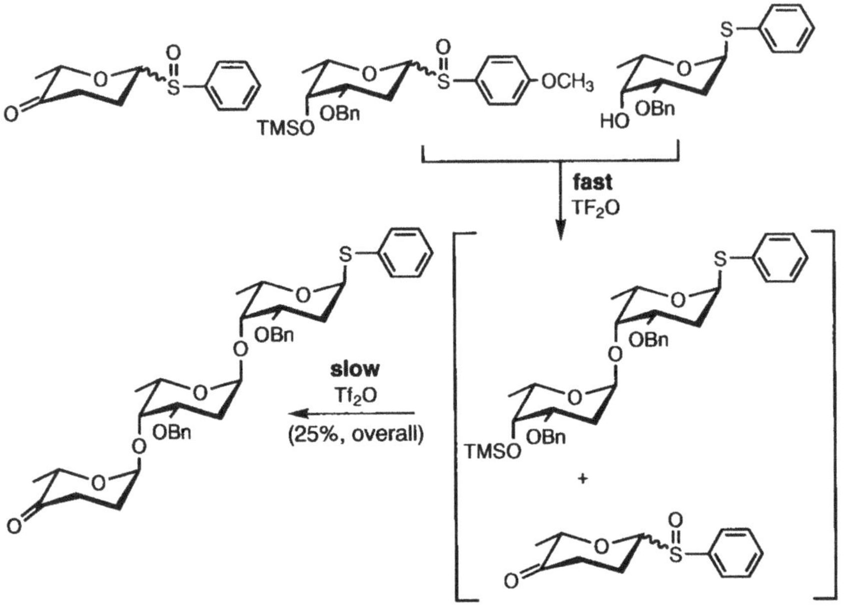

Scheme 4.17   One-pot two-step glycosylation using sulfoxides of different reactivity.

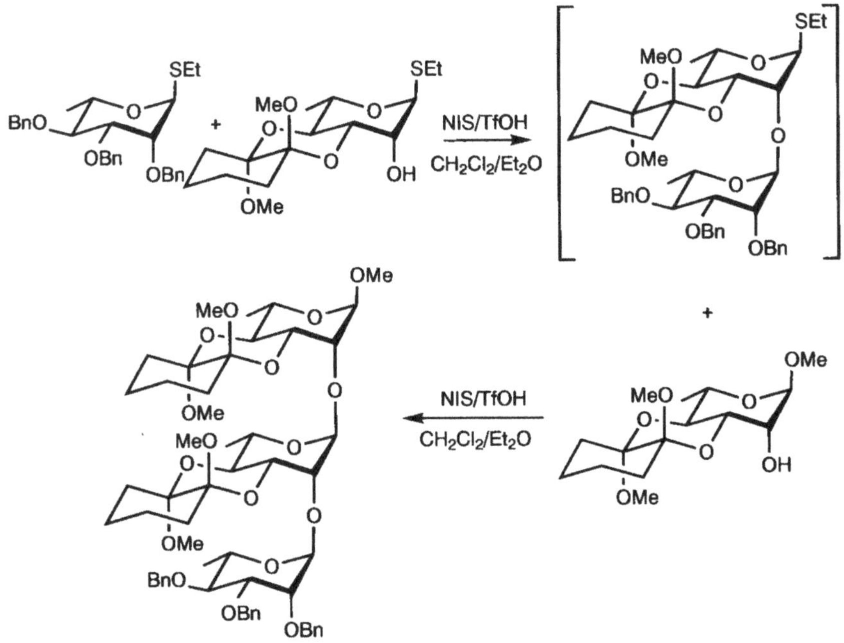

Scheme 4.18   One-pot sequential glycosylations.

strategy, a benzylated activated thioglycosyl donor was chemoselectively coupled with the less reactive cyclohexane-1,2-diacetal (CDA) protected thioglycosyl acceptor to give a disaccharide. Next, a second acceptor and more activator were added to the reaction mixture, which resulted in the clean formation of a trisaccharide. The lower reactivity of the CDA protected thioglycoside reflects the torsial strain inflicted upon the developing cyclic oxonium ion, the planarity of which is opposed by the cyclic protecting group. The one-pot two-step glycosylation strategy allows the construction of several glycosidic bonds without time-consuming work-up and purification steps. It should, however, be realised that this type of reaction will only give satisfactory results when all the glycosylations are high yielding and highly diastereoselective. For example, by exploiting neighbouring-group participation, it is relatively easy to form selectively 1,2-*trans* glycosides. Also, in general, mannosides give very high α selectivities. Other types of glycosidic linkages may impose problems.

4.9 Solid-phase oligosaccharide synthesis

Inspired by the success of solid-phase peptide and oligonucleotide syntheses, in the early 1970s several research groups attempted to develop methods for solid-supported oligosaccharide synthesis.[53] However, since the early methods for glycosidic bond formation were rather restricted, the success of these solid-phase strategies was limited and only simple di- and trisaccharides could be obtained.

In 1987, van Boom and co-workers reported[54] the solid-supported synthesis of a D-galactofuranosyl heptamer. Their synthetic approach is illustrated in Scheme 4.19. A selectively protected L-homoserine was linked to the Merrifield polymer chloromethyl polystyrene (PS = polystyrene) to give a derivatised polymer. The loading capacity of the polymer was $0.5\,\mathrm{mmol\,g^{-1}}$ of resin. After acid hydrolysis of the trityl ether, the galactofuranosyl chloride was coupled with the immobilised homoserine residue under Koenigs–Knörr conditions to give a polymer-linked homoserine glycoside. It was observed that the coupling reaction had not gone to completion and, to limit the formation of shorter fragments, the unreacted hydroxyl groups were capped by acetylation with acetic anhydride in the presence of pyridine and N,N-dimethylaminopyridine (DMAP). Elongation was performed as follows: the levulinoyl (Lev) group was removed by treatment with a hydrazine/pyridine/acetic acid mixture and the released alcohol was coupled with the chloride. Again unreacted hydroxyl groups were capped by acetylation. After repeating this procedure five times ($n = 6$) the heptasaccharide was released from

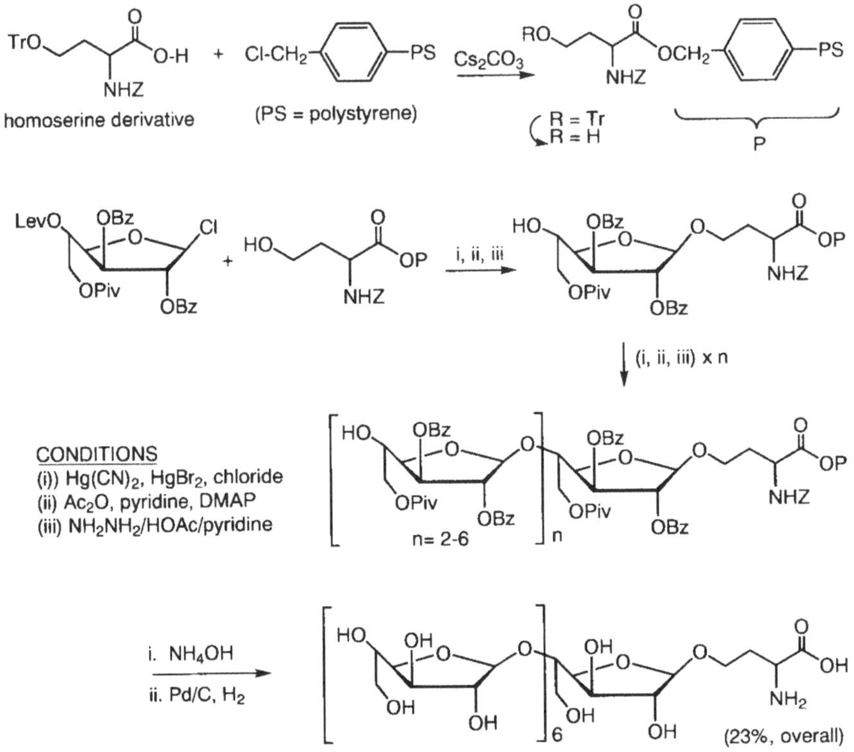

Scheme 4.19 Solid-phase oligosaccharide synthesis with glycosyl donors in solution.

the resin by basic hydrolysis. Under these conditions the benzoyl (Bz) and pivaloyl (Piv) protecting groups were also removed. Finally, cleavage of the benzyloxycarbonyl (Z) group by hydrogenolysis over Pd-C gave the target compound in an overall yield of 23%.

Several other strategies have been described for solid-phase oligosaccharide synthesis (Figure 4.4). For example, it has been shown that anomeric sulfoxides are efficient donors in solid-phase oligosaccharide assembly. In this approach, a glycosyl acceptor is immobilised on tentagel through an anomeric thiophenyl linkage.[55] The latter linkage can easily be cleaved by hydrolysis with a thiophilic reagent such as mercury trifluoroacetate. These synthetic tools were employed for the synthesis of a saccharide library. A similar thioglycosyl linker attached to polystyrene or controlled pore glass has also been used in combination with trichloroacetimidate donors.[56] Also, photo-cleavable linkers have been reported which are stable to a wide range of reaction conditions but which can be cleaved by irradiation with ultraviolet (UV) light.[57,58]

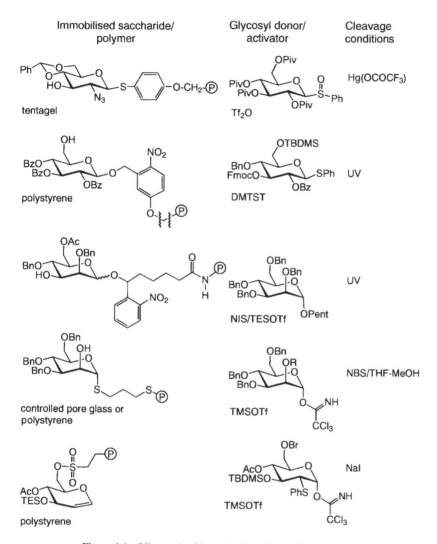

Figure 4.4 Oligosaccharide synthesis on insoluble polymers.

2-Deoxy-glycosides have been prepared on solid support by employing a sulfonate-based linker that can be cleaved by nucleophilic attack.[59]

A feature all these procedures have in common is that the anomeric centre of a saccharide is linked to the solid support and that the glycosyl donors are added to the growing chain. Danishefsky et al. reported[60] an inverse approach which used the incoming sugars as glycosyl acceptors and the immobilised saccharide as a donor. The basics of the strategy involves attachment of a glycal to a polymer support, and epoxidation to provide a 1,2-anhydro derivative. This polymer-bound glycosyl donor is then treated with a solution of a protected glycal, acting as a glycosyl acceptor, to give a polymer bound disaccharide. Reiteration of this reaction sequence provides larger oligosaccharides, which ultimately are retrieved from the support (Scheme 4.20). The commercially available 1% divinylbenzene–styrene copolymer was employed and the glycal was attached to this resin by using a diisopropylsilyl ether linker. Such a linker is stable under the employed reaction conditions but can be cleaved by fluoride-ion treatment. In previous studies, a diphenyldichlorosilane linker was used; however, it was shown that this linker was inferior to the diisopropylsilyl linker. Lithiation of the copolymer followed by quenching with diisopropyldichlorosilane provided silyl chloride modified resin.

The silylated polymer was reacted with a solution of partially protected galactal in dichloromethane and Hünig's base to give the corresponding saccharide-linked polymer construct. The loading of the solid support

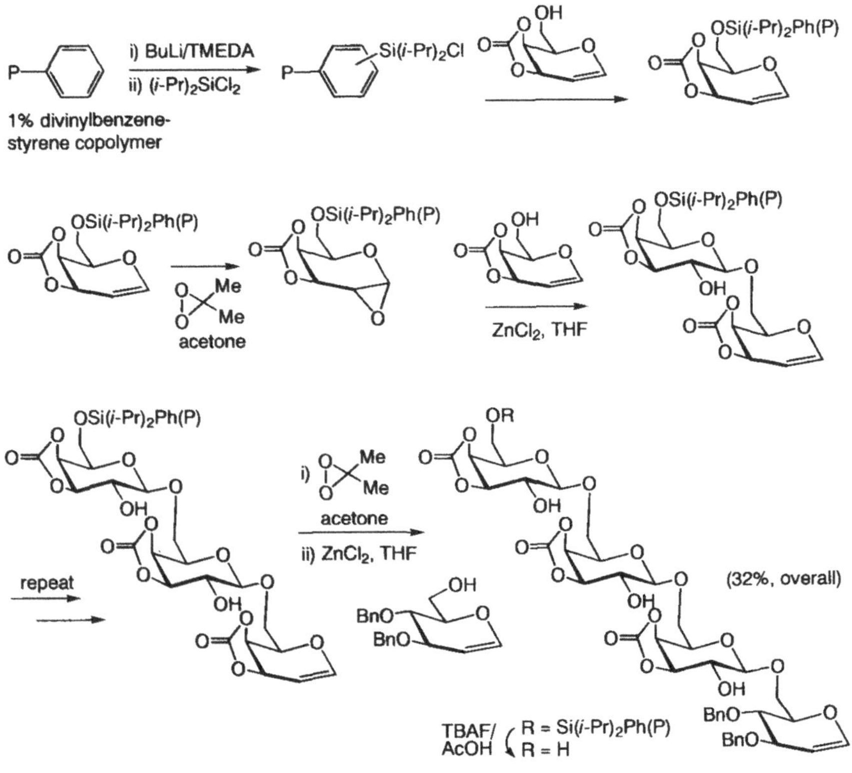

Scheme 4.20 Solid-phase oligosaccharide synthesis using an immobilised glycosyl donor.

was $0.9\,mmol\,g^{-1}$ of resin. The double bond of the polymer-bound glycal was activated by epoxidation with 3,3-dimethyldioxirane and the epoxide, thus obtained, was reacted with a tetrahydrofuran solution of a galactal acceptor in the presence of $ZnCl_2$ to give the polymer-bound dimer. The glycosidation procedure required a sixfold to tenfold excess of solution-based glycosyl acceptor and two to three equivalents of promoter. However, in some reactions less acceptor and shorter reaction times were used. It should also be noted that no glycosylation at the 2-position was observed. Twice repetition of this two-step procedure (epoxidation, glycosylation) provided a polymer-bound tetrasaccharide, which was released from the solid support by treatment with tetra-*n*-butylammonium fluoride (TBAF). The method allowed the preparation of the tetrasaccharide in a 74% overall yield. An advantageous aspect of this solid-supported approach is that no capping step is required because any unreacted epoxide will hydrolyse in the washing procedure. However, in the case of a very difficult glycosylation step, most of the solid-supported linked glycosyl donor may decompose, lowering the overall yield. In traditional solid-phase synthetic procedures, excess donor can be used to achieve acceptable yields in difficult glycosylation reactions.

Oligosaccharides can also be assembled on a solid support by using a two-directional approach.[61] In this strategy, an immobilised saccharide first acts as a glycosyl acceptor and in the next step acts as a glycosyl donor. This approach requires glycosyl donors that can be orthogonally activated, and it was found that thioglycosides and anomeric trichloro-acetimidates provide such a set (Scheme 4.21). A succinoyl-based linker was used and attachment to the solid support was achieved by amide bond formation. Cleavage was easily accomplished by base-mediated hydrolysis of the ester linkage. In an alternative approach, the immobilised material first acts as a donor and then as an acceptor. The two-directional strategy makes very good use of the immobilised saccharides and requires no protecting-group manipulations for the synthesis of a trisaccharide. In addition, it proves to be very efficient for the preparation of trisaccharide libraries.

The rate of reactions on a solid support is generally reduced compared with solution-based methods. Krepinsky and co-workers addressed this problem by using a polymer-supported solution synthesis of oligosaccharides.[62] This strategy is based on the fact that polyethyleneglycol (PEG) polymer supported saccharides are soluble under conditions of glycosylation but insoluble during the work-up procedure (PEG is soluble in many organic solvents but can be precipitated in ether). As can be seen in Figure 4.5, several different linkers have been employed for attachment of a saccharide to the PEG. A polyethyleneglycol-based polymer has also been used in combination with an inverse procedure using glycosyl fluorides and thioglycosides as acceptors and donors.[63]

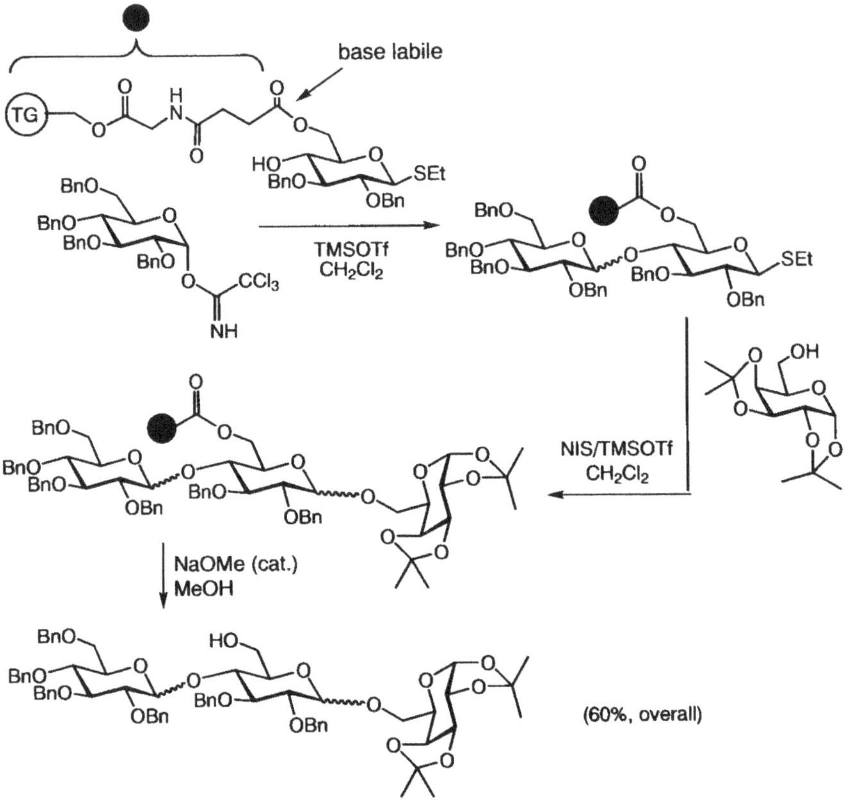

Scheme 4.21   Two-directional solid-phase oligosaccharide synthesis.

4.10   Enzymatic glycosylation strategies

The need for increasingly efficient methods for oligosaccharide synthesis has stimulated the development of enzymatic approaches.[64] The enzymatic methods bypass the need for protecting groups since the enzymes control both the regioselectivity and stereoselectivity of a glycosylation. Two fundamentally different approaches for enzymatic oligosaccharide synthesis have been developed: (1) the use of glycosyl transferases and (2) the application of glycosyl hydrolases.

Glycosyl transferases are essential enzymes for oligosaccharide biosynthesis. These enzymes can be classified as enzymes of the Leloir pathway and those of the non-Leloir pathway. The glycosyltransferases of the Leloir pathway are involved in the biosynthesis of most N-linked

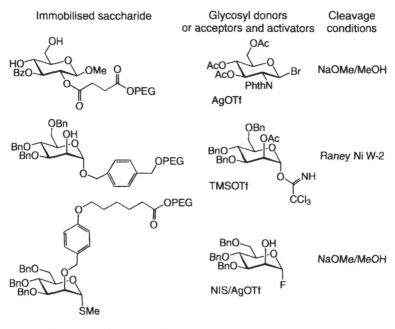

Figure 4.5 Oligosaccharide synthesis on polyethyleneglycol (PEG).

and O-linked glycoproteins in mammalians and utilise sugar nucleotide monophosphates or diphosphates as glycosyl donors. In contrast, glycosyltransferases from the non-Leloir pathway use sugar phosphates as substrates.

Glycosyl transferases are highly regioselective and stereoselective enzymes and have been successfully applied in enzymatic synthesis of oligosaccharides. Glycosyl transferases can be isolated from milk and serum and are most commonly purified by affinity column chromatography using immobilised sugar nucleotide diphosphates. Several glycosyl transferases have been cloned and overexpressed and are now readily available in reasonable quantities. However, the number of easily available glycosyl transferases is still very limited. Furthermore, these enzymes are highly substrate-specific and therefore the possibilities of preparing analogues are limited.

Nature employs glycosyl hydrolases for the degradation of oligosaccharides. However, the reverse hydrolytic activity of these enzymes can also be exploited in glycosidic bond formation. This method allows the preparation of several disaccharides and trisaccharides. Glycosyl hydrolases are much more readily available than glycosyl transferases but are generally less stereoselective and the transformations are lower yielding.

Combined approaches have been developed to overcome problems associated with chemically and enzymatically based methods. In such an approach, glycosidic linkages that are very difficult to introduce chemically are introduced enzymatically and vice versa. The enzymatic approach has proven to be extremely valuable for the introduction of neuramic acid units in a synthetically prepared oligosaccharide.

4.10.1   Glycosyl transferases

β-(1-4)-Galactosyltransferase is one of the most commonly used enzymes for glycosidic bond synthesis and catalyses the transfer of galactose from uridine diphosphate-Galp (UDP-Galp) to the 4-position of a GlcNAcp residue.[65] For example, N-acetyl lactosamine was obtained by condensation of galactosyl-UDP with N-acetyl glucosamine in the presence of β-(1–4)-galactosyltransferase, and this reaction proceeded with complete absolute stereoselectivity and regioselectivity (Scheme 4.22). The UDP that is formed during the reaction is an inhibitor of the enzyme and

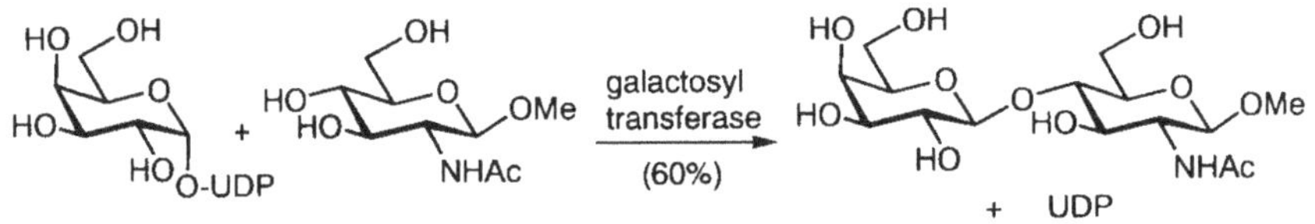

Scheme 4.22   Glycosyl transferases in stereoselective glycosidic bond synthesis.

decreases the yield of the transformation. To accelerate the enzymatic glycosylation it is desirable to remove the UDP. Indeed, when the galactosylation is performed in the presence of calf intestinal phosphatase, an enzyme that degrades UDP, the yield of the disaccharide is increased from 60% to 83%.[66] The convenience of the transformation can be further improved by in situ enzymatic synthesis of UDP-Gal from the inexpensive UDP-glucose (UDP-galactose 4'-epimerase). The β-(1–4)-galactosyltransferase tolerates some modifications in the glycosyl donor and acceptor.[64] For example, D-glucose, D-xylose, N-acetylmuramic acid and myo-inositol have been utilised as a glycosyl acceptor. Furthermore, derivatisation at the 3 and 6 positions of the glycosyl acceptor are tolerated and, for example, the disaccharide α-L-Fucp-(1–6)-GlcNAcp is a substrate for the enzyme. Fucosyl transferases,[67] sialyl transferases,[68] N-acetylglucosaminyltransferases[69] and mannosyl transferases[70] are other glycosyl transferase enzymes that have been applied for glycosidic bond synthesis.

The nucleotide diphosphates can be obtained by different approaches. Chemical synthesis provides an attractive route to these activated substrates but the methods developed are rather laborious. Alternatively, nucleotide diphosphates can be obtained by enzymatic synthesis. However, the most elegant way to utilise nucleotide phosphates is by *in situ* regeneration. An example of such an approach is illustrated in Scheme 4.23.[71] A catalytic amount of UDP-Gal*p* was used to initiate the

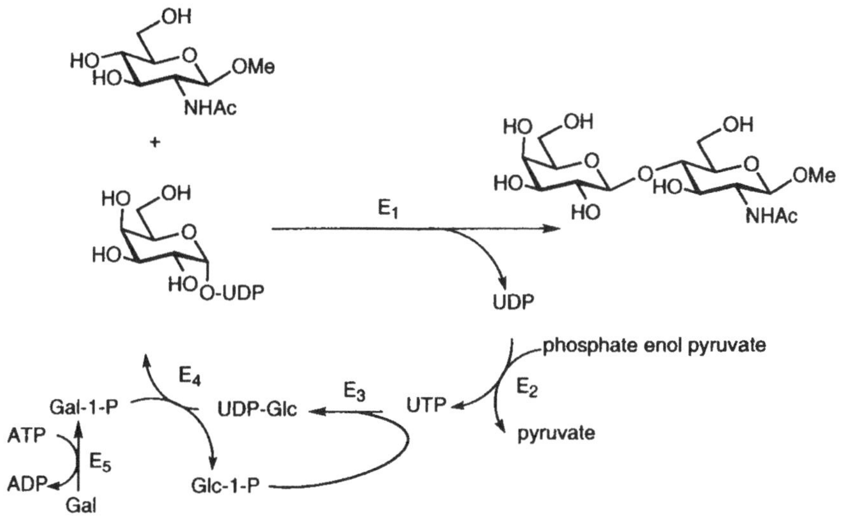

Scheme 4.23 The use of glycosyl transferases in combination with *in situ* co-factor regeneration. E_1 = galactosyl tranferase; E_2 = pyruvate kinase; E_3 = UDP-Glc-pyrophosphorylase; E_4 = galactose-1-phosphate uridyl transferase; E_5 = galactokinase.

glycosylation of *N*-acetylglucosamine to give lactosamine and UDP. Next, UDP is converted by a multienzymatic process into UDP-Gal. The reaction requires a stoichiometric amount of galactose-6-phosphate. Other co-factor regeneration processes have been developed.[72] Apart from using catalytic amounts of expensive nucleoside diphosphate, *in situ* co-factor regeneration offers the advantage of having low concentrations of uridine diphosphate in the reaction. As the latter compound is an inhibitor of the enzyme, under the *in situ* regeneration conditions, the transformation will proceed to completion.

Wong, Paulson and co-workers have employed glycosyl transferases for the preparation the tetrameric SLe^x ligand.[73] They overexpressed galactosyl, fucosyl and neuramyl transferases and used *in situ* regeneration of the nucleotide diphosphate. This methodology makes it possible to prepare multigram quantities of the SLe^x tetrasaccharide.

4.10.2 Glycosyl hydrolases

Glycosidic bonds have been prepared using glycosidases and these transformations have been accomplished under kinetic (transglycosylation) and thermodynamic (reverse hydrolysis) controlled conditions.

The reverse hydrolysis is based on a shift of the reaction equilibrium, normally in favour of the hydrolysis of glycosidic linkages in aqueous medium, towards synthesis. Thermodynamic considerations indicate that this shift in equilibrium can be achieved by either increasing the substrate concentration or decreasing the water content of the reaction solvent. The best results have been obtained by using an alcoholic solution containing a small proportion of water.[74] For example, β-D-allyl glucoside was isolated at a yield of 62% when glucose in an aqueous solution of allyl alcohol (10% water) was treated with almond β-D-glucosidase at 50°C for 48 h (Scheme 4.24). Galactosides and mannosides have been prepared by

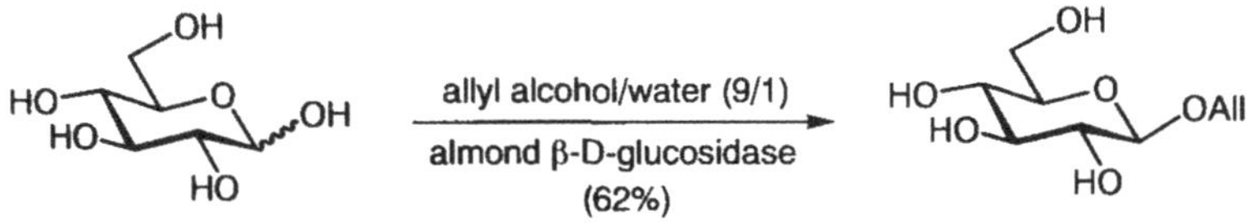

Scheme 4.24 Reverse activity of glycosyl hydrolases under thermodynamic conditions.

employing galactosidases and mannosidases, respectively. Furthermore, the methodology has been used for the preparation of benzyl, pentenyl, 6-hydroxyhexyl and 2-(trimethylsilyl)ethyl glycosides. The amount of water to be used depends on the enzyme and alcohol and in general between 8% and 20% water gives the highest rate of conversion. In general, the more hydrophobic the alcohol, the more water is needed in the medium to maintain the enzyme in an adequately hydrated state.

Kinetically controlled glycosylation is based on trapping a reactive intermediate with a glycosyl acceptor to form a new glycosidic linkage. The reactive intermediates can be obtained by a transglycosylation of disaccharides and oligosaccharides, aryl glycosides and glycosyl fluorides.

An example of a transglycosylation is given in Scheme 4.25.[75] Reaction of *p*-nitrophenyl α-fucopyranoside with methyl galactoside in the presence of α-L-fucosidase gave a mixture of two disaccharides. Thus, the enzyme hydrolyses the aryl glycosidic linkage to form an activated glycosyl – enzyme intermediate. This intermediate can now react with a wide range of exogenous nucleophiles including saccharides, alcohols and water. Surprisingly, despite the enormous amount of water that is present in the reaction medium, the glycosyl-enzyme intermediate reacts preferentially with a galactosyl acceptor. This feature is probably because

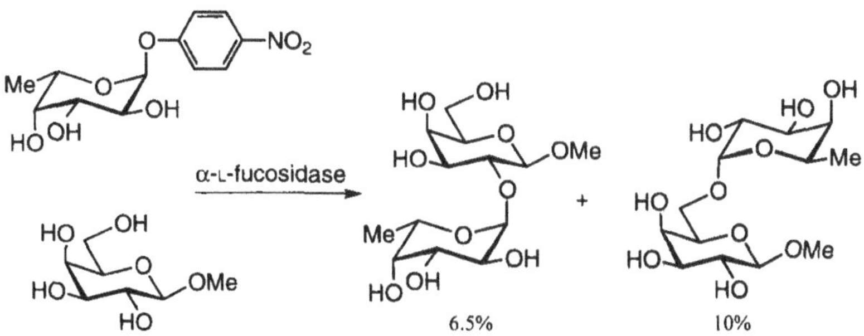

Scheme 4.25 Glycosylation with glycosyl hydrolases under kinetic conditions.

of recognition of the glycosyl acceptor by the enzyme. Many factors determine the outcome of a glycosidase-mediated glycosylation. For example, the product ratio of the fucosidase catalysed reaction could be influenced by selecting different reaction solvents.

The reaction time of a transglycosylation may also be an important determinant of the product outcome. For example, the (1–4) linked disaccharide (N,N'-diacetylchitobiose) was the initial product formed when a mixture of N-acetyl-D-glucosamine and p-nitrophenyl β-D-2-N-acetyl glucosamide was treated with β-galactosidase from *Aspergillus oryzae* (Scheme 4.26).[76] The only other transfer product observed was the corresponding (1–6) linked disaccharide formed at 10% of the total

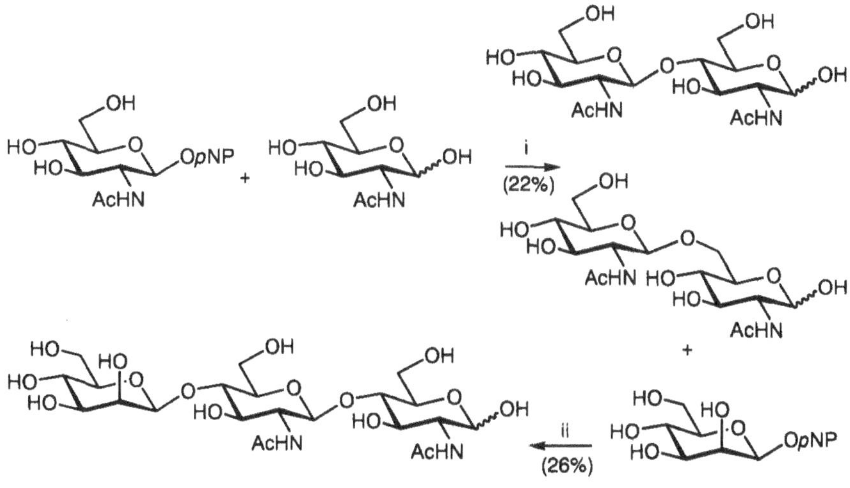

Scheme 4.26 Glycosyl hydrolases in oligosaccharide synthesis. (i) = N-acetylhexosaminidase from *A. oryzae*; (ii) = β-mannosamine from *A. orzyae*.

transfer products. After the donor had been consumed, the ratio between the disaccharides continued to evolve with a relative increase in the concentration of the initial minor product. At the point of disappearance of the glycosyl donor the ratio of (1–4) to (1–6) transfer products was 9:1. At the extreme during continued evolution of the disaccharide mixture, this ratio was reversed to 8:92. This result can be explained if it is assumed that during the second stage of the reaction, the accumulated (1-4) product acts as a glycosyl donor, transferring an N-acetyl-β-D-glucosaminyl residue to the C(4) and C(6) hydroxyl of N-acetyl-D-glucosamine increasing the overall relative and absolute concentration of the (1–6) linked product. At the same time, both are undergoing hydrolysis to the monosaccharide, the (1–4) disaccharide at a faster rate than the (1–6) disaccharide. After purification by charcoal-celite column chromatography, the disaccharide could be isolated in a yield of 22%. When the disaccharide mixture (9:1) was treated with N-acetylhexosaminidase from *Canavalia ensiformis* (Jack bean), the (1–6) isomer was selectively hydrolysed. In the presence of a mannosyl transferase from *A. oryzae*, a mannosyl unit from *p*-nitrophenyl β-D-mannopyranoside could be transferred to a disaccharide to give a trisaccharide in a yield of 26%. This trisaccharide is part of the core trisaccharide of N-linked glycoproteins.[77]

References

1. For reviews see (a). G. Wolf and G. Röhle, 1974, *Angew. Chem., Int. Ed. Engl.*, 13; (b) H. Paulsen, 1982, *ibid*, 21, 155; (c) H. Paulsen, 1990, *ibid*, 29, 823; (d) R.R. Schmidt, 1986, *ibid*, 25, 212; (e) R.R. Schmidt, 1989, *Pure Appl. Chem.*, 61, 1257; (f) P. Sinaÿ, 1991, *ibid*, 63, 519; (g) K.C. Nicolaou, T.J. Caulfield and R.D. Croneberg, 1991, *ibid*, 63, 555; (h) A. Vasella, 1991, *ibid*, 63, 507, (i) B. Fraser-Reid, U.E. Udodong, Z. Wu, H. Ottoson, J.R. Merritt, S. Rao, C. Roberts and R. Madsen, 1992, *Synlett*, 12, 927; (j) P. Fugedi, P.J. Garegg, H. Lohn and T. Norberg, 1987, *Glycoconjugate J.*, 4, 97; (k) K. Toshima and K. Tatsuta, 1993, *Chem. Rev.*, 93, 1503; (l) G.-J. Boons, 1996, *Tetrahedron*, 1095; (m) G.-J. Boons, 1996, *Contemporary Organic Synthesis*, 173; (n) P.J. Garegg, 1997, *Adv. Carbohydr. Chem. Biochem.*, 52, 179; (o) M.S. Shimizu, H. Togo and M. Yokoyama, 1998, *Synthesis*, 799.
2. (a) N.K. Kochetkov and A.F. Bochkov, 1971, *Recent Dev. Chem. Nat. Carbon Comp.*, 4, 17; (b) N.K. Kochetkov, A.F. Bochkov and T.A. Sokolovskaja, 1971, *Carbohydr. Res.*, 16, 17; (c) A.F. Bochkov and N.K. Kochetkov, 1975, *ibid*, 39, 355; (d) V.I. Betaneli, M.V. Ovchinnikov, L.V. Backinowsky and N.K. Kochetkov, 1976, *ibid*, 76, 252; (e) L.V. Backinowsky, Y.E. Tsvetkov, N.F. Balan, N.E. Byramova and N.K. Kochetkov, 1980, *ibid*, 85, 209.
3. P. Sinaÿ, 1978, *Pure Appl. Chem.*, 50, 1437.
4. (a) W. Koenigs, E. Knörr, 1901, *Ber*, 34, 957; (b) K. Igarashi, 1977, *Adv. Carbohydr. Chem. Biochem.*, 34, 243.
5. T. Mukaiyama, Y. Murai and S. Shoda, 1981, *Chem. Lett.*, 431.
6. U. Spohr and R.U. Lemieux, 1988, *Carbohydr. Res.*, 174, 211.

7. M.L. Wolfrom and W. Groebke, 1963, *J. Org. Chem.*, 28, 2986.
8. K.C. Nicolaou, R.E. Dolle, D.P. Papahatjis and J.L. Randall, 1984, *J. Am. Chem. Soc.*, 106, 4189.
9. (a) G.H. Posner and S.R. Haines, 1985, *Tetrahedron Lett.*, 26, 5; (b) W. Rosenbrook, D. A. Riley and P.A. Lartley, 1985, *Tetrahedron Lett.*, 26, 3.
10. A. Marra and P. Sinaÿ, 1990, *Carbohydr. Res.*, 195.
11. S. Sato, M. Mori and Y. Ito, 1986, *Carbohydr. Res.*, 155, C6.
12. D. Kahne, S. Walker, Y. Cheng and D. van Engen, 1989, *J. Am. Chem. Soc.*, 111, 6881.
13. (a) R.U. Lemieux and J.L. Hayimi, 1965, *Can. J. Chem.*, 43, 2162; (b) R.U. Lemieux, K. B. Hendriks, R.V. Stick and K. James, 1975, *J. Am. Chem. Soc.*, 97, 4056.
14. (a) G.M. Bedault and G.G.S. Dutton, 1974, *Carbohydr. Res.*, 37, 309; (b) H. Paulsen and O. Lockhoff, 1981, *Chem. Ber.*, 114, 3102; (c) H. Paulsen, W. Kutsschker and O. Lockhoff, 1981, *Chem. Ber.*, 114, 3233; (d) P.J. Garegg and P. Ossowski, 1983, *Acta Chem. Scand.*, B37, 249.
15. (a) C.A.A. van Boeckel, T. Beetz and S.F. van Aelst, 1984, *Tetrahedron*, 40, 4097; (b) C. A.A. van Boeckel and T. Beetz, 1985, *Recl. Trav. Chim. Pays-Bas*, 104, 174; (c) C.A.A. van Boeckel and T. Beetz, 1985, *Recl. Trav. Chim. Pays-Bas*, 104, 171.
16. A. Toepfer and R.R. Schmidt, 1990, *Carbohydr. Res.*, 202, 193.
17. (a) V.K. Srivastava and C. Schuerch, 1980, *Carbohydr. Res.*, 79, C13; (b) V.K. Srivastava and C. Schuerch, 1981, *J. Org. Chem.*, 46, 1121.
18. (a) D. Crich and S. Sun, 1996, *J. Org. Chem.*, 61, 4506; (b) D. Crich and S. Sun, 1997, *J. Org. Chem.*, 62, 1198; (c) D. Crich and S. Sun, 1997, *J. Am. Chem. Soc.*, 119, 11217.
19. (a) N.K. Kochetkov, E.M. Klimov and N.N. Malysheva, 1989, *Tetrahedron Lett*, 30, 5459; (b) N.K. Kochetkov, E.M. Klimov, N.N. Malysheva and A.V. Demchenko, 1991, *Carbohydr. Res.*, 212, 77; (c) N.K. Kochetkov, E.M. Klimov, N.N. Malysheva and A.V. Demchenko, 1992, *Carbohydr. Res.*, 232, C1.
20. (a) R.L. Halcomb and S.J. Danishefsky, 1989, *J. Am. Chem. Soc.*, 111, 6661; (b) R.G. Dushin and S.J. Danishefsky, 1992, *J. Am. Chem. Soc.*, 114, 3471; (c) V. Behar and S.J. Danishefsky, 1994, *Angew. Chem. Int. Ed. Engl.*, 33, 1468; (d) S.J. Danishefsky, V. Behar, J.T. Randolph and K.O. Lloyd, 1995, *J. Am. Chem. Soc.*, 117, 5701; (e) S.J. Danishefsky and M. Bilodeau, 1996, *Angew. Chem. Int. Ed. Engl.*, 35, 1380.
21. C.M. Timmers, G.A. van der Marel and J.H. van Boom, 1993, *Rec. Trav. Chim. Pays Bas*, 112, 609.
22. W.J. Sanders and L.L. Kiessling, 1994, *Tetrahedron Lett.*, 35, 7335.
23. (a) J. Pougny and P. Sinaÿ, 1976, *Tetrahedron Lett.*, 4073; (b) R.U. Lemieux and R. M. Ratcliffe, 1979, *Can. J. Chem.*, 57, 1244; (c) R.R. Schmidt and J. Michael, 1985, *J. Carbohydr. Chem.*, 4, 141.
24. (a) G. Stork and G. Kim, 1992, *J. Am. Chem. Soc.*, 114, 1087; (b) G. Stork and J.J.L. Clair, 1996, *J. Am. Chem. Soc.*, 118, 247.
25. (a) F. Barresi and O. Hindsgaul, 1991, *J. Am. Chem. Soc.*, 113, 9376; (b) F. Barresi and O. Hindsgaul, 1992, *Synlett*, 759; (c) F. Barresi and O. Hindsgaul, 1994, *Can. J. Chem.*, 72, 1447.
26. (a) Y. Ito and T. Ogawa, 1994, *Angew. Chem.*, 106, 1843; (b) A.D. Dan, T. Ito and T. Ogawa, 1995, *Tetrahedron Lett.*, 36, 7487.
27. (a) M. Bols, 1992, *J. Chem. Soc., Chem. Commun.*, 913; (b) M. Bols, 1993, *Acta Chem. Scand.*, 47, 829; (c) M. Bols, 1993, *J. Chem. Soc., Chem. Commun.*, 791.
28. (a) M. Bols and H.C. Hansen, 1994, *Chem. Lett.*, 1049; (b) T. Ziegler and R. Lau, 1995, *Tetrahedron Lett.*, 36, 1417; (c) R. Lau, G. Schüle, U. Schwaneberg and T. Ziegler, 1995, *Liebigs Ann.*, 1745; (d) S. Valverde, A.M. Gómez, A. Hernández, B. Herradón and J. Cristóbal López, 1995, *J. Chem. Soc., Chem. Commun.*, 2005.
29. J. Banoub, P. Boullanger and D. Lafont, 1992, *Chem. Rev.*, 92, 1167.

30. J. Debenham, R. Rodebaugh and B. Fraser-Reid, 1997, *Liebigs Ann./Recueil*, 791.
31. M.P. DeNinno, 1991, *Synthesis*, 583.
32. (a) Y. Ito and T. Ogawa, 1988, *Tetrahedron Lett.*, 29, 1061; (b) Y. Ito, T. Ogawa, M. Numata and M. Sugimoto, 1990, *Carbohydr. Res.*, 101, 165; (c) A. Hasagawa, T. Ohki, H. Nagahama, M. Ishida and M. Kiso, 1991, *Carbohydr. Res.*, 212, 277; (d) A. Hasagawa, H. Nagahama, T. Ohki, K. Hotta, M. Ishida and M. Kiso, 1991, *J. Carbohydr. Chem.*, 10, 493; (e) W. Biberg and H. Lönn, 1992, *Tetrahedron Lett.*, 33, 115.
33. (a) T.J. Martin and R.R. Schmidt, 1992, *Tetrahedron Lett.*, 33, 6123; (b) H. Kondo, Y. Ichikawa and C.-H. Wong, 1992, *J. Am. Chem. Soc.*, 114, 8748.
34. A. Marra and P. Sinaÿ, 1990, *Carbohydr. Res.*, 195, 303.
35. A. Demchenko and G.-J, Boons, 1998, *Tetrahedron Lett.*, 39, 3065.
36. (a) K. Okomoto, T. Kondo and T. Goto, 1988, *Tetrahedron*, 44, 1291; (b) Y. Ito, M. Numata, M. Sugimoto and T. Ogawa, 1989, *J. Am. Chem. Soc.*, 111, 8508; (c) T. Ercégovic and G. Magnusson, 1995, *J. Org. Chem.*, 60, 3378; (d) V. Martichonok and G.M. Whitesides, 1996, *J. Am. Chem. Soc.*, 118, 8187; (e) J.C. Catro-Palomino, Y.E. Tsvetkov and R.R. Schmidt, 1998, *J. Am. Chem. Soc.*, 120, 5434.
37. (a) J. Thiem and W. Klaffke, 1990, *Topics in Current Chemistry*, 285, 285; (b) A. Kirschning, A.F.-W. Bechthold and J. Rohr, 1997, *Topics in Current Chemistry*, 188, 2.
38. (a) J. Thiem, H. Karl and J. Schwenter, 1978, *Synthesis*, 696, (b) J. Thiem and M. Gerken, 1982, *J. Carbohydr. Chem.*, 1, 229; (c) J. Thiem and B. Schöttmer, 1987, *Angew. Chem. Int. Ed. Engl.*, 26, 555; (d) S.J. Danishefsky, D.M. Armistead, F.E. Wincott, H.G. Selnick and R. Hungate, 1989, *J. Am. Chem. Soc.*, 111, 2976; (e) Y. Ito and T. Ogawa, 1987, *Tetrahedron Lett.*, 28, 2723; (f) M. Perez and J.-M. Beau, 1989, *Tetrahedron Lett.*, 30, 75; (g) P. Avecchia, M. Trumtel, A. Veyieres and P. Sinaÿ, 1989, *Tetrahedron Lett.*, 30, 2533; (h) K.S. Suzuki, G.A. Sullikowski, R.W. Friesen and S.J. Danishefsky, 1990, *J. Am. Chem. Soc.*, 112, 8895; (i) J. Thiem and W. Klaffke, 1989, *J. Org. Chem.*, 54, 2006.
39. (a) K.C. Nicolaou, T. Ladduwahetty, J.L. Randall and A. Chucholowski, 1986, *J. Am. Chem. Soc.*, 108, 2466; (b) Y. Ito and T. Ogawa, 1987, *Tetrahedron Lett.*, 28, 2723; (c) M. Perez and J.-M. Beau, 1989, *Tetrahedron Lett.*, 30, 75; (d) K.C. Nicolaou, C.W. Hummel, N.J. Bockovisch and C.-H. Wong, 1991, *J. Chem. Soc., Chem. Commun.*, 870; (e) H.M. Zuurmond, P.A.M. van der Klein, G.A. van der Marel and J.H. van Boom, 1992, *Tetrahedron Lett.*, 33, 2063; (f) G. Grewel, N. Kaila and R.W. Franck, 1992, *J. Org. Chem.*, 57, 2084; (g) H.M. Zuurmond, P.A.M. van der Klein, G.A. van der Marel and J.H. van Boom, 1993, *Tetrahedron*, 49, 6501; (h) J.C. Castro-Palomino and R.R. Schmidt, 1998, *Synlett*, 501.
40. K. Toshiba, Y. Nozaki, H. Inokuchi, M. Nakata, K. Tatsuta and M. Kinoshita, 1993, *Tetrahedron Lett.*, 34, 1611.
41. (a) R.R. Schmidt and M. Stumpp, 1983, *Liebigs Ann. Chem.*, 1249; (b) K.-H. Jung, M. Hoch and R.R. Schmidt, 1989, *ibid*, 1099; (c) R. Bommer, W. Kinzy and R.R. Schmidt, 1991, *ibid*, 425; (d) A. Toepfer and R.R. Schmidt, 1992, *Tetrahedron Lett.*, 33, 5161; (e) R. Windmüller and R. R. Schmidt, 1994, *Tetrahedron Lett.*, 35, 7927.
42. (a) A. Demchenko and G.-J. Boons, 1997, *Tetrahedron Lett.*, 38, 1629.
43. B. Fraser-Reid, U.E. Udodong, Z. Wu, H. Ottoson, J.R. Merritt, S. Rao, C. Roberts and R. Madsen, 1992, *Synlett*, 12, 927.
44. (a) G.H. Veeneman and J.H. van Boom, 1990, *Tetrahedron Lett.*, 31, 275; (b) G.H. Veeneman, S. H. van Leeuwen and J.H. van Boom, 1990, *Tetrahedron Lett.*, 31, 1331.
45. R.W. Friesen and S.J. Danishefsky, 1989, *J. Am. Chem. Soc.*, 111, 6656.
46. S. Hashimoto, H. Sakamoto, T. Honda and S. Ikegami, 1997, *Tetrahedron Lett.*, 38, 5181.
47. (a) M.I. Barrena, R. Echarri and R. Catillón, 1996, *Synlett*, 675; (b) L. Green, B. Hinzen, S.J. Ince, P. Langer, S.V. Ley and S.L. Warriner, 1998, *Synlett*, 440.

48. (a) G.J. Boons, P. Grice, R. Leslie, S.V. Ley and L.L. Yeung, 1993, *Tetrahedron Lett.*, 34, 8523.; (b) G.-J. Boons, R. Geurtsen and D. Holmes, 1995, *Tetrahedron Lett.*, 36, 6325.
49. (a) D. Kahne, S. Walker, Y. Cheng and D. Van Engen, 1989, *J. Am. Chem. Soc.*, 111, 6881; (b) S. Raghavan, D. Kahne, *J. Am. Chem. Soc.*, 115 (1993) 1580.
50. (a) S.V. Ley and H.W.M. Priepke, 1994, *Angew. Chem. Int. Ed. Engl.*, 33, 2292; (b) P. Grice, S.V. Ley, J. Pietruszka, H.W.M. Priepke and E.P.E. Walther, 1995, *Synlett*, 781.
51. (a) H. Yamada, T. Harada, H. Miyazaki and T. Takahashi, 1994, *Tetrahedron Lett.*, 35, 3979; (b) H. Yamada, T. Harada and T. Takahashi, 1994, *J. Am. Chem. Soc.*, 116, 7919.
52. G.-J. Boons and T. Zhu, 1997, *Synlett*, 809.
53. (a) J.M. Frechet and C. Schuerch, 1972, *J. Am. Chem. Soc.*, 94, 604; (b) J.M. Frechet and C. Schuerch, 1971, *J. Am. Chem. Soc.*, 93, 492; (c) J.M. Frechet and C. Schuerch, 1972, *Carbohydr. Res.*, 22, 399; (d) R. Eby and C. Schuerch, 1975, *Carbohydr. Res.*, 39, 151; (e) R. Guthrie, A.D. Jenkins and J. Stehlicek, 1971, *J. Chem. Soc. C*, 2690; (f) R. Guthrie, A.D. Jenkins and G.A.F.J. Roberts, 1973, *J. Chem .Soc., Perkin Trans. 1*, 2414; (g) G. Excoffier, D. Gagnaire, J.P. Utille and M. Vignon, 1972, *Tetrahedron Lett.*, 13, 5065; (h) U. Zehavi and A.J. Patchornik, 1973, *J. Am. Chem. Soc.*, 95, 5673; (i) S.H.L. Chiu and L. Anderson, 1976, *Carbohydr. Res.*, 50, 227.
54. G.H. Veeneman, S. Notermans, R.M.J. Liskamp, G.A. van der Marel and J.H. van Boom, 1987, *Tetrahedron Lett.*, 28, 6695.
55. (a) L. Yan, C.M. Taylor, R. Goodnow and D. Kahne, 1994, *J. Am. Chem. Soc.*, 116, 6953; (b) R. Ling, L. Yan, J. Loebach, M. Ge, Y. Uozumi, K. Sekanina, N. Horan, J. Gildersleeve, C. Thompson, A. Smith, K. Biswas, W.C. Still and D. Kahne, 1996, *Science*, 274, 1520.
56. (a) J. Rademann and R.R. Schmidt, 1996, *Tetrahedron Lett.*, 37, 3989; (b) A. Heckel, E. Mross, K.-H. Jung, J. Rademann and R.R. Schmidt, 1998, *Synlett*, 171; (c) J. Rademann, A. Geyer and R.R. Schmidt, 1998, *Angew. Chem. Int. Ed.*, 37, 1241.
57. (a) K.C. Nicolaou, N. Winssinger, J. Pastor and F DeRoose, 1997, *J. Am. Chem. Soc.*, 119, 449; (b) K.C. Nicolaou, N. Watanabe, J. Li, J. Pator and N. Winssinger, 1998, *Angew. Chem. Int. Ed.*, 37, 1559.
58. R. Rodebaugh, B. Fraser-Reid and H.M. Geysen, 1997, *Tetrahedron Lett.*, 38, 7653.
59. J.A. Hunt and W. R. Roush, 1996, *J. Am. Chem. Soc.*, 118, 9998.
60. (a) S.J. Danishefsky, K.F. McClure, J.T. Randolph and R.B. Ruggeri, 1993, *Science*, 260, 1307; (b) S.J. Danishefsky, J.T. Randolph and K.F. McClure, 1995, *J. Am. Chem. Soc.*, 117, 5712; (c) S.J. Danishefsky, K.F. McClure, J.T. Randolph and R.B. Ruggeri, 1995, *Science*, 269, 202.
61. T. Zhu and G.-J. Boons, 1998, *Angew. Chem. Int. Ed. Engl.*, 37, 1898.
62. (a) S.P. Douglas, D.M. Whitfield and J.J. Krepinsky, 1991, *J. Am. Chem. Soc.*, 113, 5095; (b) S.P. Douglas, D.M. Whitfield and J.J. Krepinsky, 1995, *J. Am. Chem. Soc.*, 117, 2116; (c) R. Verduyn, P.A.M. van der Klein, M. Douwers, G.A. van der Marel and J.H. van Boom, 1993, *Recl. Trav. Chim. Pays-Bas*, 112, 464; (d) O.T. Leung, D.M. Whitfield, S.P. Douglas, H.Y.S. Pang and J.J. Krepinsky, 1994, *New J. Chem.*, 18, 349; (e) A.A. Kandil, N. Chan, P. Chong and M. Klein, 1992, *Synlett*, 555.
63. Y. Ito, O. Kanie and T. Ogawa, 1996, *Angew. Chem. Int. Ed. Engl.*, 35, 2510.
64. (a) C.-H. Wong, R.L. Halcomb, Y. Ichikawa and T. Kajimoto, 1995, *Angew. Chem. Int. Ed. Eng.*, 34 , 412; (b) C.-H. Wong, R.L. Halcomb, Y. Ichikawa and T. Kajimoto, 1995, *Angew. Chem. Int. Ed. Eng.*, 34, 521.
65. T.A. Beyer, J.E. Sadler, J.I. Rearick, J.C. Paulson and R.L. Hill, 1981, *Adv. Enzymol.*, 52, 23.
66. (a) C. Unverzagt, H. Kunz and J. Paulson, 1990, *J. Am. Chem. Soc.*, 112, 9308; (b) C. Unverzagt, S. Kelm and J. Paulson, 1994, *Carbohydr. Res.*, 251, 285.
67. (a) U.B. Gokhale, O. Hindsgaul and M.M. Palcic, 1990, *Can. J. Chem.*, 68, 1063; (b) M.M. Palcic, A.P. Venot, R.M. Ratcliffe and O. Hindsgaul, 1989, *Carbohydr. Res.*, 190, 1.

68. (a) C. Auge and C. Gautheron, 1988, *Tetrahedron Lett.*, 29, 789; (b) S. Sabesan and J.C. Paulson, 1986, *J. Am. Chem. Soc.*, 108, 2068; (c) J. Thiem and W. Treder, 1986, *Angew. Chem. Int. Ed. Engl.*, 25, 1096.

69. (a) K.J. Kaur, G. Alton and O. Hindsgaul, 1991, *Carbohydr. Res.*, 210, 145; (b) O. Hindsgaul, K.J. Kaur, U.B. Gokhale, G. Srivastava, G. Alton and M.M. Palcic, 1991, *ACS Symp. Ser.*, 466, 38.

70. (a) W. McDowell, T.J. Grier, J.R. Rasmussen and R.T. Schwartz, 1987, *Biochem. J.*, 248, 523; (b) P. Wang, G.-J. Shen, Y.-F. Wang, Y. Ichikawa and C.-H. Wong, 1993, *J. Org. Chem.*, 58, 3985.

71. C.-H. Wong, S.L. Haynie and G.M. Whitesides, 1982, *J. Org. Chem.*, 47, 5416.

72. (a) C.-H. Wong, R. Wang and Y. Ichikawa, 1992, *J. Org. Chem.*, 57, 4343; (b) L. Elling, M. Grothus and M.-R. Kula, 1993, *Glycobiology*, 3, 349.

73. Y. Ichikawa, Y.-C. Lin, D.P. Dumas, G.-J. Shen, E. Garcia-Junceda, M.A. Williams, R. Bayer, C. Ketcham, L.E. Walker, J.C. Paulson and C.-H. Wong, 1992, *J. Am. Chem. Soc.*, 114, 9283.

74. (a) G. Vic and D.H.G. Crout, 1995, *Carbohydr. Res.*, 279, 315; (b) G. Vic, J.J. Hastings and D.H.G. Crout, 1996, *Tetrahedron Asym.*, 7, 1973.

75. S.C.T. Svensson and J. Thiem, 1990, *Carbohydr. Res.*, 200, 391.

76. S. Singh, J. Packwood and D.H.G. Crout, 1994, *J. Chem. Soc., Chem. Commun.*, 2227.

77. S. Singh, M. Scigelova and D.H.G. Crout, 1996, *J. Chem. Soc., Chem. Commun.*, 993.

5 The chemistry of *O*- and *N*-linked glycopeptides

G.-J. Boons

5.1 Introduction

For a long time, carbohydrates and peptides were considered as separate classes of natural products. However, it is now well established that most proteins of eukaryotic cells are glycoproteins; they are covalently link-ed to saccharide portions. Various functional roles for carbohydrate moieties of glycoproteins have been postulated,[1,2] and glycosylation strongly affects physiochemical properties as well as biological functions. Uptake, distribution, excretion and proteolytic stability, immunological masking of peptide epitopes, as well as the initiation and control of protein folding are examples of the important influences of glycosyla-tion.[3] Protein-bound carbohydrates have been identified as ligands that are involved in cell–cell interactions and are recognition sites for viruses and bacteria. Several studies have revealed that saccharide attachment to peptides, which are not glycosylated in nature, can influence pharmaco-logical and pharmacokinetic properties of peptides functioning as enzyme inhibitors, neuropeptides and hormones.

Glycopeptide linkages can be divided into two principal groups: (1) the *N*-linked group, bearing an *N*-glycosidic linkage to the site chain of L-asparagine and (2) the more diverse *O*-linked group, bearing an *O*-glycosidic linkage to L-serine, L-threonine, 4-hydroxy-L-proline, or δ-hydroxy-L-lysine (Scheme 5.1). In living systems, post-translational construction of the *N*-linked glycoprotein begins while the nascent protein is still on the ribosome.[4] The context for this process is well defined,[5] and *N*-linked glycoproteins show even less structural diversity in the linkage region than their *O*-linked counterparts, which are assembled in the trans-Golgi apparatus and its membranous extensions. Typically, the *N*-linked glycoproteins contain a GlcNAc*p*–β-(1 → 4) GlcNAc*p*–β(1 → N) linkage to Asn–Xxx–Ser or Asn–Xxx–Thr (chitobiose → Asn linkage), whereas the *O*-linked glycoproteins display at least four distinct types of linkages, which vary both in terms of the carbohydrate attachment, as well as in the identity of the amino acid carrier. Unlike serine and threonine, the hydroxyproline and hydroxylysine residues are created after protein assembly by post-translational modification of the assembled protein prior to glycosylation.[2d]

Despite the tremendous amount of research that has been committed to unveil the biosynthetic pathways and biological effects of protein

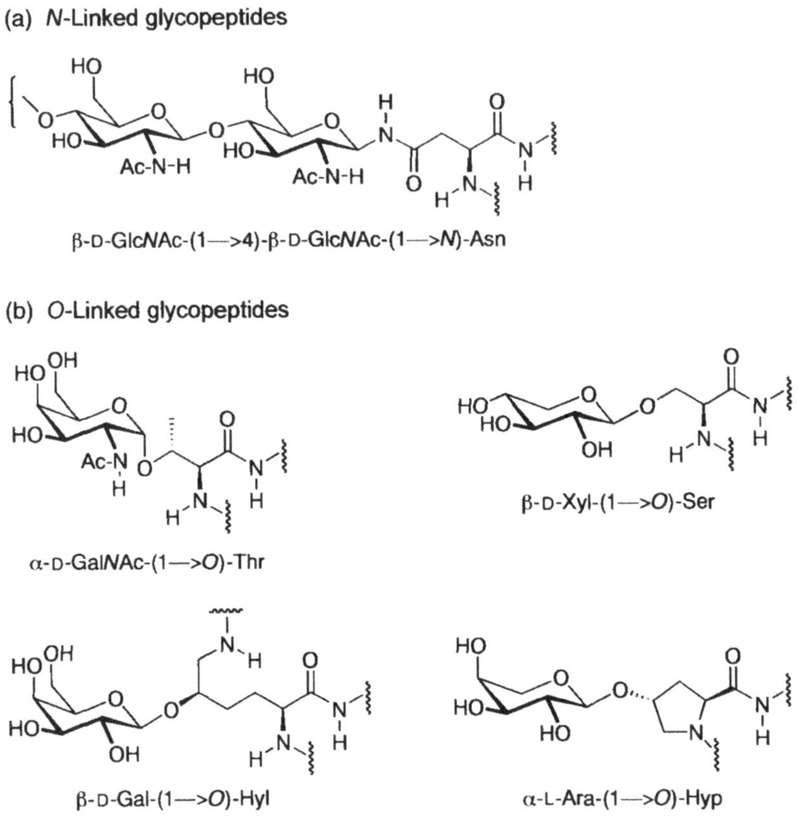

(a) *N*-Linked glycopeptides

β-D-GlcNAc-(1—>4)-β-D-GlcNAc-(1—>*N*)-Asn

(b) *O*-Linked glycopeptides

α-D-GalNAc-(1—>*O*)-Thr

β-D-Xyl-(1—>*O*)-Ser

β-D-Gal-(1—>*O*)-Hyl

α-L-Ara-(1—>*O*)-Hyp

Scheme 5.1   *N*- and *O*-linked glycopeptides.

glycosylation, even for the more extensively studied classes of glycopro-teins the biological role of the saccharide moiety of these compounds are not well understood[6] and the synthesis and structural analysis of simpler glycopeptides is required.[7]

5.2   Strategies for the chemical synthesis of glycopeptides

A short introduction to peptide synthesis will be given before the synthesis of glycopeptides is discussed. Proteins are assembled by a nucleic acid templated reaction from a menu comprising 19 L-α-amino acids of general structure **1** and the amino acid L-proline (**2**) (Figure 5.1). Nature synthesises proteins by a stepwise assembly of amino acid building blocks. Since these compounds are readily available in

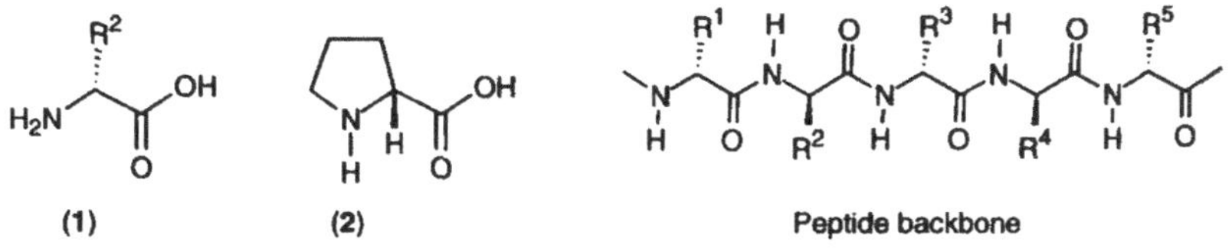

Figure 5.1   Amino acid and peptide structures. $1 =$ general structure; $2 = $ L-proline.

homochiral form, this is also the most obvious way for the chemist to proceed in the laboratory. The condensation of two amino acids will give a complex mixture of compounds (different dimers and oligomers). In order to obtain a particular dipeptide, the amino group of one amino acid and the carboxyl group of the other amino acid both need to be blocked selectively (Scheme 5.2a).

The most commonly used amino protecting groups are the *t*-butoxycarbonyl (Boc) and fluorenylmethoxycarbonyl (Fmoc) groups. The Boc group is stable to catalytic hydrogenolysis and to basic and nucleophilic reagents. However, the Boc group is conveniently removed by treatment with the moderately strong acid trifluoroacetic acid (TFA). The Fmoc group is very stable to acidic conditions, but is cleaved with mild basic reagents, and 20% piperidine in dimethylformamide (DMF) is routinely used. The carboxyl acid is frequently protected as a benzyl or *t*-butyl ester. These groups can be removed by catalytic hydrogenolysis or by TFA, respectively. Alternatively, the carboxylic acid can be used to immobilise an amino acid to a solid support. Some functional groups of side chains may interfere with amide bond formation and therefore need protection. The amino and guanidino groups of lysine and arginine, the carboxylates of aspartic and glutamic acids, and the nucleophilic thiol and imidazole groups of cysteine and hystadine, respectively, are usually protected.

There are many methods of carboxylate activation. A coupling reaction should allow complete coupling without affecting the rest of the molecule. *N*,*N*-Dicyclohexylcarbodiimide (DCC or DDCI) is a widely used coupling reagent. It provides a high level of activation and therefore all the reactive side chains usually need to be blocked. Extensive racemisation may take place with particular activated amino acids. Addition of 1-hydroxybenzotriazole (HOBt) to the coupling reaction will greatly reduce the danger of racemisation. In order to synthesise a tripeptide, either the amino protecting group or the carboxylic acid protecting group of a dimer has to be removed without effecting the other protecting groups. It is common practice to remove the amino protecting group. The carboxyl of an amino acid monomer will be activated and coupled with the dipeptide to give a tripeptide. This process is repeated until the whole

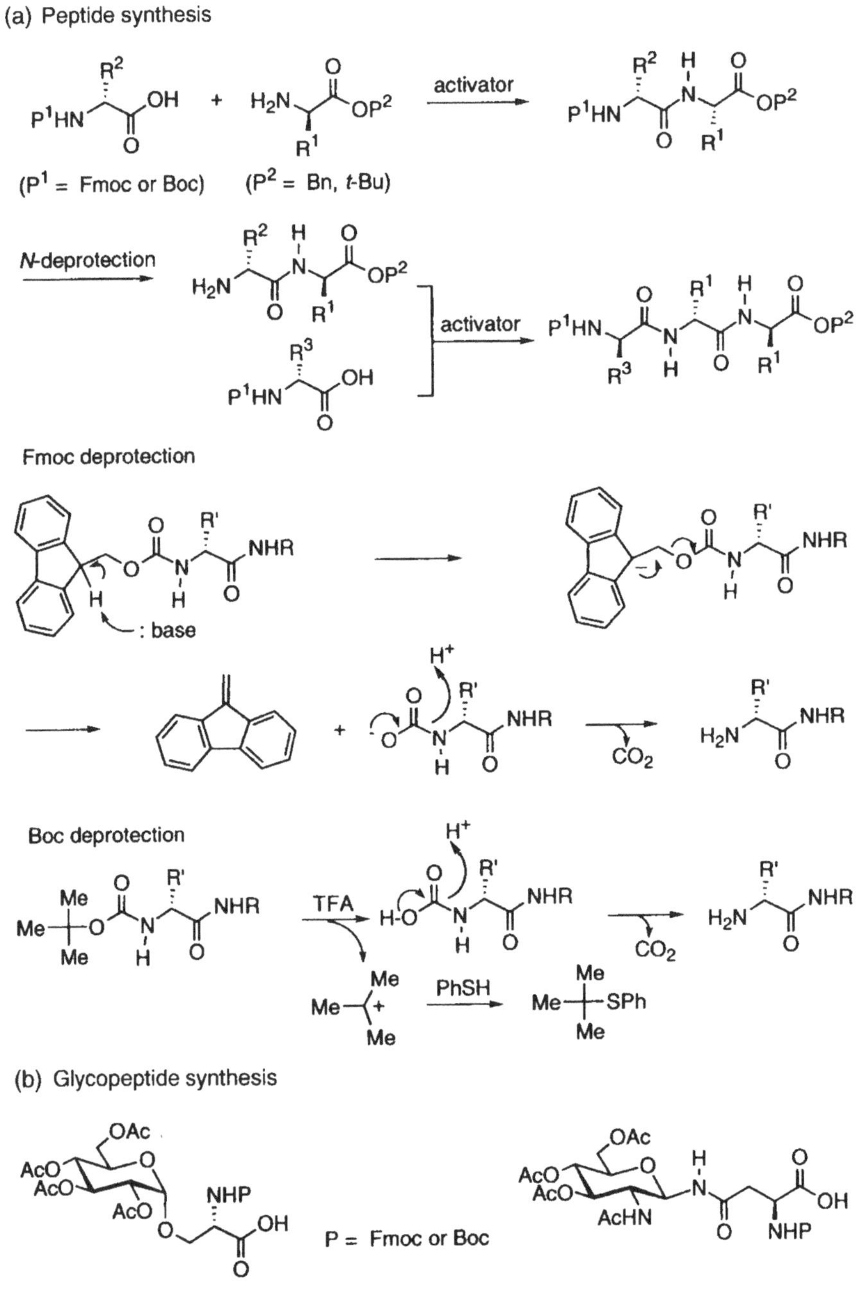

(a) Peptide synthesis

Fmoc deprotection

Boc deprotection

(b) Glycopeptide synthesis

P = Fmoc or Boc

Scheme 5.2 General strategies for peptide and glycopeptide synthesis.

peptide is assembled. Finally, all protecting groups are cleaved to give the target compound.

Two approaches can be considered for the synthesis of glycopeptides. The convergent strategy entails the glycosylation of properly protected oligopeptides. The success of O-glycosylation of peptides is limited because of the low solubility of peptides under the conditions commonly employed for glycosylation and the low reactivity of the side-chain hydroxyls of serine and threonine. Condensation of a glycosylamine and a peptide containing an aspartic acid residue has been accomplished in the synthesis of N-linked glycopeptides.[8] The major problem of this reaction is the intramolecular aspartimide formation with the C-terminal amide group upon activation of the aspartic side chains.

The second method is based on the preparation of glycosylated amino acids, which are used in the stepwise synthesis of glycopeptides (Scheme 5.2b). In this strategy, an oligosaccharide is prepared and coupled to a properly protected amino acid and this building block used in glycopeptide assembly. The stepwise approach is more reliable than block synthesis and can be performed in solution or on solid support. However, it should be realised that many methods and reaction conditions that are commonly used for peptide synthesis cannot be applied for the preparation of glycopeptides. These limitations arise from the chemical lability of glycopeptides. For example, the O-serine and threonine glycosidic linkages are base sensitive and can undergo β elimination (Scheme 5.3).[9] Strong basic conditions can also racemise the stereogenic centres of peptides. Furthermore, glycosidic bond linkages are sensitive under strong acidic conditions. In early synthetic studies

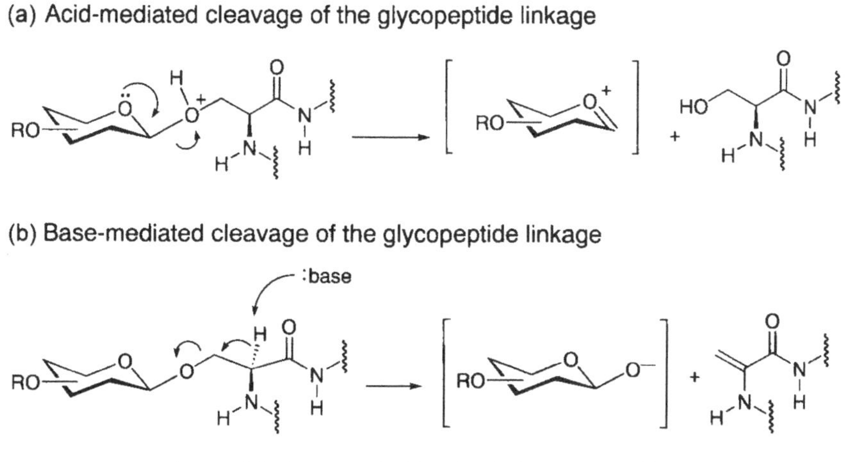

(a) Acid-mediated cleavage of the glycopeptide linkage

(b) Base-mediated cleavage of the glycopeptide linkage

Scheme 5.3   Chemical lability of O-linked glycopeptides.

unnecessary precautions were taken to prevent β elimination but systematic studies have shown that the O-glycopeptide linkages are stable to conditions used for Fmoc cleavage (piperidine) and to catalytic NaOMe in methanol, conditions that are used for cleavage of acetyl protecting groups. These reaction conditions will not racemise chiral centres in the peptide backbone. Glycosidic bonds are cleaved by strong acids such as HF but saccharides are stable to short treatment with TFA and therefore many side-chain protecting groups that are cleaved under these conditions can be used for glycopeptide synthesis. However, care has to be taken when a saccharide contains a fucosidic linkage since this type of glycosidic bond is more labile than normal glycosidic linkages.

5.3 Protecting groups in glycopeptide synthesis

The reaction conditions applied for the protecting group manipulations should be compatible with the glycopeptide structure and in general strong acidic and basic conditions should be avoided. As discussed in Chapter 2, benzyl ethers are often employed as permanent protecting groups during oligosaccharide synthesis; however, this type of protection is rarely used in the syntheses of glycopeptides that contain cysteine residues. The problems of using benzyl groups arise from poisoning of the Pd catalyst by the sulfur atom of cysteine. Thus, benzyl-protecting groups can be used for the assembly of oligosaccharide moieties but have to be replaced by other protecting groups prior to the synthesis of such glycopeptides. Often acetyl protecting groups are used for the protection of the oligosaccharide. This protecting group can be removed by treatment with a catalytic amount of sodium methoxide in methanol or with saturated ammonia in methanol. These reaction conditions do not affect the glycopeptide structure. The use of O-acetates has an additional advantage in that glycosidic linkages are stabilised by this electron-withdrawing protecting group. Thus, it is often advantageous to remove a glycopeptide from a solid support with acid and follow this by cleavage of the acetyl protecting groups.

The fluorene-methoxy-carbonyl (Fmoc)[10] group is often employed as a temporary amino protecting group on glycosylated amino acid building blocks. Cleavage of the Fmoc can best be accomplished by the weak base morpholine as it gives a cleaner reaction product than when piperidine or 1,8-diazabicyclo[5.4.0]undec-7-ene (DBU) is used. The t-butyloxycarbonyl (Boc) group,[11] which can be cleaved with TFA, has been used as an amino protecting group in solution-phase glycopeptide synthesis. The Boc methodology often cannot be used for solid-phase synthesis because the linkers used for these approaches require strong acidic conditions for cleavage (e.g. HF) and glycosidic bonds frequently do not survive this treatment.

Allyl, benzyl, phenylacyl and t-butyl esters are commonly used for protection of the carboxyl group. These esters can be removed without affecting the Fmoc group, and the resulting deprotected glycosylated amino acids can be used in glycopeptide synthesis.

5.4 Chemical synthesis of serine O-glycoside derivatives

The Koenigs–Knörr methodology gives low yields when serine or threonine residues of properly protected amino acid or peptides fragments are glycosylated.[12–15] The disappointing yields are probably because of the low reactivity of serine and threonine hydroxyls. One suggested explanation for the poor reactivity of serine-containing peptides (serine amides) is that unfavourable intramolecular hydrogen bonding (N—H ← :OH) reduces the nucleophilicity of the serine hydroxyl.[16a] Thus, by protecting the serine or threonine amine as an imine moiety, the unfavourable hydrogen bonding pattern can be reversed (C=N: → H—O), leading to *enhanced* reactivity of the hydroxyl group in the Koenigs–Knörr reaction. By running side-by-side reactions under identical conditions, it was demonstrated that the nature of the amino protecting group plays an important role in the course of the Koenigs–Knörr reaction with tetra-O-acetyl-glucosyl bromide and various serine esters (Scheme 5.4).[16b] Competition experiments showed

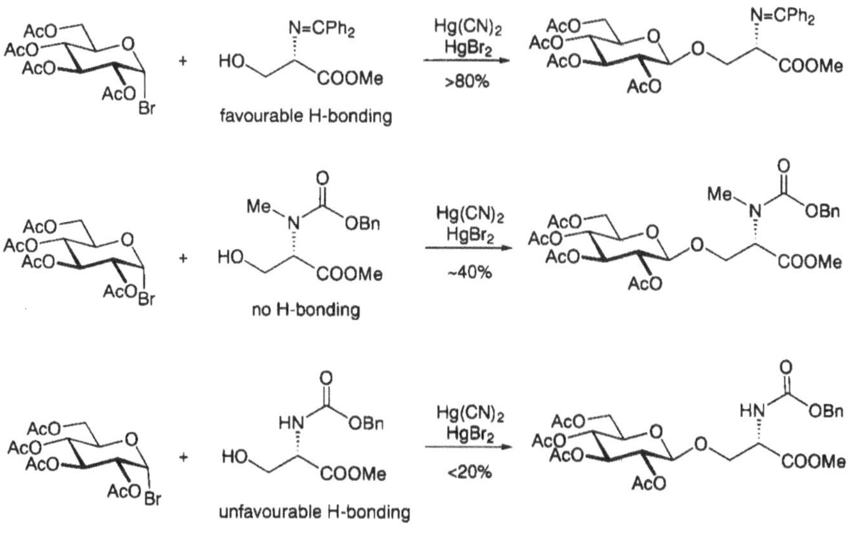

Scheme 5.4 The effect of hydrogen bonding in the Koenigs–Knörr reaction.

that the benzophenone Schiff base esters of serine are more nucleophilic than their benzyloxycarbonyl protected (Cbz, Z-protected) counterparts.[17] It is interesting to note that Z-protected γ-hydroxy-proline derivatives, which are presumably not capable of intramolecular H-bonding because of geometrical constraints, glycosylate more easily than their unconstrained β-hydroxy analogues.[18] To provide useful synthetic methods for solid-phase synthesis, this approach to serine glycosides requires hydrolysis or hydrogenolysis of the Schiff base nitrogen and subsequent N-acylation (Fmoc), thus adding to the number of steps required to incorporate the glycoside-bearing residue into a peptide.

Despite the reduced nucleophilicity of N-acylated serine derivatives, direct glycosylation of Fmoc protected serine and threonine esters is feasible, and has become a popular approach to O-linked glycopeptides (Scheme 5.5).[19] In this approach, the pentafluorophenyl (Pfp) ester has a

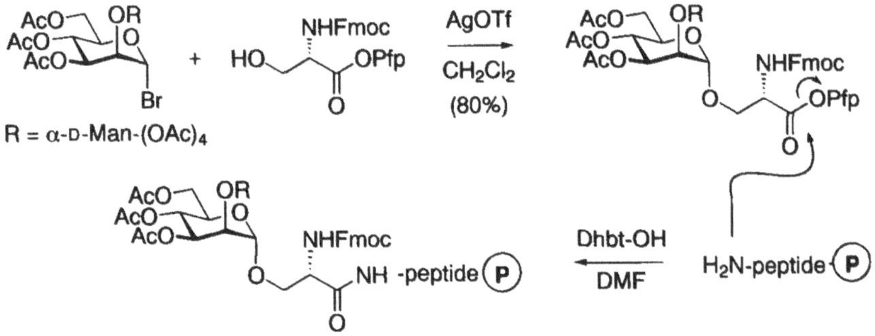

Scheme 5.5 Pentafluorophenyl (Pfp) ester as a protecting and activating group.

dual role; first, it acts as a protecting group during the glycosylation process and subsequently it functions as an activating group during peptide bond formation. Potential problems with the Pfp ester glycosides include hydrolysis of the activated ester, during isolation and chromatography of the amino acid precursor, and modest reactivity of the Pfp esters during the peptide coupling reaction. It is noteworthy that the use of racemisation suppressors such as hydroxybenztriazole[20,21] accelerate peptide bond formation.[19c]

Fmoc protected serine and hydroxyproline can be glycosyled with 1,2-trans-glycosyl peracetates, using Lewis acids (BF$_3$ · Et$_2$O or SnCl$_4$) as promoters (Scheme 5.6).[22] Thus, in this approach, the carboxylic acid of the amino acids is unprotected. Surprisingly, coupling of the glucosyl penta-O-acetate with an Fmoc protected serine derivative gave a lower yield of coupling product. The yields reported for this approach are modest (34%–65%), but the starting materials are readily available, and the products can be used directly in solid-phase glycopeptide assembly.

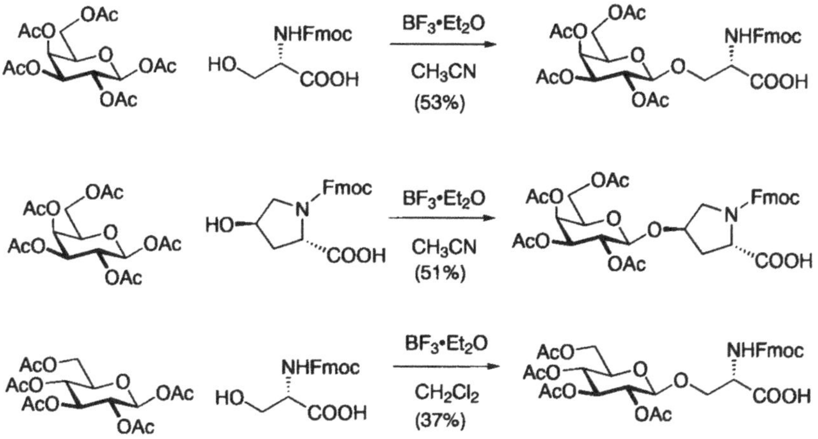

Scheme 5.6   Direct glycosylation of Fmoc protected amino acids.

Perhaps more troublesome than the moderate yields is the necessity for reversed-phase high-performance liquid chromatography (HPLC) to purify the Fmoc protected amino acid glycosides. It has been reported,[23] that the carboxylic acid moiety of an amino acid can be protected as a phenacyl ester and after glycosylation be removed with HOAc/Zn° (Scheme 5.7). Thus, a thioglycosyl donor, having a nonparticipating functionality at C(2), was reacted with the Fmoc protected phenacyl ester

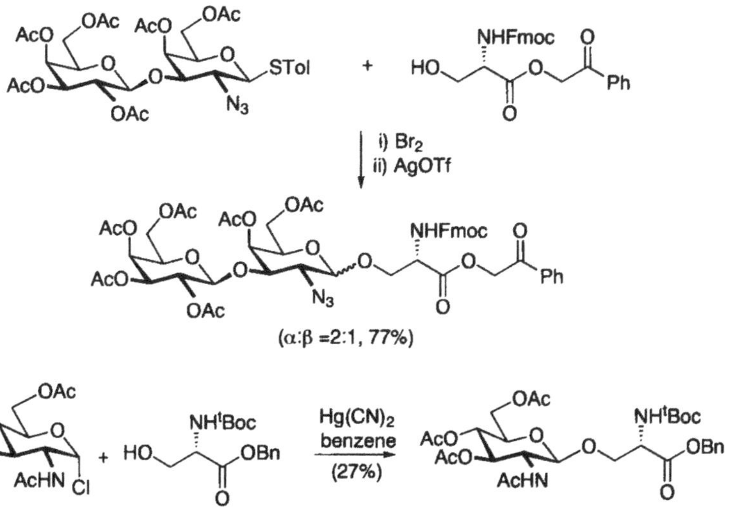

Scheme 5.7   Glycosylation of Fmoc and 'Boc esters.

to provide a mixture of anomers ($\alpha : \beta = 2:1$). Subsequent reduction of the azide moiety with thioacetic acid, followed by acylation, led to the desired 2-deoxy-2-N-acetyl glycosides. It is noteworthy that azides can conveniently be reduced under a variety of conditions to give an amino functionality and the azido group has proven to be an attractive nonparticipating masking functionality for an amino group in glycoconjugate and oligosaccharide synthesis (see Chapters 2 and 4). Furthermore, this approach gives glycosylated amino acids in higher yields but is more laborious than the method employing Pfp esters.

An even earlier Helferich-type glycosylation of the Boc and benzyl protected serine derivative with a glycosyl donor, having a participating N-acetyl functionality at C(2), gave a coupling product in modest yield. The benzyl ester of the product could subsequently be removed by hydrogenolysis.[24] The first linkage type [α-GalNAc–(1 → O-Ser] is by far the most abundant type found in naturally occurring O-linked glycoproteins,[2d] and the second type [β-GlcNAc–(1 → O-Ser] has been found only within nuclear glycoproteins.[25] It should be noted that the glycosyl donor containing an N-acetyl moiety reacts via an oxazolidine intermediate leading to *trans*-β-glycosides and has the advantage that N-deprotection and subsequent reacylation is not necessary. However, oxazolidines are rather unreactive glycosyl donors and give poor results with unreactive acceptors. 2-Amino-glycosyl donors having N-trichloro-ethoxycarbonyl (Troc) as protection for the amino functionality have also been used in glycosylations with serine and threonine derivatives as acceptors.[16b,c, 26] This participating group favours β glycosides, as does the classical N-phthalimide. It is important that the N protection of the glycosyl donor is orthogonal to the N protection of the amino acid acceptor, and it is essential that any N deprotection performed after peptide assembly does not destroy the newly formed peptide bonds.

Although very reactive sulfoxides, derived from thioglycosides (Scheme 5.8), can be useful in the synthesis of highly congested glycosidic linkages,[27] their use with serine and threonine acceptors led to anomeric mixtures.[28] Use of the classical Koenigs–Knörr procedure with a silver catalyst (AgClO$_4$) in conjunction with an azido sugar bromide and an imine protected acceptor gave superior results.[16c] This reaction system has provided high α stereoselectivities ($\alpha : \beta \geq 10 : 1$) as well as excellent overall chemical yields of α glycosides.

The trichloroacetimidate methodology has been successfully applied to the synthesis of sialyl and asialyl T$_N$ epitopes (Scheme 5.9).[29] The use of this methodology can lead to predominant orthoester formation with 2-O-acetyl protected glycosyl donors, but acceptable results can be obtained with 2-azido-glycosyl donors. The complexity of the tetrasaccharide donor probably contributes to the poor chemical yield of this

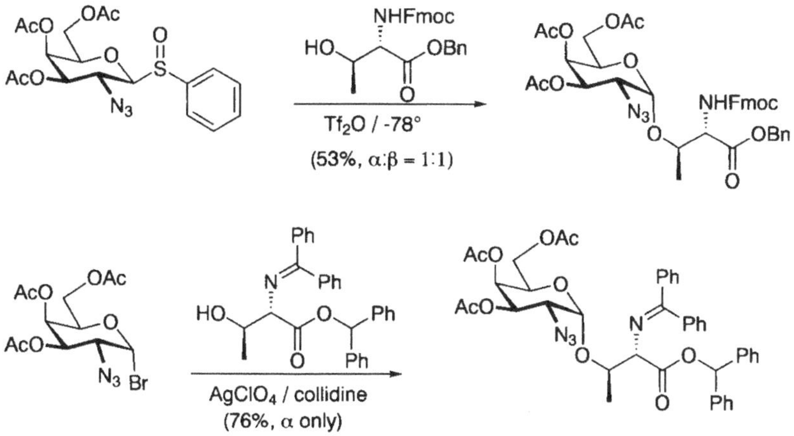

Scheme 5.8 α-Gal*N*Ac: important *O*-linked core region.

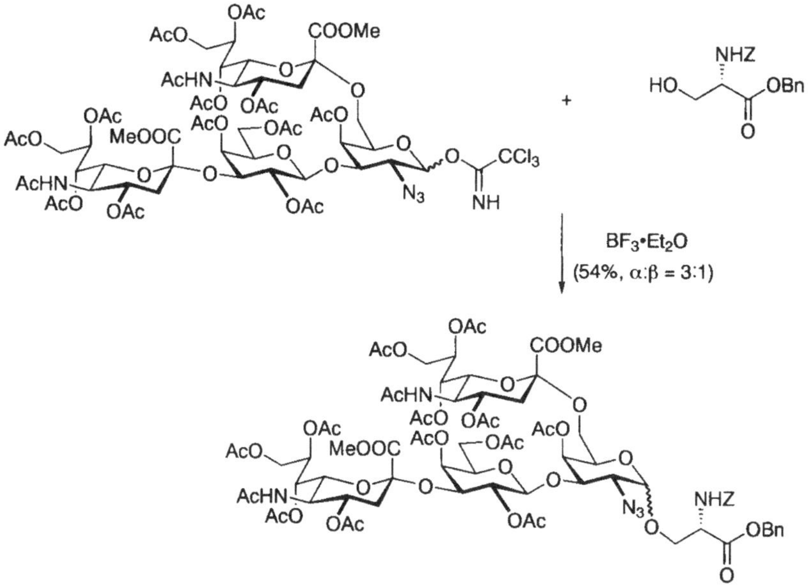

Scheme 5.9 Trichloroacetimidate methodology.

glycosylation, as well as the modest ($\alpha : \beta < 3 : 1$) anomeric selectivity. The principal value of the trichloroacetimidate approach is that fairly complex oligosaccharides may be assembled from saccharide building blocks in a convergent manner prior to glycosylation of the serine or threonine moiety. The thioglycoside approaches also share this advantage and the anomeric thio moiety can function as an efficient anomeric protecting group prior to glycosylation.

Recently, a 2-azido-galactosyl fluoride was coupled with an allyl/Fmoc protected serine derivative and, in this case, the coupling product was obtained with high α selectivity.[30] After a protecting group interconversion step, the derivative obtained was used as a glycosyl acceptor, and subsequent glycosyl donors were added to give a glycotetraosyl peptide structure which represented a predominant substructure in complex glycan-glycoproteins present in human blood group A ovarian mucin.

5.5   The synthesis of N-glycopeptides

The preparation of N-glycopeptides requires a completely different strategy because an anomeric amide has to be introduced instead of a glycosidic linkage. The most commonly employed methods for the preparation of N-glycopeptides proceeds through a glycosyl azide which is reduced to a labile intermediate glycosylamine which is subsequently condensed with an appropriately protected aspartic acid derivative.[31] For example, treatment of 2-acetamido-3,4,6-tri-O-acetyl-2-deoxy-α-D-glucopyranosyl chloride with silver azide gave the corresponding β-D-glucopyranosyl azide (Scheme 5.10). Hydrogenation of the anomeric azide with Pd-C resulted in the formation of glucopyranosyl amine as a mixture of anomers. Coupling of the anomeric mixture of amines with an aspartic acid derivative in the presence of dicyclohexylcarbodiimide (DCC) yielded β-linked glycopeptides in a reasonably good yield. Surprisingly, α anomers are never observed in this condensation. Other activating reagents that have been used to introduce the N-glycosidic linkage include ethyl-2-ethoxy-1,2-dihydroquinoline-1-carboxylate (EEDQ),

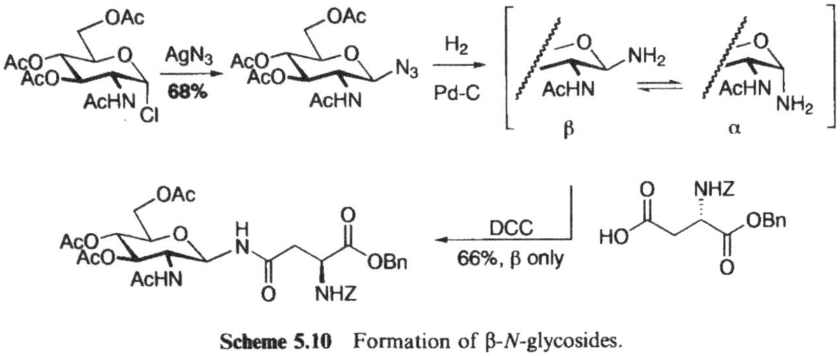

Scheme 5.10   Formation of β-N-glycosides.

2-*i*-butoxy-carbonyl-1,2-dihydroquinoline (IIDQ) or the water-soluble carbodiimide 1-ethyl-3-(3'-dimethylaminopropyl)-carbodiimide (EDC).

It has been shown that glycosyl azides can be prepared in higher yields by using phase-transfer conditions.[32] Furthermore, an anomeric amino group can be introduced by reaction of 2-acetamido-2-deoxy-glucose with ammonium hydrogen carbonate.[33] In the latter approach, the anomeric amino group was protected as an Fmoc group which aided purification. Treatment of anomeric acetates with $TMSN_3$ represents another interesting entry to glycosyl azides.[34]

5.6 Solution-phase and solid-phase glycopeptide synthesis

Properly protected glycosyl amino acids have been used for incorporation into peptide chains using standard methods for peptide coupling. The Fmoc/Boc protecting group combination is most commonly applied using the coupling reagent 2-ethoxy-1-ethoxycarbonyl-2,2-dihydroquino-line for the incorporation of a glycosyl amino acid.[35] In addition, pentafluorophenyl esters,[36] 3,4-dihydroxy-4-oxo-1,2,3-benzotriazine (Dhbt)[37] or DCC/HOBt[38] have been successfully used. The reactivity of the Pfp esters can be increased by the addition of HOBt. Solution-phase techniques have made it possible to synthesise glycopeptides that contain 10–12 amino acid residues and one or more glycosyl side chains. An example of the synthesis of the divalent Lewis^x containing hexapeptide is given in Scheme 5.11. Thus, a trisaccharide β-linked to *N*-Boc protected α-allyl-aspartate was obtained as detailed in Section 5.5. The Boc protecting group of the *N*-glycosyl asparagine derivative was cleaved by treatment with HCl in diethylether, and the deprotected amino functionality was condensed with Boc-glycine. Next, the Boc group of the glycosylated dipeptide was removed using standard conditions. The second glycosylamino acid was prepared by cleavage of the allyl ester of the starting *N*-glycosyl asparagine derivative using Wilkinson's catalyst. This derivative was coupled with the glycosylated dipeptide employing a water-soluble carbodiimide (EDC) to give a tripeptide that has two saccharide attachments. The *N*-terminal Boc group of glycotripeptide was replaced by a biological compatible acetyl group followed by cleavage of the allyl ester. Next, the carboxyl function of this product was easily condensed with the amino group of a tripeptide to give a glycohexapeptide. Finally, the *t*-butyl ester of this compound was cleaved by formic acid and the acetyl groups of the saccharide moiety were removed by treatment with NaOMe in methanol.

Solid-phase-based methods offer a faster approach to the synthesis of more complex glycopeptides and also enable multiglycopeptide

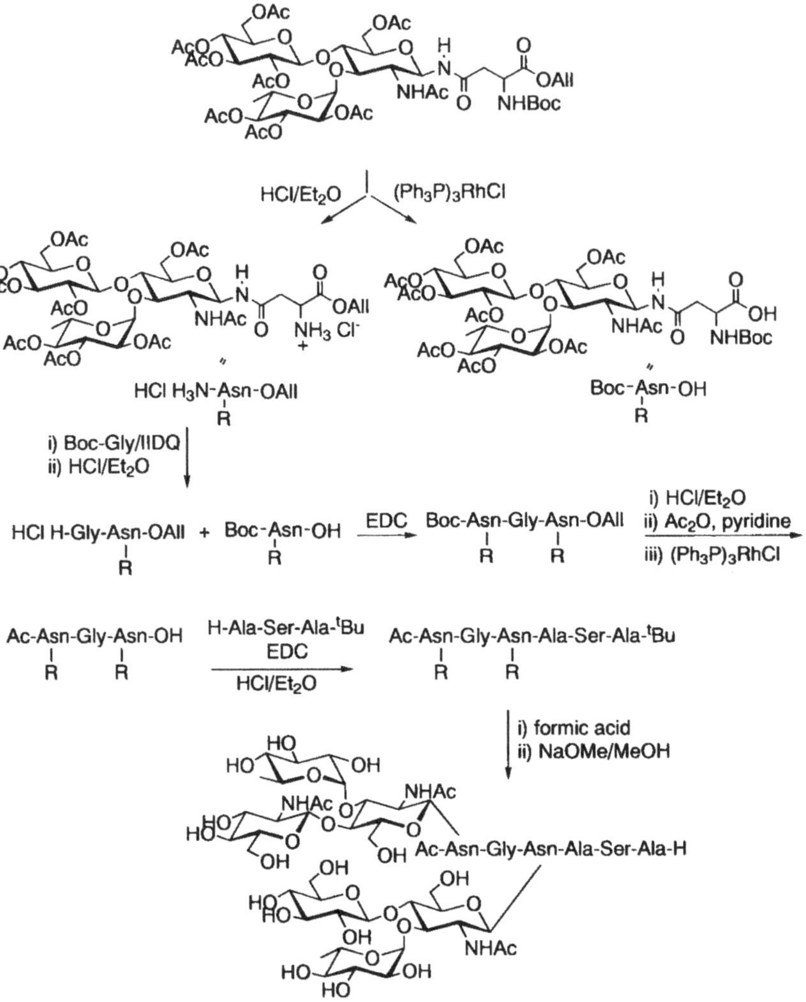

Scheme 5.11 Solution-phase glycopeptide synthesis.

synthesis[39] and the construction of glycopeptide libraries.[40] Only a few standard methods for solid-phase-based synthesis of peptides are applicable to the preparation of glycopeptides. This restriction is a result of the acid lability of glycosidic linkages. Therefore, solid-phase synthesis that involves cleavage of the product from the resin with HF is not applicable. The linkers summarised in Figure 5.2 have been successfully applied and allow cleavage under neutral or mild acidic conditions. However, a persistent problem in the glycosidic linkage between the

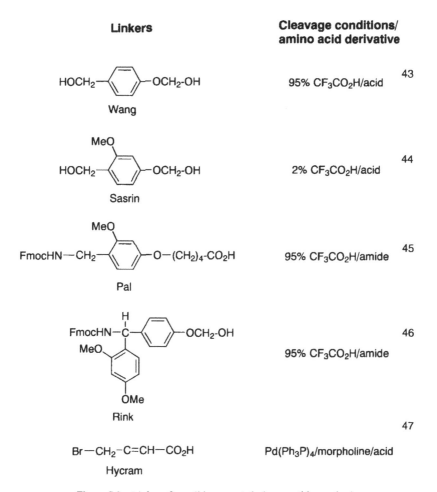

Figure 5.2 Linkers for solid-supported glycopeptide synthesis.

hydroxyl of serine and threonine is that they can be readily cleaved via β elimination under basic conditions (Scheme 5.3).

Polystyrene polymers cross-linked with divinyl benzene are commonly used for solid-phase glycopeptide synthesis; however, for continuous-flow synthesis the kieselguhr-supported polydimethyl acrylamide resin is preferred.[41] Recently, the use of the PEGA resin (polyethylene glycol–polyacrylamide copolymer) was described.[42] The high swelling capacity and disaggregating properties of this resin allows high rates of mass transfer, resulting in fast reactions.

A successful approach for the synthesis of a diglycosylated decapeptide is depicted in Scheme 5.12.[43] Fmoc protected and pentafluorophenyl

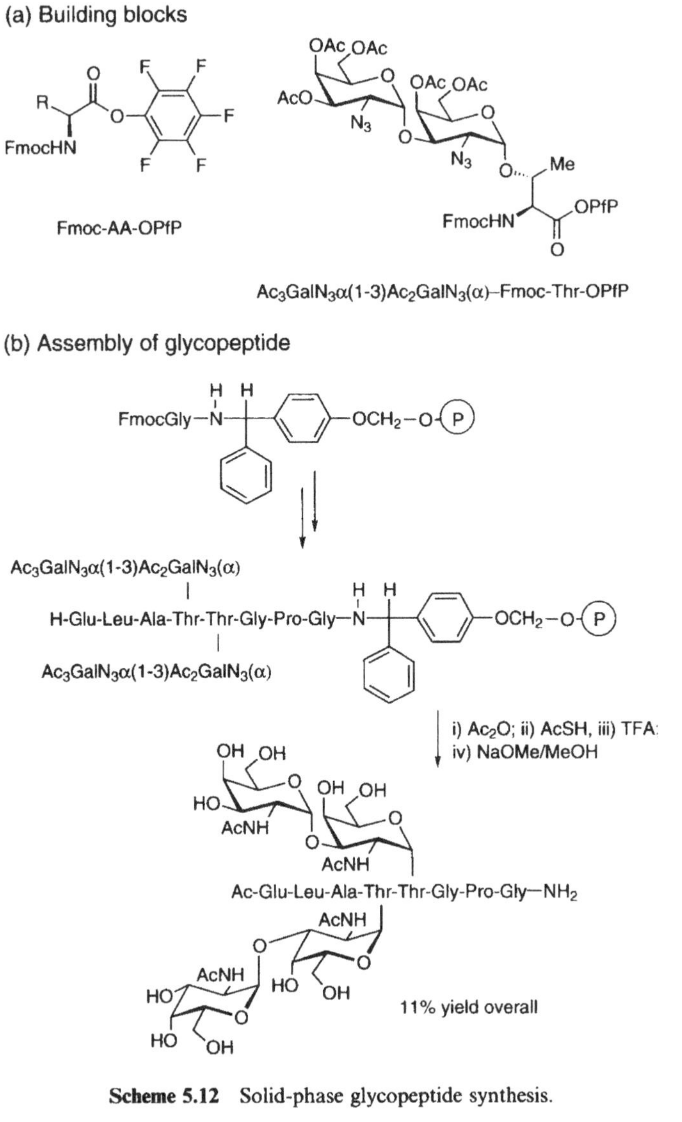

(a) Building blocks

Fmoc-AA-OPfP

Ac$_3$GalN$_3\alpha$(1-3)Ac$_2$GalN$_3$(α)–Fmoc-Thr-OPfP

(b) Assembly of glycopeptide

Ac$_3$GalN$_3\alpha$(1-3)Ac$_2$GalN$_3$(α)
|
H-Glu-Leu-Ala-Thr-Thr-Gly-Pro-Gly—N
|
Ac$_3$GalN$_3\alpha$(1-3)Ac$_2$GalN$_3$(α)

i) Ac$_2$O; ii) AcSH, iii) TFA:
iv) NaOMe/MeOH

Ac-Glu-Leu-Ala-Thr-Thr-Gly-Pro-Gly—NH$_2$

11% yield overall

Scheme 5.12 Solid-phase glycopeptide synthesis.

activated amino acids and a glycosylated threonine derivative were used for the assembly which was performed on a Rink linker. After disassembly of the product, the azides were reduced with thioacetic acid and the revealed amino functionality was acetylated with acetic anhydride. The product was cleaved from the solid support by treatment with TFA, resulting in amide formation at the C terminus. Finally, the acetyl groups were cleaved with sodium methoxide in methanol. The electron-withdrawing acetyl group of the saccharide stabilised the glycosidic linkages during the acid-mediated release from the solid support.

5.7   Enzyme-mediated glycopeptide synthesis

Proteases have proven to be useful catalysts for the stereoselective and racemisation-free coupling of peptide fragments. Wong *et al.* applied proteases for the preparation of glycopeptides, and a range of *N*- and *O*-glycosyl amino acids and peptides were coupled under kinetically controlled conditions using the serine proteases subtilusin BPN′ and thiosubtilusin (Scheme 5.13).[48] The latter protease is genetically engineered and favours aminolysis over hydrolysis. It can be used in anhydrous DMF as well as in aqueous solutions and is more stable then the wild protease. Coupling of the *N*-protected *O*-glycopeptide with dipeptides in the presence of thiosubtilisin in an aqueous solution at 50°C gave the glycopeptides in reasonable yields. Thus, the enzyme catalyses the transesterification of the methyl ester of the glycopeptide with the amine of the dipeptide. To combine the use of glycosyl transferases and proteases in the synthesis of glycopeptides, the glycopeptide was galactosylated using a galactosyl transferase. The galactosylation involved regeneration of UDP-Gal (see Chapter 4). It was also found that *N*-protected dipeptide esters with either a peracetylated or an unprotected β-GlcNac moiety are suitable substrates for subtilusin BPN′.

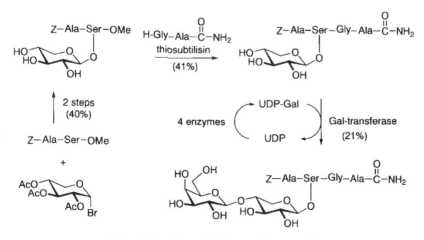

Scheme 5.13   Enzymatic glycopeptide assembly.

The enzymatic synthesis of glycopeptides does not require protection of the functional groups of the amino acid side chains and sugar hydroxyls, because of the high stereoselectivity and regioselectivity of proteases. However, the substrate selectivity of these enzymes may limit a wider range of applications.

References

1. (a) N.J. Maeji, Y. Inoue and R. Chujo, 1987, *Biopolymers*, 26, 1753; (b) M. Hollosi, A. Perczel and G.D. Fasman, 1990, *Biopolymers*, 29, 1549; (c) H. Paulsen, R. Busch, V. Sinnwell and A. Pollex-Krüger, 1991, *Carbohydr. Res.*, 214, 227; (d) K. Dill, R.E. Hardy, M.E. Daman, J.M. Lacombe and A.A. Pavia, 1982, *Carbohydr. Res.*, 108, 31; (e) R. Shogren, T.A. Gerken and N. Jentoft, 1989, *Biochemistry*, 28, 5525; (f) T.A. Gerken, K.J. Butenhof and R. Shogren, 1989, *Biochemistry*, 28, 5536.
2. (a) A. Gottschalk, 1960, *Nature (London)*, 186, 949; (b) H. Eylar, 1965, *J. Theor. Biol.*, 10, 89; (c) P.J. Winterburn and C.F. Phelps, 1972, *Nature (London)*, 236, 147; (d) J. Montreuil, 1980, *J. Adv. Carbohydr. Chem. Biochem.*, 37, 157.
3. (a) T. Feizi, 1989, in *Ciba Foundation Symposium 145*, Wiley, Chichester, 62; (b) C.M. Starr and J.A. Hanover, 1992, in *Nuclear Trafficking* (C.M. Feldherr, ed.), Academic Press, San Diego, CA, 175; (c) E.F. Hounsell, M.J. Davies and D.V. Renouf, 1996, *Glycoconjugate J.*, 13, 19; (d) H.C. Joao, I.G. Scragg and R.A. Dwek, 1992, *FEBS Lett.*, 307, 343; (e) P.A. Gleeson, R.D. Teasdale and J. Burke, 1994, *Glycoconjugate J.*, 11, 381; (f) R.U Lemieux, 1989, *Chem. Soc. Rev.*, 18, 347; (g) M.I. Horowitz (ed.), 1982, *The Glycoconjugates*, Vol. I–V, Academic Press, New York; (h) R.U. Margolis and R.K. Margolis (eds.), *Neurobiology of Glycoconjugates*, Plenum Press, New York; (i) H.J. Allen and E.C. Kisailus (eds.), 1992, *Glycoconjugates: Composition, Structure, and Function*, Marcel Dekker, New York; (j) P. Kovác (ed.) 1994, *Synthetic Oligosaccharides: Indispensible Probes for the Life Sciences, ACS Monograph Series 560*, American Chemical Society, Washington, DC; (k) D.G. Large and C.D. Warren, 1997, *Glycopeptides and Related Compounds*, Marcel Dekker, New York.
4. S.C. Hubbard and R.J. Ivatt, 1981, *Annu. Rev. Biochem.*, 50, 555.
5. (a) A. Abbadi, M. Mcharfi, A. Aubry, S. Premilat, G. Boussard and M. Marraud, 1991, *J. Am. Chem. Soc.*, 113, 2729; (b) B. Imperiali, K.L. Shannon, M. Unno and K.W. Rickert, 1992, *J. Am. Chem. Soc.*, 114, 7944; (c) K.W. Rickert and B. Imperiali, 1995, *Chemistry and Biology*, 2, 751.
6. (a) J.C. Paulson, 1989, *TIBS*, 272; (b) A. Varki, 1993, *Glycobiol.*, 218, 1; (c) R.A. Dwek, 1996, *Chem. Rev.*, 96, 683.
7. (a) J. Kihlberg and M. Elofsson, 1997, *Current Medicinal Chemistry*, 4, 85; (b) H. Kunz, 1987, *Angew. Chem. Int. Ed. Engl.*, 26, 294–308; (c) R.L. Halcomb and C.H. Wong, 1993, *Curr. Opin. Struct. Biol.*, 3, 694; (d) H. Krotkiewski, 1988, *Glycoconjugate J.*, 5, 35; (e) H.G. Garg and R.W. Jeanloz, 1986, *Adv. Carbohydr. Chem. Biochem.*, 43, 135; (f) M. Meldal, 1994, *Curr. Opin. Struct. Biol.*, 4, 710.
8. (a) S.T. Cohen-Anisfeld and P.T. Lansbury Jr, 1993, *J. Am. Chem. Soc.*, 115, 10531; (b) D. Vetter, D. Tumelty, S.K. Singh and M.A. Gallop, 1995, *Angew. Chem. Int. Ed. Eng.*, 34, 60; (c) J. Offer, M. Quibell and T.J. Johnson, 1996, *J. Chem. Soc., Perkin Trans. I*, 175.
9. J. Kihlberg and T. Vuljanic, 1993, *Tetrahedron Lett.*, 34, 6135.
10. (a) L.A. Carpino, 1970, *J. Am. Chem. Soc.*, 92, 5748; (b) L.A. Carpino, 1987, *Accounts Chem. Res.*, 20, 401.
11. (a) L. Moroder, A. Hallett, A. Wünsch, O. Keller and G. Wersin, 1976, *Hoppe-Seylers Zeitschrift Physiol. Chem.*, 357, 1651; (b) D.S. Tarbell, Y. Yamamoto and B.M. Pope, 1972,

Proc. Natl. Acad. Sci. USA., 69, 730; (c) G. Perseo, S. Piani and R. de Catiglione, 1983, *Int. J. Peptide Protein Res.*, 21, 227; (d) G.W. Anderson and A.C. McGregor, 1987, *J. Am. Chem. Soc.*, 79, 6180.

12. (a) F. Micheel and H. Köchling, 1958, *Chem. Ber.*, 91, 673; (b) J.K.N. Jones, M.B. Perry, B. Shelton and D.J. Walton, 1961, *Can. J. Chem.*, 39, 1005; (c) V.A. Derevitskaya, M.G. Vafina and N.K. Kochetkov, 1967, *Carbohydr. Res.*, 3, 377.

13. (a) H.G. Garg and R.W. Jeanloz, 1975, *Proc. Am. Peptide Symp.*, 4, 379; (b) H.G. Garg and R.W. Jeanloz, 1977, *Proc. Am. Peptide Symp.*, 5, 477; (c) H.G. Garg and R.W. Jeanloz, 1979, *Carbohydr. Res.*, 76, 85.

14. M. Hollosi, E. Kollat, I. Laczko, K.F. Medzihradszky, J. Thurin and L. Otvos, 1991, *Tetrahedron Lett.*, 32, 1531.

15. A.A. Pavia, 1985, *Proc. Am. Peptide Symp.*, 9, 469.

16. (a) L. Szabo, Y.S. Li and R. Polt, 1991, *Tetrahedron Lett.*, 32, 585; (b) R. Polt, L. Szabo, J. Treiberg, Y.S. Li and V.J. Hruby, 1992, *J. Am. Chem. Soc.*, 114, 10249; (c) L. Szabo, J. Ramza, C. Langdon and R. Polt, 1995, *Carbohydr. Res.*, 274, 11.

17. P.J. Garegg, P. Konradsson, I. Kvarnstrom, T. Norberg, S.C.T. Svensson and B. Wigilius, 1985, *Acta Chem. Scand. B.*, 39, 569.

18. E. Bardaji, J.L. Ttorres, P. Clapes, F. Albericio, G. Barany and G. Valencia, 1990, *Angew. Chem. Int. Ed. Engl.*, 29, 291.

19. (a) A.M. Jansson, M. Meldal and K. Bock, 1990, *Tetrahedron Lett.*, 31, 6991; (b) M. Meldal, 1994, in *Neoglycoconjugates: Preparation and Applications* (Y.C. Lee and R.T. Lee, eds.), Academic Press, San Diego, CA, 145.

20. W. Konig and R. Geiger, 1970, *Chem. Ber.*, 103, 788.

21. L.A. Carpino, 1993, *J. Am. Chem. Soc.*, 115, 4397.

22. (a) M. Elofsson, B. Walse and J. Kihlberg, 1991, *Tetrahedron Lett.*, 32, 7613; (b) L. A. Salvador, M. Elofsson and J. Kihlberg, 1995, *Tetrahedron*, 51, 5643.

23. (a) B. Lüning, T. Norberg and J. Tejbrant, 1989, *J. Chem. Soc., Chem. Commun.*, 17, 1267; (b) B. Lüning, T. Norberg and J. Tejbrant, 1989, *Glycoconjugate J.*, 6, 5.

24. S. Lavielle, N.C. Ling, R. Saltman and R.C. Guillemin, 1981, *Carbohydr. Res.*, 89, 229.

25. (a) R.S. Haltiwanger, W.G. Kelly, E.P. Roquemore, M.A. Blomberg, L.Y.D. Dong, L. Kreppel, T.Y. Chou and G.W. Hart, 1992, *Biochem. Soc. Trans.*, 20, 264; (b) T.Y. Chou, C.V. Dang and G.W. Hart, 1995, *Proc. Natl. Acad. Sci. USA*, 92, 4417.

26. K. Higashi, K. Nakayama, T. Soga, E. Shioya, K. Uoto and T. Kusama, 1990, *Chem. Pharm. Bull.*, 38, 3280.

27. L. Yan and D. Kahne, 1996, *J. Am. Chem. Soc.*, 118, 9239.

28. A.H. Andreotti and D. Kahne, 1993, *J. Am. Chem. Soc.*, 115, 3352.

29. H. Iijima and T. Ogawa, 1989, *Carbohydr Res.*, 186, 107.

30. W.M. Macindoe, H. Ijima, Y. Nakahara and T. Ogawa, 1995, *Carbohydr. Res.*, 269, 227.

31. (a) G.S. Marks, R.D. Marshall and A. Neuberger, 1963, *Biochem. J.*, 85, 274; (b) C. H. Bolton and R.W. Jeanloz, 1963, *J. Org. Chem.*, 28, 3228.

32. J. Thiem and T. Wiemann, 1990, *Angew. Chem., Int. Ed. Engl.*, 29, 80.

33. L.M. Likhosherstov, O.S. Novikova, V.A. Derevitskaya and N.K. Kochetkov, 1986, *Carbohydr. Res.*, 146, C1.

34. H. Kunz, W. Sager, D. Schanzerbach and M. Decker, 1991, *Liebigs Ann. Chem.*, 649.

35. H. Paulsen and K. Adermann, 1989, *Liebigs Ann. Chem.*, 751.

36. L. Kisfaludy and I. Schön, 1983, *Synthesis*, 325.

37. (a) W. König and R. Geiger, 1970, *Chem. Ber.*, 103, 2034; (b) E. Atherton, J.L. Holder, M. Meldal, R.C. Sheppard and R.M. Valerio, 1988, *J. Chem. Soc., Perkin Trans. 1*, 2887.

38. (a) S.S. Wang, 1973, *J. Am. Chem. Soc.*, 95, 1328; (b) H. Paulsen, G. Merz and U. Weichert, 1988, *Angew. Chem. Int. Ed. Engl.*, 27, 1365; (c) H. Paulsen, G. Merz, S. Peters and U. Weichert, 1990, *Liebigs Ann. Chem.*, 1165.

39. (a) S. Peters, T. Bielfeldt, M. Meldal, K. Bock and H. Paulsen, 1991, *Tetrahedron Lett.*, 32, 5067; (b) S. Peters, T. Bielfeldt, M. Meldal, K. Bock and H. Paulsen, 1992, *J. Chem. Soc., Perkin Trans 1*, 1163; (c) S. Peters, T. Bielfeldt, M. Meldal, K. Bock and H. Paulsen, 1994, *Liebigs Ann. Chem.*, 369.
40. (a) M.A. Gallop, R.W. Barrett, W.J. Dower, S.P.A. Fodor and E.M. Gordon, 1994, *Med. Chem.*, 37, 1233; (b) M.A. Gallop, R.W. Barrett, W.J. Dower, S.P.A. Fodor and E.M. Gordon, 1994, *Med. Chem.*, 37, 1385.
41. A. Atherton, E. Brown, R.C. Sheppard and A. Rosevear, 1981, *J. Chem. Soc., Chem. Commun.*, 1151.
42. M. Meldal, 1992, *Tetrahedron Lett.*, 33, 3077.
43. S. Rio-Anneheim, H. Paulsen, M. Meldal and K. Bock, 1995, *J. Chem. Soc., Perkin Trans. I*, 1071.
44. (a) M. Mergler, R. Tanner, J. Gosteli and P. Grogg, 1988, *Tetrahedron Lett.*, 29, 4005; (b) M. Mergler, R. Nyfeler, R. Tanner, J. Gosteli and P. Grogg, 1988, *Tetrahedron Lett.*, 29, 4009; (c) B. Lüning, T. Norberg and J. Tejbrant, 1989, *J. Chem. Soc., Chem. Commun.*, 1267; (d) B. Lüning, T. Norberg, C. Rivera-Baeza and J. Tejbrant, 1991, *Glycoconjugate J.*, 8, 450.
45. (a) F. Albericio, N. Kneib-Cordonier, S. Biavcalana, L. Gera, R.I. Masada, D. Hudson and G. Barany, 1990, *J. Org. Chem.*, 55, 3730; (b) T. Bielfeldt, S. Peters, M. Meldal, K. Bock and H. Paulsen, 1992, *Angew. Chem. Int. Ed. Engl.*, 31, 857.
46. H. Rink, 1987, *Tetrahedron Lett.*, 28, 3787.
47. H. Kunz and B. Dombo, 1988, *Angew. Chem. Int. Ed. Engl.*, 27, 711.
48. C.-H. Wong, M. Schuster, P. Wang and P. Sears, 1993, *J. Am. Chem. Soc.*, 115, 5893.

Part B

Natural Product Synthesis from Monosaccharides

In Part B, we analyse and discuss some recent enantiospecific natural product syntheses that have utilised monosaccharides as chiral starting materials. The coverage is not meant to be exhaustive or extensive but, rather, illustrative of the way in which monosaccharides are currently being used in asymmetric natural product synthesis. For more exhaustive coverage of this field, the reader is referred to several fine monographs and reviews.[1-8]

1. B. Fraser-Reid and R.C. Anderson, 1980, *Fortschr. Chem. Org. Naturst,* Springer-Verlag, Wien, 39, 1.
2. A. Vasella, 1980, *Modern Synthetic Methods 1980* (R. Scheffold, ed.), Salle and Sauelander, Aurau, 174.
3. S. Hanessian, 1983, *Total Synthesis of Natural Products: The Chiron Approach,* Pergamon Press, Oxford.
4. T.D. Inch, 1984, *Tetrahedron,* 40, 3161.
5. N.K. Kochetkov, A.F. Sviridov, M.S. Ermolenko et al., 1984, *Carbohydrates in the Synthesis of Natural Compounds,* USSR.
6. K.J. Hale, 1993, Monosaccharides: Use in the Asymmetric Synthesis of Natural Products in *Rodd's Chemistry of Carbon Compounds,* Vol. IE/F/G, Second Supplements to the 2nd edn. (M. Sainsbury, ed.), Elsevier, Amsterdam, 315.
7. R.J. Ferrier (Senior Reporter), *Carbohydrate Chemistry, Specialist Periodical Reports of the Royal Society of Chemistry,* Cambridge.
8. R.J. Ferrier and S. Middleton, 1993, *Chem. Rev.,* 93, 2779.

6 (−)-Echinosporin

K.J. Hale

6.1 Introduction

(−)-Echinosporin (XK-213) is a novel antitumour antibiotic isolated[1] from the fermentation broths of *Streptomyces echinosporus*. It is of great pharmacological interest on account of its significant anticancer effects in several rodent tumour models. (−)-Echinosporin was recently synthesised in enantiomerically pure form by A.B. Smith and coworkers starting from L-ascorbic acid.[2] Their synthesis is described below.

6.2 Smith's retrosynthetic analysis of (−)-echinosporin

Because the lactone ring in (−)-echinosporin is very highly strained, and susceptible to ring-opening by nucleophilic addition, Smith and coworkers opted to defer its construction until the final step of their total synthesis. Two strategic possibilities appeared to exist for lactone ring-closure (Scheme 6.1). Either the hemiacetal hydroxyl of 1 could be converted to a reactive leaving-group L, and cyclisation initiated by internal attack of the carboxylic acid group (path A), or, alternatively, the carboxylic acid unit could be activated and ring-closure effected by path B. If the latter strategy were followed, a reactive α-epoxy-lactone would almost certainly form, whose carbonyl would be inappropriately positioned for subsequent intramolecular attack by the hemiacetal OH. Recognition of this fact led Smith to formulate a synthetic end-game based on path A. Specifically, his plan was to prepare dihydroxyacid 1 and implement a Mitsunobu lactonisation[3] to obtain the target (Scheme 6.2). Compound 1 was disassembled by protecting its carboxylic acid as a methyl ester, and ring-opening its enolised hemiacetal. This afforded the enolate-aldehyde 3 (Scheme 6.2), which was then subjected to an intramolecular aldol transform to obtain 4. Compound 4 was an attractive intermediate for a forward synthesis, since it could potentially be derived from a [2 + 2]-photocycloaddition between 7 and 2-cyclopentene-1-one (Scheme 6.2). In such an addition, the cyclopentenone would add to the less-hindered exo-face of 7 (see Section 6.4), which would help guarantee the 6,5-*cis*-ring-junction stereochemistry. A correct regiochemical outcome was predictable on electronic grounds. Obviously, in the synthetic direction, two key transformations would be needed to

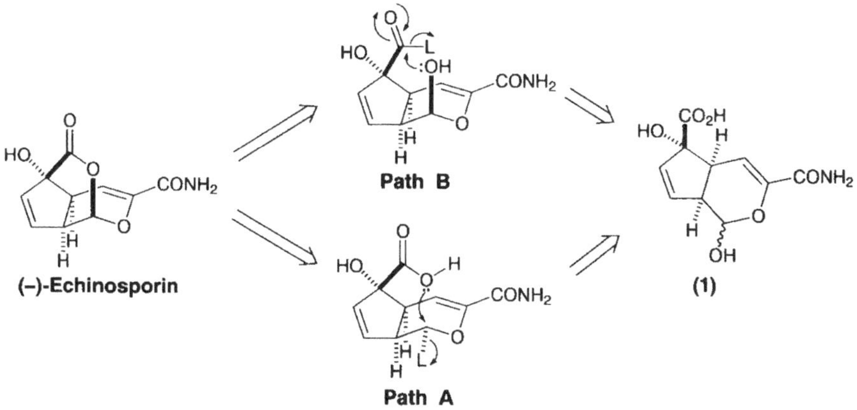

Scheme 6.1   Strategic possibilities for lactonisation in (−)-echinosporin.

convert **4** into **1**, namely, a base-mediated retro-aldol reaction, and a hemiacetalisation. The retro-aldol process could be expected to occur spontaneously, as a result of considerable strain in the cyclobutane ring of **4**. Hemiacetal formation could also be predicted to be facile. Smith concluded his retrosynthetic analysis by correlating the C(2)-stereocentre of **7** with the C(4)-stereocentre of vitamin C (L-ascorbic acid). With this as background, we now examine Smith's synthesis of (−)-echinosporin in detail.

6.3   Smith's (−)-echinosporin synthesis

Smith's route[2] to (−)-echinosporin began with the regioselective protection of L-ascorbic acid with 2,2-dimethoxypropane (Scheme 6.3). This afforded the 5,6-*O*-isopropylidene acetal **8** as the sole reaction product, in 77% yield. The 2,3-acetal was not formed in this reaction because of excessive ring strain. Compound **9** was accessed by oxidative degradation of **8** with basic hydrogen peroxide.[4] Hydroxyl protection and esterification next followed to form **10**. After partial reduction of the methyl ester to the aldehyde, ring-closure to D-threose **11** was initiated by removing the isopropylidene group with pyridinium-*p*-toluene sulfonate (PPTS) in methanol. *O*-Isopropylidenation of **11** provided **12** in high yield. The *cis*-fused 1,2-acetal emerged exclusively from this reaction, due to significant ring strain in the alternative *trans*-fused 1,2- and 2,3-acetals that could have formed. The next three steps of the synthesis were Swern oxidation, *N*-tosylhydrazone formation, and base-mediated Bamford–Stevens reaction[5] to prepare glycal **7**. Although Bamford–Stevens reactions are

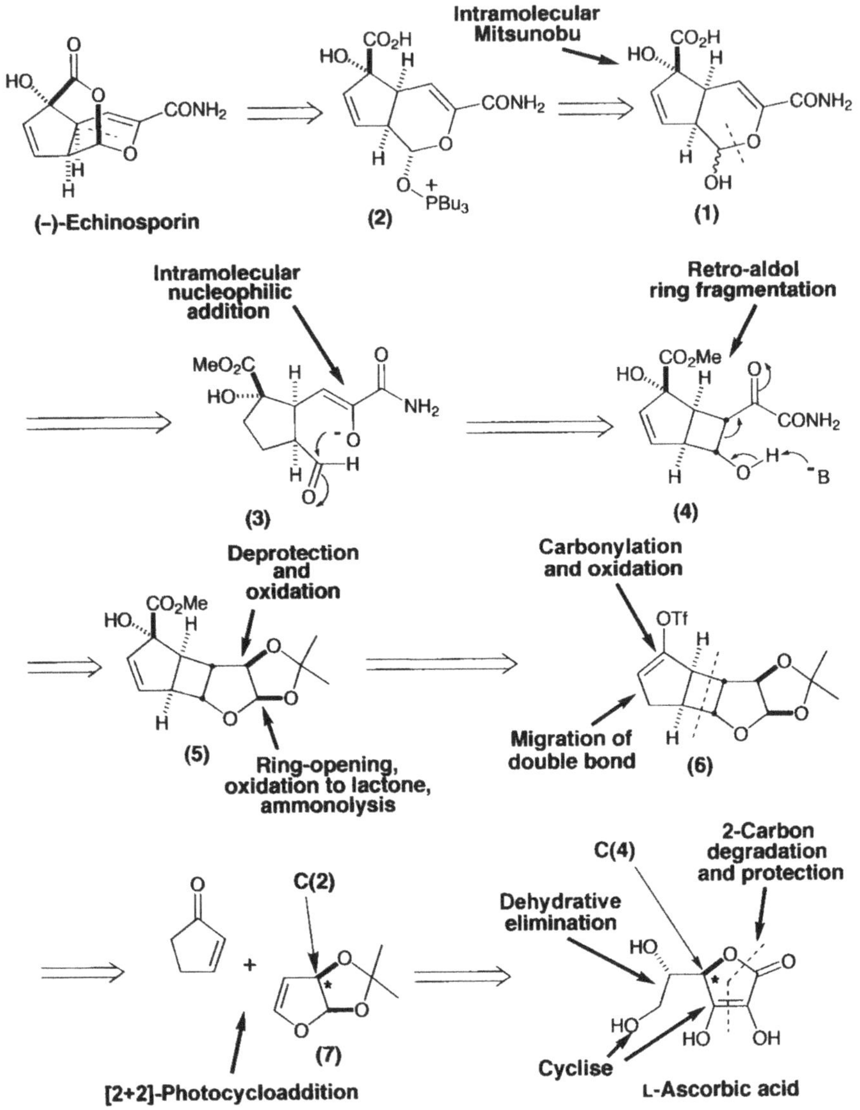

Scheme 6.2 Smith's retrosynthetic analysis of (−)-echinosporin.

sometimes low-yielding, for compound **13** the process proved highly efficient.

The key [2 + 2]-photocycloaddition reaction between **7** and cyclopentenone afforded a mixture of three photoadducts **14**, **20**, and **21**, in which the head-to-tail products **14** and **20** predominated (Scheme 6.4). The major isomer **14** was isolated in 50% yield. Smith next set about converting **14** into enol-triflate **6**. The requisite enolate was generated under kinetically-controlled conditions, by reacting **14** with lithium

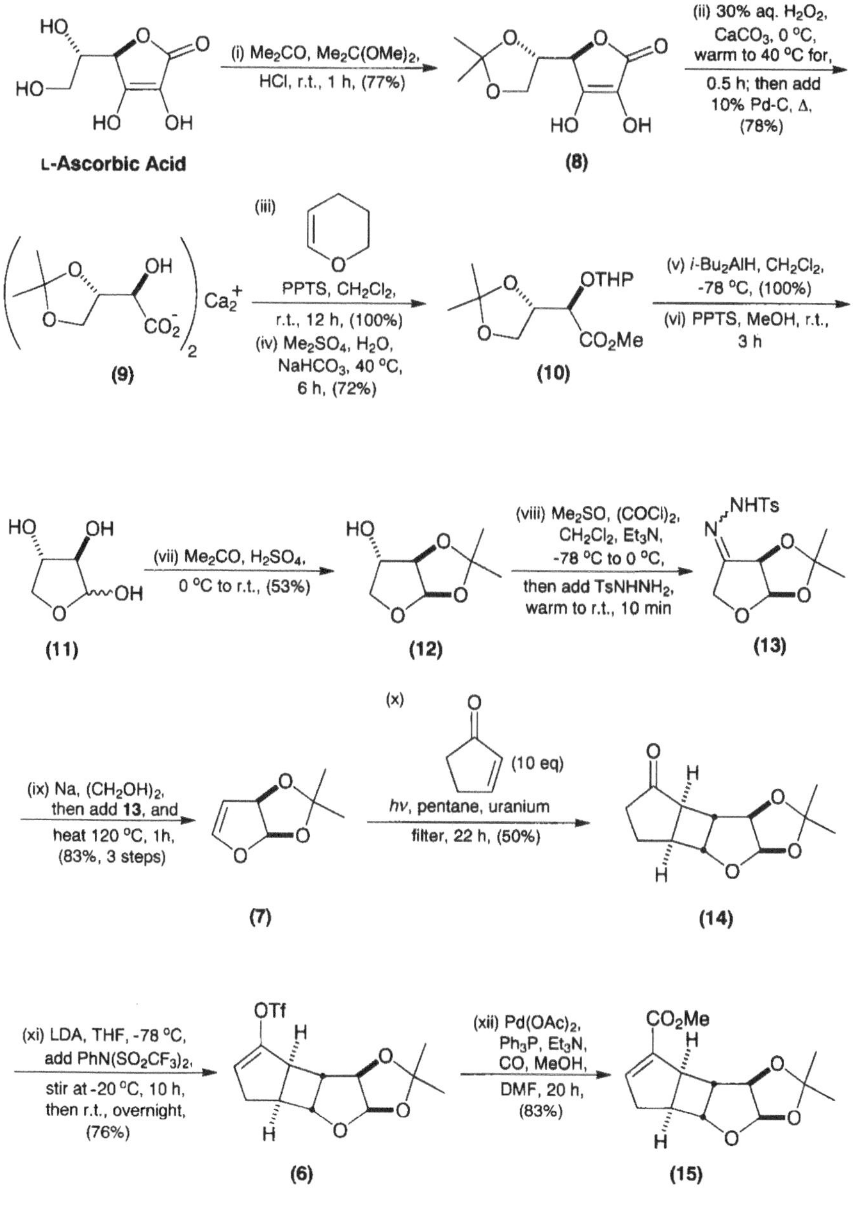

Scheme 6.3 Smith's total synthesis of (−)-echinosporin from L-ascorbic acid.

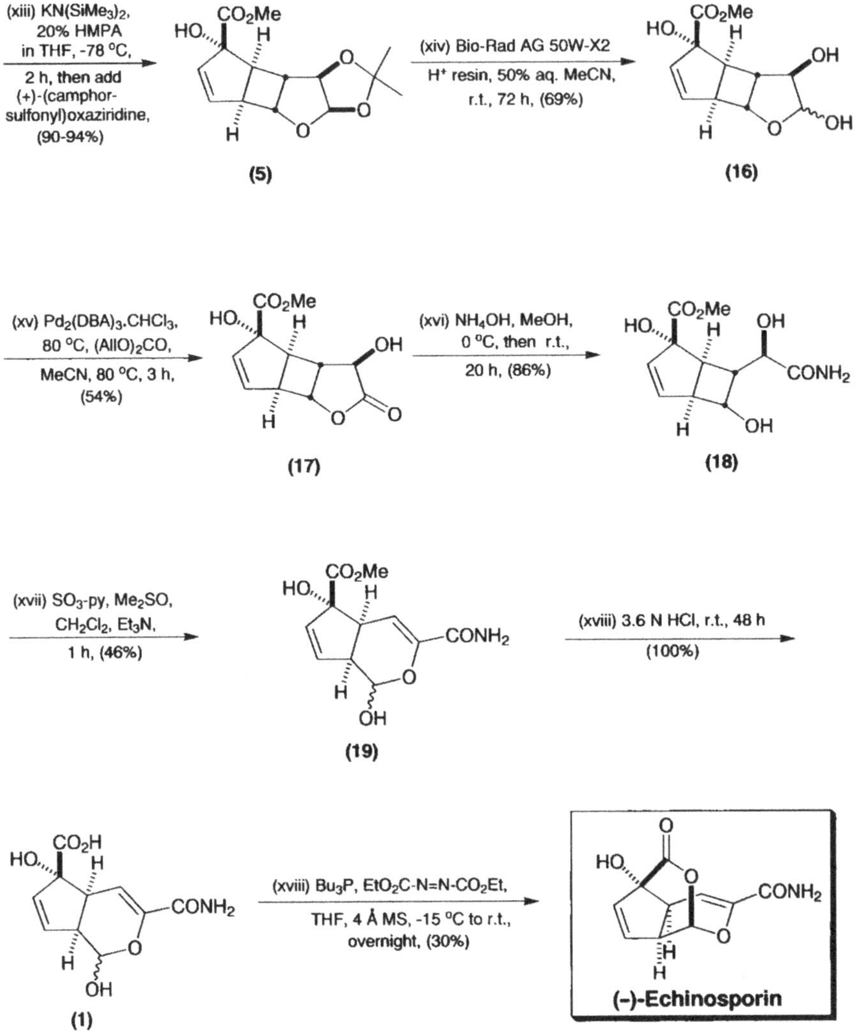

Scheme 6.3 (Continued).

diisopropylamide (LDA) in tetrahydrofuran (THF) at low temperature. Being a very bulky base, LDA preferentially abstracted the more sterically accessible α-carbonyl proton from **14**, to produce the less stable enolate of the two that could have potentially formed. Trapping of this enolate with *N*-phenyltriflimide[6] afforded the desired product **6**. It proved a relatively straightforward task to transform **6** into the α,β-unsaturated ester **15** via palladium(0)-catalysed carbonylation.[7] The

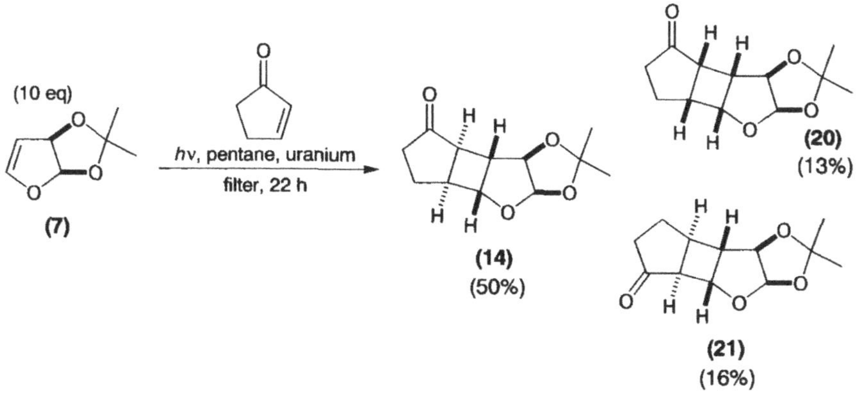

Scheme 6.4 [2 + 2]-Photocycloaddition reaction.

resulting enolate **15** was then subjected to γ-deprotonation with potassium hexamethyldisilazide. This created an extended enolate that reacted with [+]-camphorsulfonyloxaziridine[8] to produce **5** (see Section 6.4). The fact that the enolate and oxidant had 'matched' chirality clearly helped reinforce the stereoselectivity of this oxidation through double asymmetric induction. Hexamethylphosphoric triamide (HMPA) was added to the reaction mixture solely to enhance the reactivity of the enolate through coordination to the metal counterion. Typically, electrophilic hydroxylation proceeded in 90–94% yield. Smith now focused on detaching the O-isopropylidene group from **5** with acidic resin in aqueous acetonitrile. The resulting lactol **16** was then oxidised to lactone **17** with diallyl carbonate and palladium(0).[9]

Since lactones are generally more susceptible to nucleophilic attack than ordinary alkyl esters, Smith found it possible to selectively ring-open **17** with ammonium hydroxide to form **18**. A selective oxidation was then performed with the Parikh–Doering reagent [SO$_3$-py/dimethylsulfoxide (SO$_3$-py/DMSO)][10] to obtain **4**. Selectivity was possible in this reaction because of the higher acidity of the amide α-hydroxyl in **18**. This conferred much greater nucleophilicity on this OH towards the activated-DMSO reagent. Once generated, α-ketoamide **4** (see Scheme 6.2) underwent a spontaneous de Mayo-type[11] retro-aldol fragmentation to relieve ring strain. This afforded the α-ketoamide enolate **3**, which immediately cyclised to hemiacetal **19**. Because hemiacetal **19** readily decomposed when treated with strong base, Smith hydrolysed its methyl ester with 3.6 N HCl (see Scheme 6.3). This furnished **1** in quantitative yield. The finale of Smith's (−)-echinosporin synthesis was his Mitsunobu lactonisation[3] of hydroxyacid **1**. Here the less-hindered hemiacetal hydroxyl was selectively converted to a reactive glycosyl phosphonium ion, which then underwent internal displacement by the pendant acid. It provided (−)-echinosporin in 28–30% yield.

Smith's synthesis of (−)-echinosporin stands out for its good stereo- and regiocontrol at every stage. Its large number of highly chemoselective reactions, and its use of an asymmetric [2 + 2]-photocycloaddition reaction for constructing the *cis*-fused 6,5-bicyclic core are also commendable. The latter reaction provides a beautiful illustration of how properly harnessed photochemistry can be decisively deployed for the stereocontrolled synthesis of complex natural products.

6.4   Mechanistic analysis of some key reactions employed in the Smith (−)-echinosporin synthesis

The conversion of 8 to 9
After conjugate addition of the hydroperoxide anion to the chelated α,β-unsaturated lactone, the resulting enolate counterattacks the proximal β-hydroperoxide group to form an epoxide. The latter then ring-expands to create another enolate, which captures a proton from water. The product hemiacetal finally undergoes ring-opening and saponification to produce 9 (Scheme 6.5).

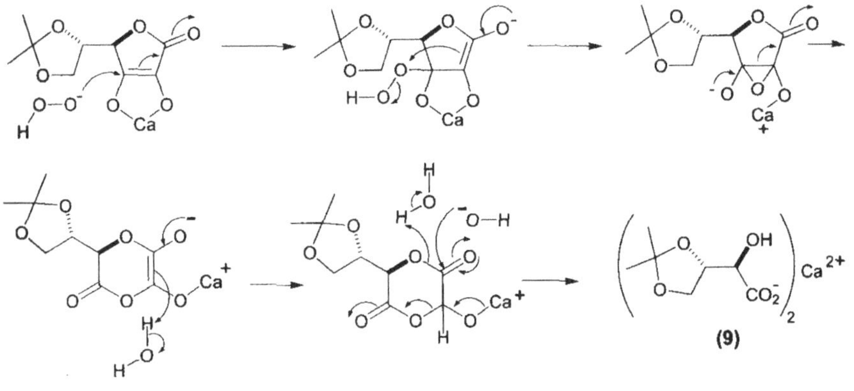

Scheme 6.5.

The conversion of 13 to 7
The classical Bamford–Stevens reaction[6] is a two-step elimination process involving a hydrazone dianion intermediate. The latter sequentially

expels *p*-toluenesulfinate anion and nitrogen to give a vinyl anion which is protonated (Scheme 6.6).

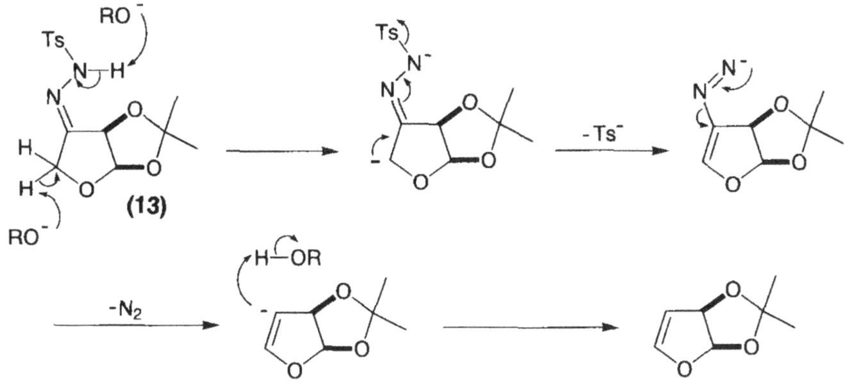

Scheme 6.6.

The conversion of 7 to 14

It is widely believed that enone–alkene photoadditions proceed through an exciplex (excited complex). For cyclopentenone and **7**, the exciplex forms from the photoexcited enone in its triplet state and the glycal in its ground-state. The regiochemistry of addition probably reflects a preferred alignment of the addends in the exciplex. Because a photoexcited enone probably has considerable diradical character, it is reasonable to assume that the more electrophilic α-keto radical would prefer to bind to the more nucleophilic portion of the alkene, since this would maximise attractive interactions within the complex. The reaction of **7** with cyclopentenone would thus favour the formation of diradical **22**, which would then ring-close to cyclobutane **14** (Scheme 6.7).

The stereochemical outcome of this photoaddition can be rationalised by the cyclopentenone adding on to the less hindered exo-face of the bicyclic dihydrofuran **7**, as shown in Figure 6.1.

The conversion of 6 into 15

Initially the Pd(0) complex oxidatively adds to enol triflate **6** to form a vinyl–Pd(II) species. Carbon monoxide then inserts into the new Pd—C σ-bond to yield a palladium(II)–acyl complex which captures methanol. The methanolysis step is formally a reductive elimination reaction in which the Pd(0) catalyst is regenerated to propagate the catalytic cycle (Scheme 6.8).[7]

Favoured reaction pathway

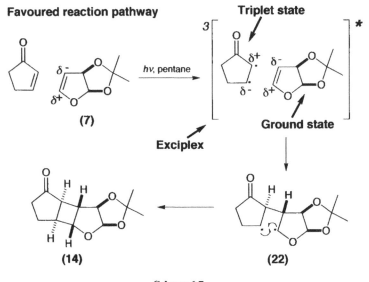

Scheme 6.7.

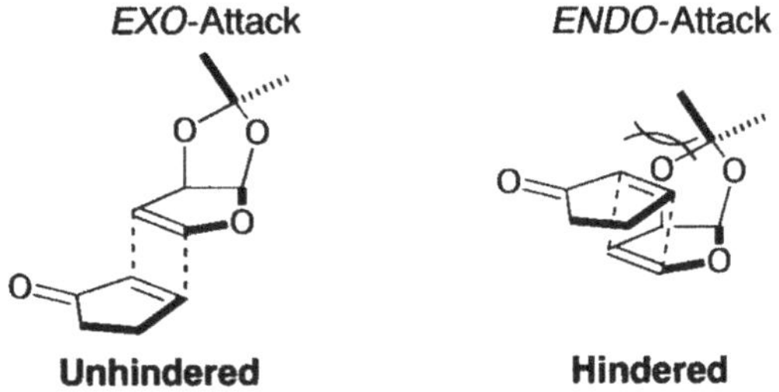

EXO-Attack   **ENDO**-Attack

Unhindered   **Hindered**

Figure 6.1   Unhindered and hindered faces of bicyclic hydrofuran.

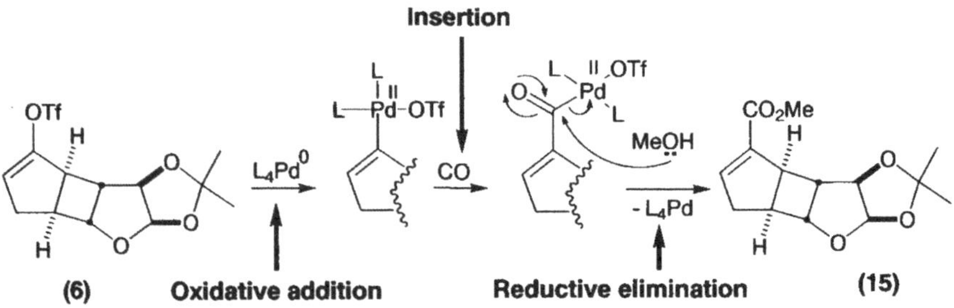

Insertion

(6)   **Oxidative addition**   **Reductive elimination**   (15)

Scheme 6.8.

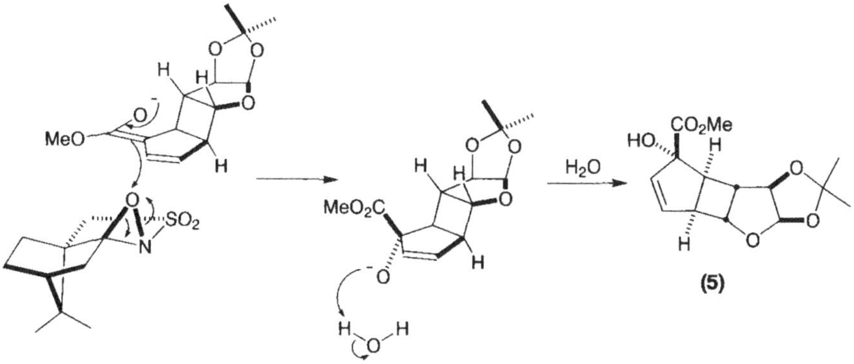

Scheme 6.9.

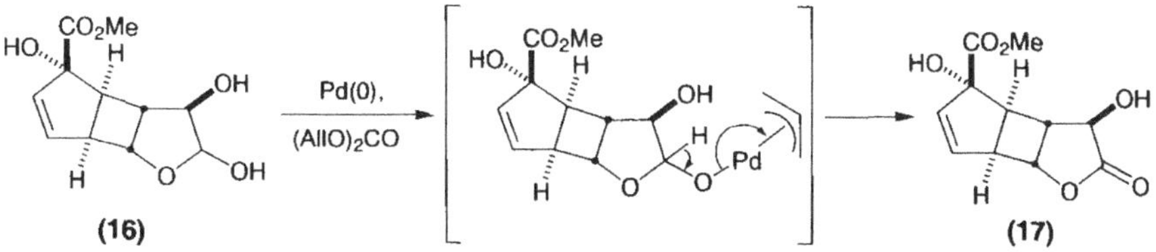

Scheme 6.10.

The conversion of 15 into 5

Kinetic enolisation of **15** generates a dienolate which reacts on its less hindered underside with the Davis [+]-camphorsulfonyloxaziridine reagent[8] as shown in Scheme 6.9.

The conversion of 16 to 17

Here the Pd(0) complex reacts with diallyl carbonate to form a π-allyl palladium alkoxide that ligates to **16**. The resulting intermediate then undergoes β-hydride elimination to produce the lactone and propene (Scheme 6.10).[9]

References

1. (a) T.N. Iida, N. Hirayama and K. Shirahata, 1981, *Abstracts of Papers*, 44th Annual Meeting of the Japan Chemical Society, 12 October, 1E-18, p. 403; (b) R. Okachi, I. Kawamoto and T. Sato (Kyowa Hakko), 1981, *Japan Kokai*, 59,777, 23 May; (c) T. Sato, I. Kawamoto, T. Oka and R. Okachi, 1982, *J. Antibiot.*, 35, 266; (d) M. Morimoto and R. Imai, 1985, *J. Antibiot.*, 38, 490; (e) N. Hirayama, T. Iida, K. Shirahata, Y. Ohashi and Y. Sasada, 1983, *Bull. Chem. Soc. Jpn.*, 56, 287.

2. (a) A.B. Smith, III, G.A. Sulikowski and K. Fujimoto, 1989, *J. Amer. Chem. Soc.*, 111, 8039; (b) A.B. Smith, III, G.A. Sulikowski, M.M. Sulikowski and K. Fujimoto, 1992, *J. Amer. Chem. Soc.*, 114, 2567.

3. (a) O. Mitsunobu, 1981, *Synthesis*, 1; (b) D.L. Hughes, 1992, *Org. React.*, 42, 335; (c) I. D. Jenkins and O. Mitsunobu, 1995, *Encyclopedia of Reagents for Organic Synthesis* (L.A. Paquette, ed.), 8, 5379.

4. C.C. Wei, S. De Bernado, J.P. Tengi, J. Borgese and M. Weigele, 1985, *J. Org. Chem.*, 50, 3462.

5. (a) R.H. Shapiro, 1976, *Org. React.*, 23, 405; (b) M.A. Gianturco, P. Friedel and V. Flanagan, 1965, *Tetrahedron Lett.*, 1847; (c) A.R. Chamberlin and S.H. Bloom, 1990, *Org. React.*, 39, 1.

6. (a) J.E. McMurry and W.J. Scott, 1983, *Tetrahedron Lett.*, 24, 979; (b) A.B. Smith, III, K.J. Hale, L.M. Laakso, K. Chen and A. Riera, 1989, *Tetrahedron Lett.*, 30, 6963.

7. S. Cacchi, E. Morera and G. Ortar, 1985, *Tetrahedron Lett.*, 26, 1109.

8. F.A. Davis, M.S. Haque, T.G. Ulatowski and J.C. Towson, 1986, *J. Org. Chem.*, 51, 2402; (b) F.A. Davis and M.S. Haque, 1986, *J. Org. Chem.*, 51, 4083; (c) F.A. Davis, T.G. Ulatowski and M.S. Haque, 1987, *J. Org. Chem.*, 52, 5288; (d) F.A. Davis, A.C. Sheppard, B.-C. Chen and M.S. Haque, 1990, *J. Amer. Chem. Soc.*, 112, 6679.

9. I. Minami and J. Tsuji, 1987, *Tetrahedron*, 43, 3909.

10. (a) J.R. Parikh and W.E. von Doering, 1967, *J. Amer. Chem. Soc.*, 89, 5505; (b) T.T. Tidwell, 1990, *Org. React.*, 39, 297.

11. (a) P. de Mayo, J.-P. Pete and M. Tchir, 1967, *J. Amer. Chem. Soc.*, 89, 5712; (b) P.E. Eaton, 1968, *Acc. Chem. Res.*, 1, 50; (c) P. de Mayo, 1971, *Acc. Chem. Res.*, 4, 41; (d) D. Termont, D. De Keukeleire and M. Vandewalle, 1977, *J. Chem. Soc. Perkin Trans.*, 1, 2349.

7 (+)-Zaragozic acid C

K.J. Hale

7.1 Introduction

Zaragozic acid C is a novel metabolite of the fungus *Leptodontium elatius*.[1] It functions as a picomolar inhibitor of squalene synthase, an enzyme thought to play a role in human cholesterol biosynthesis. It is widely believed that zaragozic acid C might function as a cholesterol-lowering agent in humans. As a result, there has been considerable synthetic interest in this and related family members. The first enantiospecific total synthesis of (+)-zaragozic acid C was achieved by Carreira and Du Bois,[2] using D-isoascorbic acid as the chiral starting material. We now analyse their synthesis, which is outstanding for its excellent planning and execution.

7.2 Carreira's retrosynthetic analysis of (+)-zaragozic acid C

Carreira's opening retrosynthetic move on zaragozic acid C disengaged its C(6) *O*-acyl side chain, and blocked its three carboxyls as acid-labile *t*-butyl esters (Scheme 7.1). The C(7)-hydroxyl was then protected as an acid-labile *t*-butyl carbonate group to provide compound **1** as a subtarget. Carreira felt that protection of this hydroxyl was necessary because previous work on *O*(6)-deacylated zaragozic acid had shown that the C(7)-OH was more susceptible to *O*-acylation, possibly for electronic reasons. The same study also revealed that the C(4) tertiary OH could be left unmasked during such a forward acylation.

Carreira's plan for introducing the tertiary alcohol group within **1** was to add a lithium acetylide reagent onto ketone **3**. Chelation of the organolithium with the C(7)-pivaloyloxy group in **3** would help guide its addition to the more hindered β-face of the ketone. Multiple oxidation of **2** was then envisaged for transforming it into **1**. Carreira next performed an imaginary reduction on **3** and manipulated its protecting groups to access **4**. This permitted unravelling of the unusual bicyclic ketal array by retrosynthetic acid hydrolysis; the result was ketone **5**. Because the ketal stereochemistry of **4** was identical to that found in the zaragozic acids, Carreira naturally assumed that this had to be the most energetically stable arrangement that the acetal could adopt. As a result, he envisaged

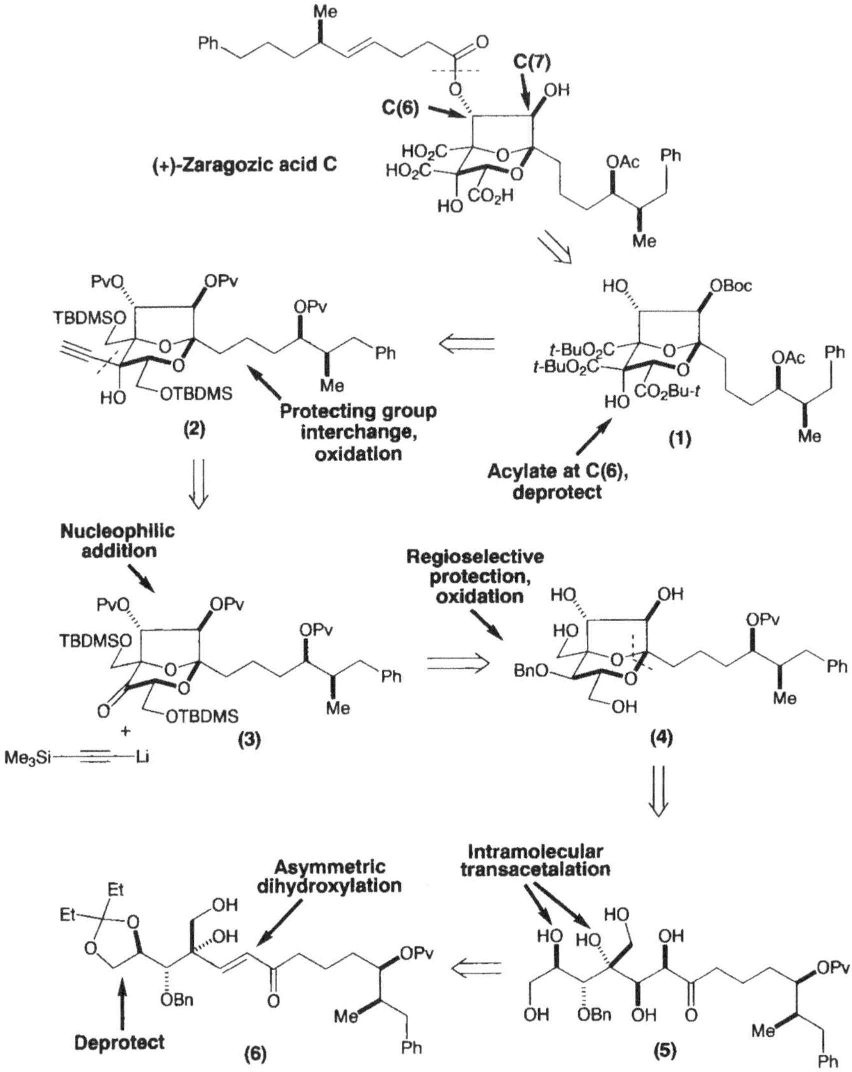

Scheme 7.1 Carreira's retrosynthetic plan for the synthesis of (+)-zaragozic acid C.

assembling it under thermodynamically-controlled (equilibrating) trans-acetalation conditions. These would favour formation of the most energetically stable acetal product. The terminal diol group in **5** was then retrosynthetically protected as an acid-labile *O*-pentanonide acetal, to allow selective clearance of the hydroxyls α and β to the ketone, and reveal enone **6** as a subtarget. Since *trans*-disubstituted enones are often good substrates for the Sharpless asymmetric dihydroxylation (AD) reaction,[3] this protocol appeared ideal for introducing this functionality in the forward direction. The selection of an enone subtarget also opened up the possibility of exploiting an alkynone precursor 7 in the synthesis (Scheme 7.2). Tactically, it is often good practice to work towards a 1,2-disubstituted alkyne in a retrosynthetic analysis, since 1,2-disubstituted alkynes can often be readily disconnected at either σ-bond into viable electrophile and acetylenic anion partnerships. The incorporation of an acetylenic diadduct into a synthetic plan can thus allow for quite dramatic retrosynthetic simplification of a target molecule. In the case of 7, Carreira's use of this transform led to the selection of 8 and 9 as subtargets. A further application of this procedure on 10 revealed ketone 11 and trimethylsilylethynylmagnesium bromide as possible reactants. Thus, in one fell swoop quite considerable molecular simplification was achieved. The final phase of the retrosynthetic planning was chirality matching of 12 to a readily available chiral starting material. Carreira appreciated that the two stereogenic centres present within 12 could be readily equated with the C(4) and C(5) hydroxyls of D-isoascorbic acid. This abundant, yet much underused, starting sugar was therefore selected as the optically active starting material for this synthesis.

7.3   Carreira's total synthesis of (+)-zaragozic acid C

Early attention focused on the need for developing a viable synthetic route to 16, the precursor required for the assembly of ketal 17 (see Scheme 7.4). Utilising technology first introduced by Cohen and co-workers at Hoffmann-La Roche,[4] D-isoascorbic acid was oxidatively degraded with basic hydrogen peroxide, and the crude product lactonised with 6N HCl (Scheme 7.3). The lactone ring of 13 was then opened up with dimethylamine, and the terminal diol regioselectively protected with 3,3-dimethoxypentane to access 12. Although in principle three acetals could have potentially arisen from this reaction, only one product was observed in practice, this being compound 12 for steric reasons. After protection of the hydroxyl as an O-benzyl ether, exposure of the product to ethoxyvinyllithium afforded ketone 11, by nucleophilic addition–elimination. The next objective was stereoselective positioning of the chiral tertiary alcohol unit within 10. This was achieved by a chelation-controlled addition of trimethylsilylethynylmagnesium bromide to ketone 11. Ozonolysis then selectively cleaved the double bond in 10 to provide ethyl ester 14. Note how the alkyne unit survived this oxidation step.

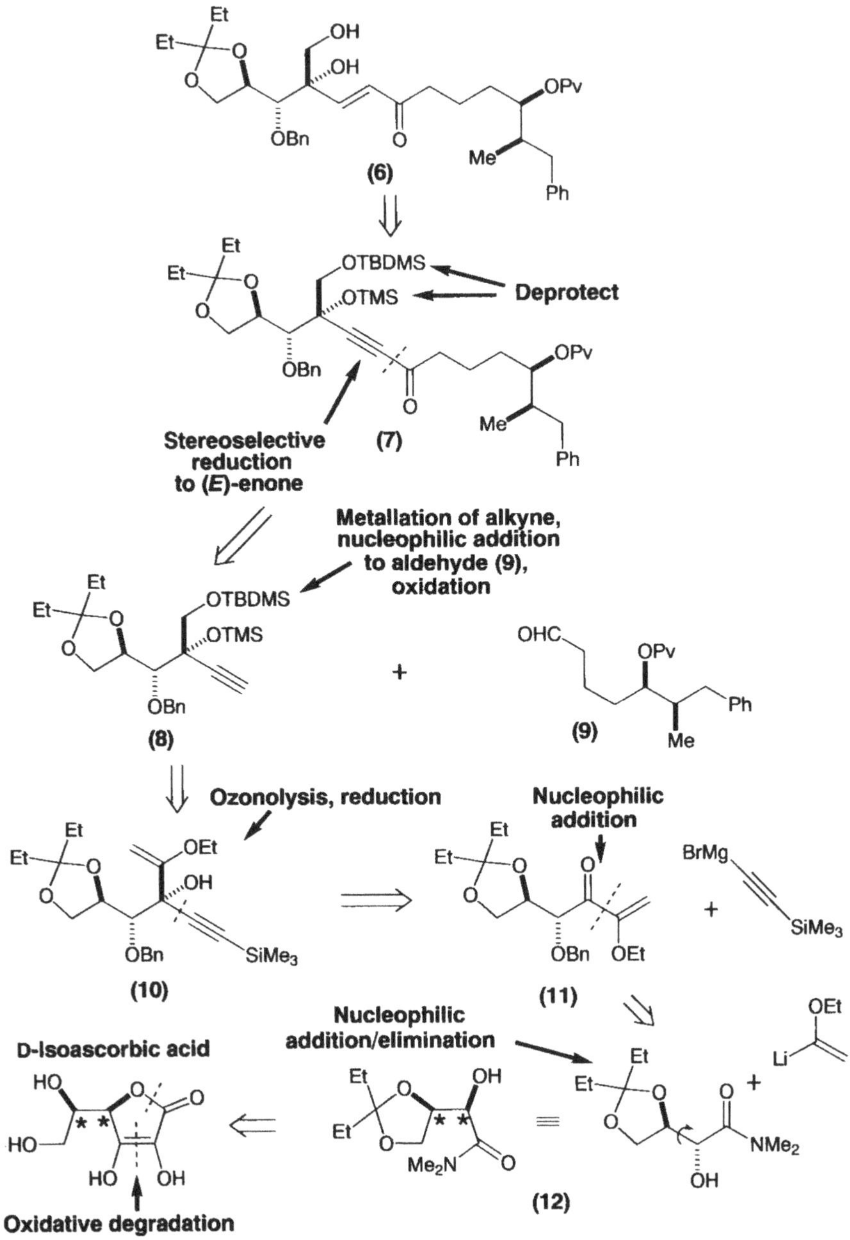

Scheme 7.2 Carreira's retrosynthetic plan for the synthesis of (+)-zaragozic acid C.

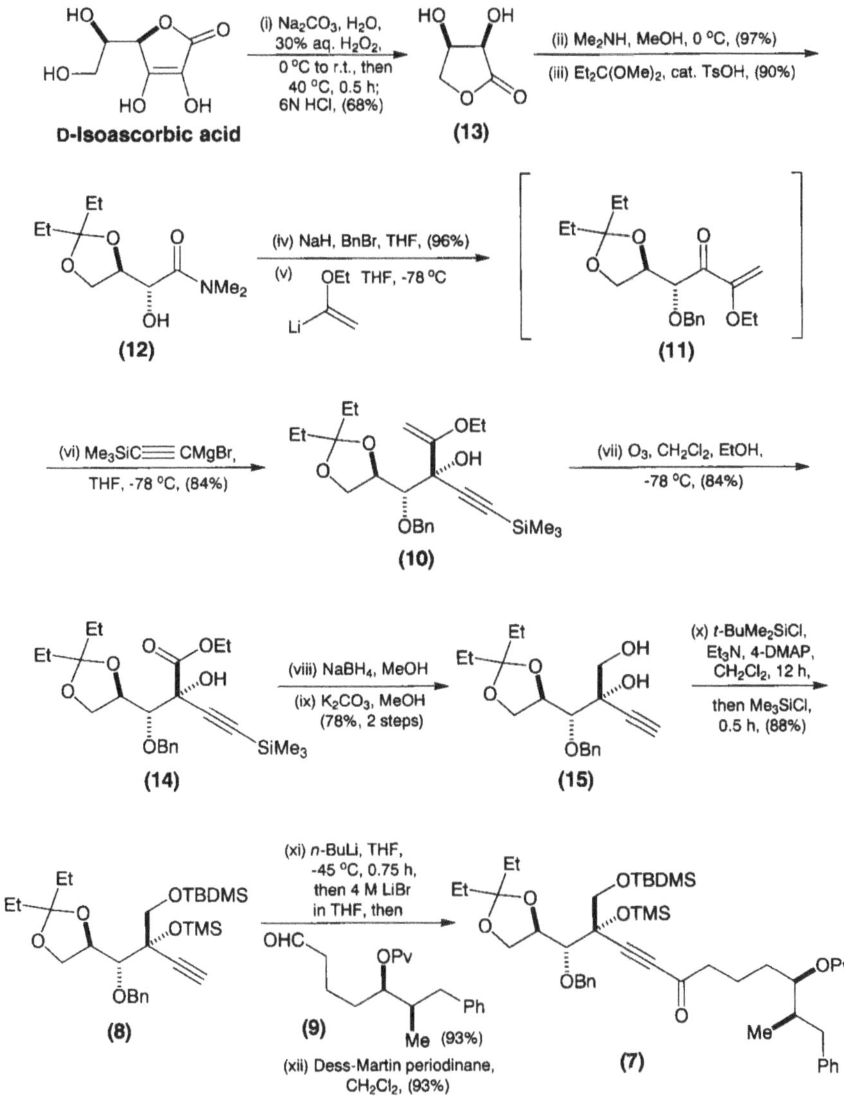

Scheme 7.3 Carreira's total synthesis of (+)-zaragozic acid from D-isoascorbic acid.

In order to side-step potential problems with hydroxyl-directed hydro-metallation of the alkyne, ester **14** was reduced with sodium borohydride in methanol. This also led to a small amount of alkyne desilylation (5%–10%). Fortunately, however, this was the next step of the synthesis. After selective protection of diol **15** to obtain **8**, the stabilised acetylenic anion was generated. This reacted with aldehyde **9** to produce a mixture of alcohols that readily underwent oxidation to alkynone **7** when exposed to the Dess–Martin reagent (Scheme 7.3).

Although $CrCl_2$ and $CrSO_4$ are the two low-valent chromium(II) salts typically used[5] to reduce ynones to *trans* enones, both reagents failed to give satisfactory yields of product when applied to **7**. However, when $[Cr(OAc)_2 \cdot H_2O]_2$ was employed for this reduction (Scheme 7.4), a 60% yield of product was obtained. Double *O*-desilylation was next performed to access diol **6**, the substrate needed for the Sharpless AD reaction.[3]

One of the most striking features of alkene **6** is its pre-existing chirality close to the double bond soon to be dihydroxylated. Given the presence of this stereogenicity, the confident prediction of product stereochemistry from the AD clearly becomes difficult. In the event, an identical mixture of products (1.7:1 in favour of **16**) emerged from the reaction of **6** with either $(DHQD)_2$-PHAL/OsO_4 or $(DHQ)_2$-PHAL/OsO_4. Such behaviour was most unexpected. Given the structural complexity of this alkene, a reliable rationalisation of this result currently seems beyond our grasp. However, such a finding clearly emphasises the need for exercising great caution when predicting the stereochemical outcome of AD reactions where there is an asymmetric centre in the vicinity of the double bond undergoing reaction.

With tetraol **16** in hand, it now proved possible to investigate the key internal ketalisation reaction needed to obtain **4** (see Scheme 7.1 for the structure of **4**). The best protocol for initiating ring-closure exposed **16** to conc. HCl in methanol (Scheme 7.4). Compound **4** was then selectively *O*-silylated with TBSCl, and *O*-pivaloylated to access compound **17**. Hydrogenolytic cleavage of the *O*-benzyl group followed next. The product alcohol was then oxidised to ketone **3** (see Scheme 7.1) under standard Swern conditions, and a chelation-controlled addition executed with lithium trimethylsilylacetylide (see Scheme 7.4). This installed the tertiary alcohol present within the target. A 6.1:1 ratio of epimers was observed in this reaction in favour of the desired addition product. The synthesis continued with a silver-nitrate-mediated cleavage of the alkynyl trimethylsilyl group to obtain **2**. Diisobutylaluminium hydride (DIBAL) was then employed to reductively cleave the pivaloyl esters from **2** to generate the corresponding triol. After *O*-acetylation, a semi-hydrogenation of the alkyne with a poisoned Pd on C catalyst afforded the desired alkene. Buffered HF-pyridine then effected global *O*-desilylation to access **18**.

The closing stages of Carreira's synthesis of (+)-zaragozic acid C (Scheme 7.4) began with the Dess–Martin oxidation[6] of **18**. An ozonolysis next yielded trialdehyde **19**, and a buffered sodium chlorite oxidation provided the corresponding triacid. Of particular interest here was

the compatibility of the tertiary alcohol unit, the bicyclic ketal, and the
acetate esters with this three-step oxidation sequence. *N,N*-Diisopropyl-
O-t-butylisourea[7] was next used to anchor the *t*-butyl esters within **20**,

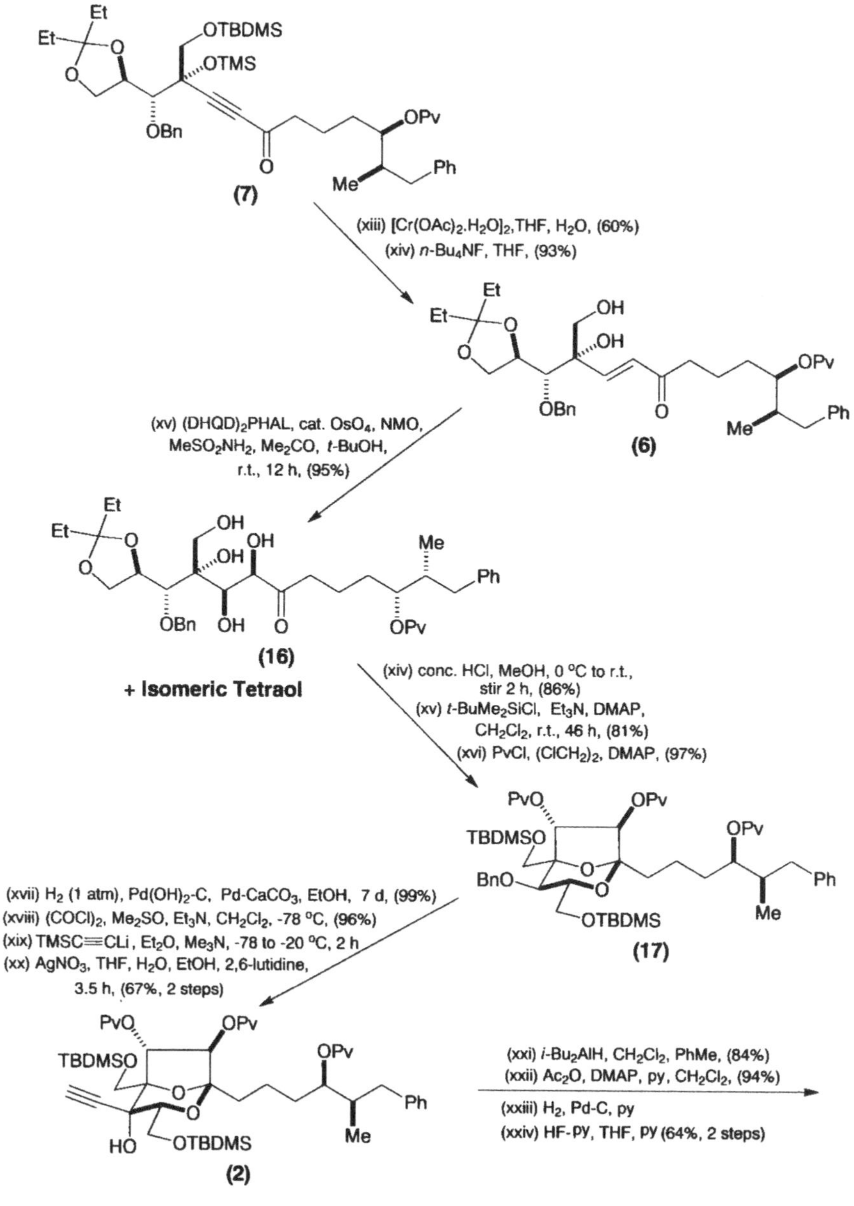

Scheme 7.4 Carreira's total synthesis of (+)-zaragozic acid from D-isoascorbic acid.

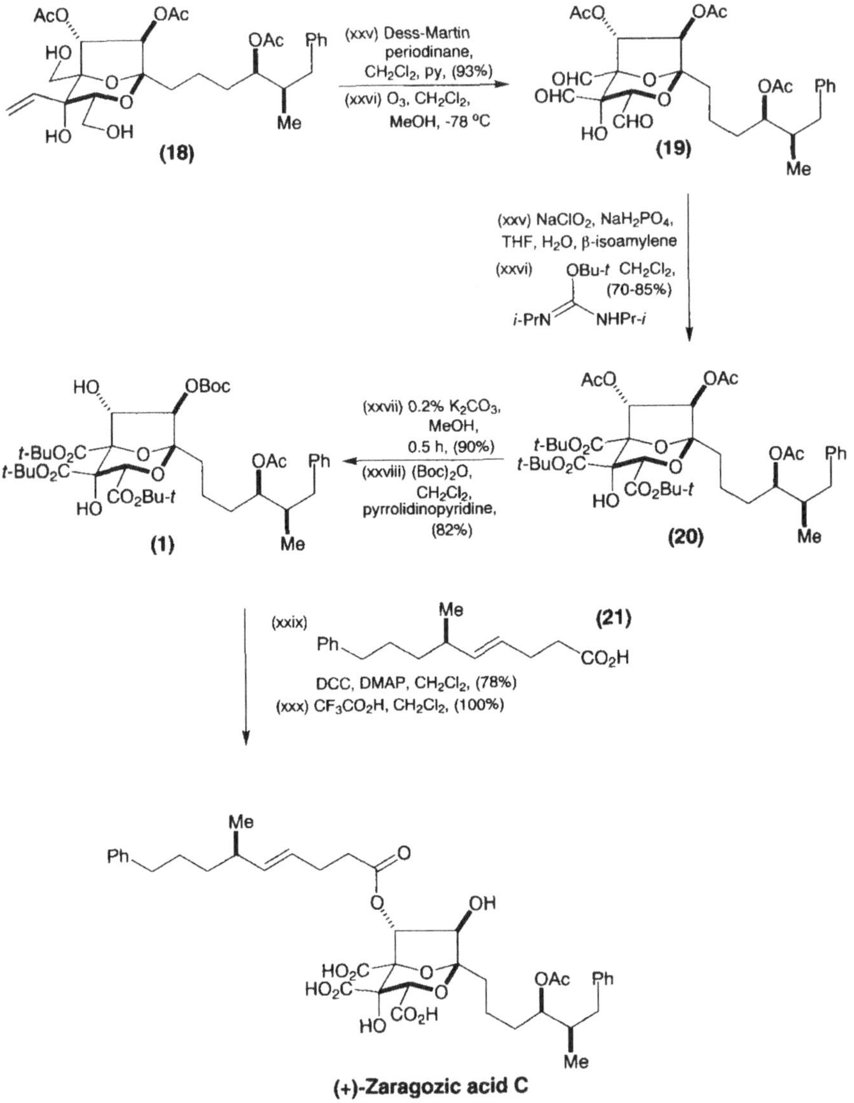

Scheme 7.4 (Continued).

under nonacidic conditions. Conventional methodology for installing *t*-butyl esters (i.e. conc. H₂SO₄ and isobutene) would almost certainly have faltered here, because of the presence of the bicyclic ketal array. Introduction of the C(6)-O-acyl side chain was accomplished in a further three steps. The first step entailed selective transesterification of the more electrophilic C(6)- and C(7)-O-acetates with potassium methoxide. A selective O-acylation of the more reactive C(7)-hydroxyl next followed with Boc_2O. Finally, esterification with acid 21 and t-butyl deprotection with trifluoroacetic acid completed this most impressive synthesis of (+)-zaragozic acid C.

7.4 Mechanistic analysis of some of the key steps in Carreira's synthesis of (+)-zaragozic acid C

The conversion of 11 to 10
The sole formation of 10 is consistent with a chelated five-membered transition state operating in this reaction, involving both the α-benzyloxy group and the carbonyl oxygen. At low temperatures, such a chelate would rigidly hold the carbonyl in the position shown (Scheme 7.5), and favour approach of the alkynyl Grignard reagent *syn* to the C—H bearing the α-benzyloxy group, that is, from the less-hindered direction.

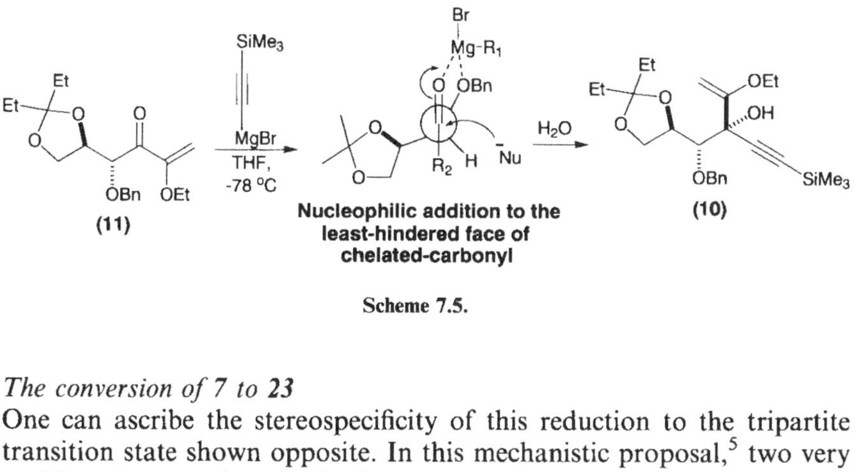

Nucleophilic addition to the
least-hindered face of
chelated-carbonyl

Scheme 7.5.

The conversion of 7 to 23
One can ascribe the stereospecificity of this reduction to the tripartite transition state shown opposite. In this mechanistic proposal,[5] two very rapid and successive single-electron transfers occur from the Cr(II) reagent to opposite sides of the triple bond. This leads to a vinyl-dichromium species 22 that subsequently protonates with retention of configuration to produce the *trans*-enone 23 (Scheme 7.6).

The conversion of 6 to 16
This is the Sharpless asymmetric dihydroxylation reaction;[3] one of the most powerful and versatile catalytic asymmetric reactions ever to be discovered. The Sharpless AD reaction owes its success to the presence of

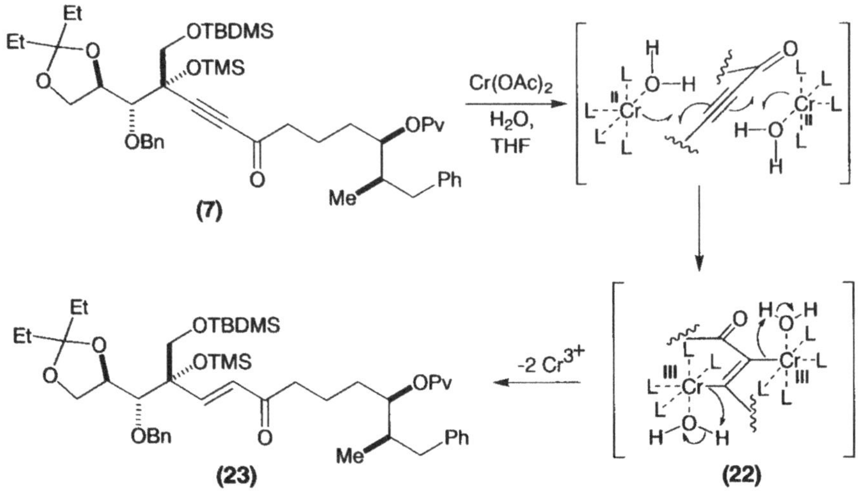

Scheme 7.6.

two *Cinchona* alkaloid ligands, (DHQD)$_2$-PHAL and (DHQ)$_2$-PHAL (Figure 7.1). These complex OsO$_4$ via their quinuclidine nitrogens to form a pair of chiral oxidants that can enantioselectively dihydroxylate all classes of substituted alkene. For 1,2-*trans*-disubstituted and trisubsti-

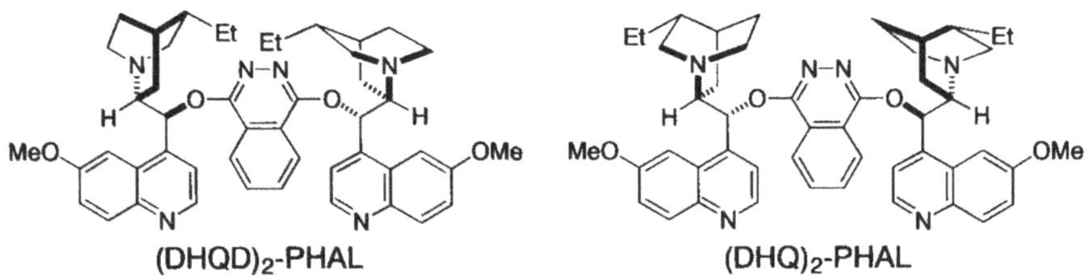

Figure 7.1 Two *Cinchona* alkaloid ligands: (DHQD)$_2$-PHAL and (DHQ)$_2$-PHAL.

tuted alkenes, the enantioselectivity of dihydroxylation is usually very high. However, for monosubstituted alkenes the product enantiomeric excesses (ees) can sometimes be lower. Sharpless has formulated a simple rule for predicting product stereochemistry in AD reactions with (DHQD)$_2$PHAL-OsO$_4$ and (DHQ)$_2$PHAL-OsO$_4$ (Scheme 7.7). The designations L, M, and S refer to substituents that are large, medium, and small, respectively. It is noteworthy that even though both oxidants are formally diastereoisomers, they still usually give rise to enantiomeric products. For the vast majority of substituted *achiral* alkenes, the

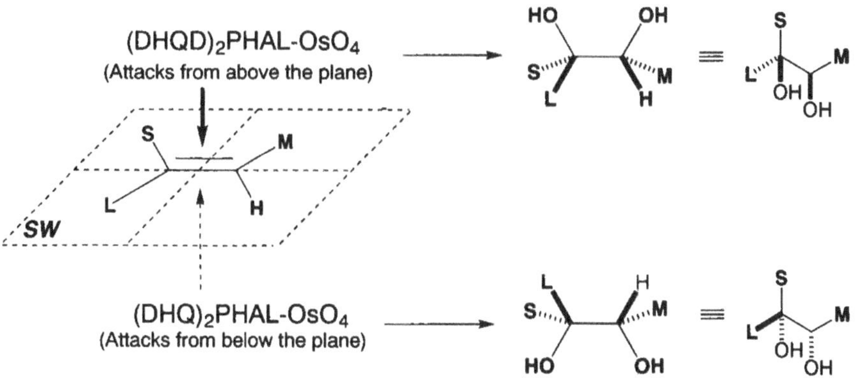

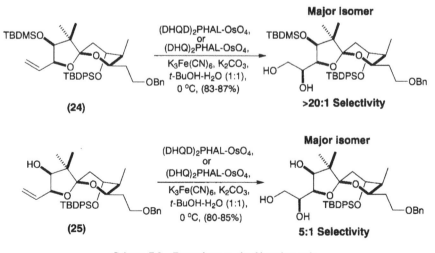

Scheme 7.7 Sharpless rule for predicting product stereochemistry.

Sharpless AD face-selection rule works very well. However, there are some categories of 1,1-disubstituted[8] and monosubstituted alkenes[9] for which violations have occurred. For *chiral* alkenes, the rule should be applied with caution, since pre-existing chirality within a substrate can also often affect the stereochemical outcome, sometimes in an unpredictable manner,[10,11] as Carreira's experience[2] with alkene **6** shows. For this substrate, an identical mixture of diols (1.7:1) was observed with either *Cinchona* alkaloid oxidant. Similar behaviour was encountered by Smith and coworkers[11] for the chiral alkenes **24** and **25** (Scheme 7.8). Smith's observations led him to propose that with some chiral substrates 'steric

Scheme 7.8 Exceptions to the Sharpless rule.

interactions may disfavor the normal AD transition states, causing both the matched and unmatched reagents to behave as bulky but effectively achiral species'.[11]

Eyring plot data gathered over a very wide temperature range by Gobel and Sharpless,[12] and $^{12}C/^{13}C$ kinetic isotope work by Corey et al.[13] and Sharpless et al.,[14] both support an AD mechanism in which the alkene reversibly coordinates to ligand-bound OsO_4, prior to participating in an irreversible [3 + 2]-cycloaddition. Although we recognise that the most favourable [3 + 2]-transition state adopted by a particular alkene will depend upon its precise structure, we have schematically depicted the AD mechanism in Figure 7.2.

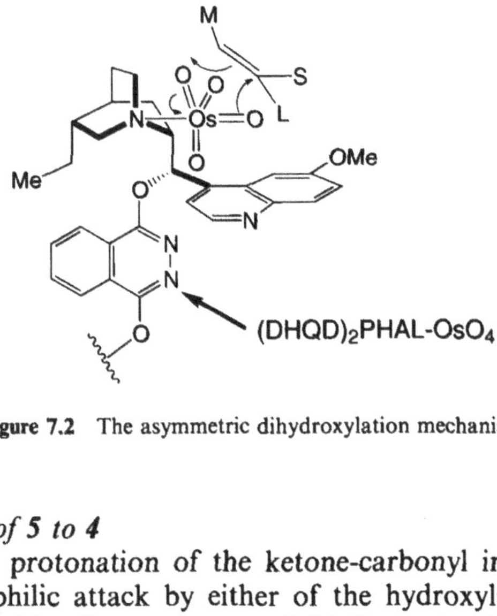

Figure 7.2 The asymmetric dihydroxylation mechanism.

The conversion of 5 to 4
It is likely that protonation of the ketone-carbonyl in **5** facilitates an internal nucleophilic attack by either of the hydroxyls that ultimately form part of the bicyclic acetal array. Initially nucleophilic addition will afford a hemiketal whose exo-hydroxyl can protonate and ultimately be lost as water. This would lead to a stabilised oxonium ion that could cyclise to **4**. In Scheme 7.9, we present one mechanism which satisfies the above discussion, but we emphasise that a myriad of mechanistic pathways can potentially operate.

The conversion of 3 into 26
Molecular models of **3** clearly indicate that the C(7)-O-pivaloyl group is appropriately positioned to engage in chelation with the alkynyllithium reagent in the manner shown overleaf. Chelation of this sort would obviously help guide in the nucleophile to the more-hindered top face of

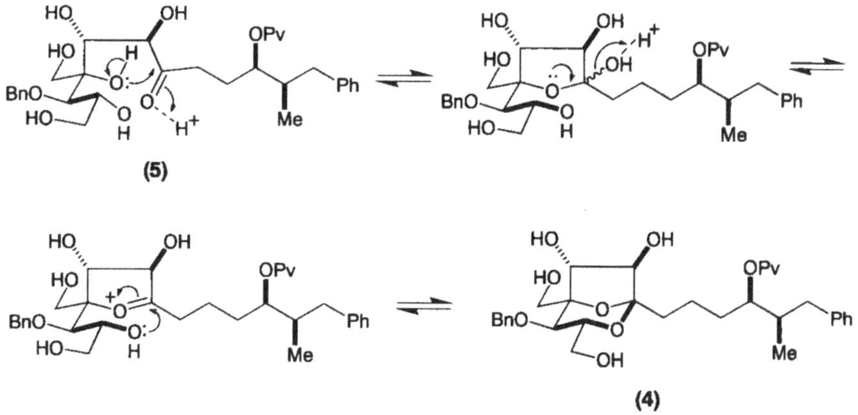

Scheme 7.9.

ketone **3**. A tetrameric organolithium is depicted in Scheme 7.10, primarily because simple ether solvents and tertiary amines are known to promote the formation of tetrameric aggregates for alkynyl-lithiums.[15]

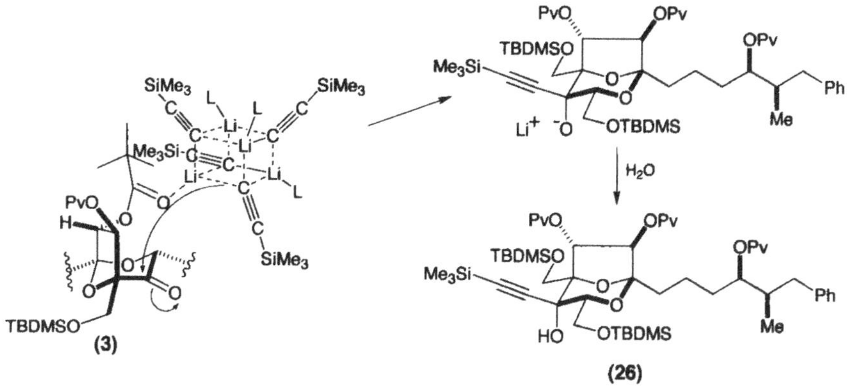

Scheme 7.10.

References

1. C. Dufresne, K.E. Wilson, D. Zink, J. Smith, J.D. Bergstrom, M. Kurtz, D. Rew, M. Nallin, R. Jenkins, K. Bartizal, C. Trainor, G. Bills, M. Meinz, L. Huang, J. Onishi, J. Milligan, M. Mojena and F. Pelaez, 1992, *Tetrahedron*, 48, 10221.
2. (a) E.M. Carreira and J. Du Bois, 1994, *J. Amer. Chem. Soc.*, 116, 10825; (b) E.M. Carreira and J. Du Bois, 1995, *J. Amer. Chem. Soc.*, 117, 8106.

3. (a) H.C. Kolb, M.S. VanNieuwenhze and K.B. Sharpless, 1994, *Chem. Rev.*, 94, 2483; (b) P.J. Walsh and K.B. Sharpless, 1993, *Synlett.*, 605.

4. N. Cohen, B.L. Banner, R.J. Lopresti, F. Wong, M. Rosenberger, Y.-Y. Liu, E. Thom and A.A. Liebmann, 1983, *J. Amer. Chem. Soc.*, 105, 3661.

5. (a) C.E. Castro and R.D. Stephens, 1964, *J. Amer. Chem. Soc.*, 86, 4358; (b) A.B. Smith, III, P.A. Levenberg and J.Z. Suits, 1986, *Synthesis*, 184.

6. (a) D.B. Dess and J.C. Martin, 1983, *J. Org. Chem.*, 48, 4156; (b) R.E. Ireland and L. Lui, 1993, *J. Org. Chem.*, 58, 2899.

7. L.J. Mathias, 1979, *Synthesis*, 561.

8. (a) K.J. Hale, S. Manaviazar and S.A. Peak, 1994, *Tetrahedron Lett.*, 35, 425; (b) A. Nelson, P. O'Brien and S. Warren, 1995, *Tetrahedron Lett.*, 2685; (c) P. O'Brien and S. Warren, 1996, *J. Chem. Soc. Perkin Trans. 1*, 2129; (d) D.J. Krysan, 1996, *Tetrahedron Lett.*, 37, 1375; (e) K.P.M. Vanhessche and K.B. Sharpless, 1996, *J. Org. Chem.*, 61, 7978.

9. (a) P. Salvadori, S. Superchi and F. Minutolo, 1996, *J. Org. Chem.*, 61, 4190; (b) D.L. Boger, J.A. McKie, T. Nishi and T. Ogiku, 1996, *J. Amer. Chem. Soc.*, 118, 2301.

10. (a) D.J. Krysan, T.W. Rockway and A.R. Haight, 1994, *Tetrahedron: Asymmetry*, 5, 625; (b) M.D. Cliff and S.G. Pyne, 1997, *J. Org. Chem.*, 62, 1023; (c) J.M. Gardiner and S.E. Bruce, 1998, *Tetrahedron Lett.*, 39, 1029.

11. M. Iwashima, T. Kinsho and A.B. Smith, 1995, *Tetrahedron Lett.*, 36, 2199.

12. T. Gobel and K.B. Sharpless, 1993, *Angew. Chem. Int. Ed. Engl.*, 32, 1329.

13. E.J. Corey, M.C. Noe and M.J. Grogan, 1996, *Tetrahedron Lett.*, 37, 4899; (b) E.J. Corey, A. Guzman-Perez and M.C. Noe, 1995, *J. Amer. Chem. Soc.*, 117, 10805.

14. A.J. DelMonte, J. Haller, K.N. Houk, K.B. Sharpless, D.A. Singleton, T. Strassner and A.A. Thomas, 1997, *J. Amer. Chem. Soc.*, 119, 9907.

15. G. Fraenkel and P. Pramanik, 1983, *J. Chem. Soc., Chem. Commun.*, 1527.

8 (+)-Neocarzinostatin

K.J. Hale

8.1 Introduction

Synthetic interest in the enediyne antitumour antibiotics has been enormous in recent years,[1] primarily because of their novel structures and their ability to cleave DNA selectively. One conspicuous member of this class is (+)-neocarzinostatin. Its most noteworthy feature is its unusual nine-membered epoxydienediyne core, which is saddled onto a highly functionalised, monoglycosylated, cyclopentene ring. The significant ring strain present within this molecule, coupled with its unique arrangement of reactive functional groups, its rare α-linked 2-methyl-aminofucosamine unit, and its exceedingly labile epoxy dienediyne motif, all add to the challenges it provides for asymmetric total synthesis. To date, only one synthetic pathway[2] to (+)-neocarzinostatin has been devised, and that is due to Myers and his co-workers, now at Harvard.

8.2 Myers' retrosynthetic planning for the synthesis of (+)-neocarzinostatin

From a tactical perspective, early retrosynthetic removal of the epoxy dienediyne core must dominate all synthetic planning for (+)-neocarzinostatin, since its presence in any intermediate will render that molecule highly susceptible to nucleophilic attack, and will jeopardise the success of all the remaining chemistry. In essence, it would be strategically advisable to install the epoxy dienediyne unit late in a forward total synthesis of (+)-neocarzinostatin.

Although early attachment of the sugar moiety could potentially overcome the thorny issue of having to glycosidate with a fully-armed aglycone, the premature presence of the sugar could necessitate implementation of a more elaborate protecting group strategy. Logistically, therefore, postponement of glycosylation until the final stages of a synthesis would probably be best.

Good retrosynthetic planning for (+)-neocarzinostatin would also make every effort to identify nine-membered carbocycle precursors that were considerably less strained. Such molecules should be far easier to construct in a forward route. Obviously, retrosynthetic functionalisation of the dienediyne array could assist in the identification of these less strained and less reactive intermediates.

Sound strategic planning would also attempt to pinpoint those features, already present in the epoxydienediyne core, which could greatly assist in assembly of the nine-membered carbocycle. Clearly, such structural elements would have to be retained in any prospective cyclisation precursor. With these things in mind, we will now examine Myers' retrosynthetic plan for this target.

Myers' first disconnection was made across the O-glycosidic bond of (+)-neocarzinostatin to furnish 1 and 2 as subtargets (Scheme 8.1). A Schmidt O-trichloroacetimidate glycosidation reaction[3] was envisaged for constructing the difficult 1,2-cis-O-glycosidic bond found in the target, since the stereochemical outcome of such glycosidations is often controllable through variation of the reaction solvent or the Lewis acid promoter. Two examples[3] of this behaviour are shown in Scheme 8.2.

In addition, Nicolaou[4] had already successfully used a Schmidt glycosylation during his synthesis of the enediyne antibiotic, calicheamycin γ^1 (Scheme 8.3). Myers therefore felt reasonably confident that success could accrue from following such an approach. Nevertheless, in his proposed union of 1 with 2, Myers was still testing the scope of the Schmidt glycosidation beyond all past boundaries, for never before had it been successfully used on a 2-methylamino glycosyl donor, nor had it been employed to attach an aglycone containing quite such a delicate array of reactive functionality. It was not at all clear, therefore, whether this methodology could match up to the requirements of the present situation. Indeed, all previous reports on the use of the Schmidt protocol for building α-linked 2-amino glycosides, had utilised 2-azido-1-O-trichloroacetimidates as glycosyl donors, and had reduced the azido function after glycosidation had been achieved.[5] Obviously, the prospects for achieving a fruitful reductive N-methylation in the present system would be remote to say the least. Hence, Myers quickly discounted this particular option.

Before departing completely from the topic of retrosynthetic glycosidation, it is perhaps worthwhile reflecting on why Myers selected O-triethylsilyl (TES) ethers as sugar protecting groups for this coupling step. As we have already seen in Nicolaou's calicheamycin work,[4] TES ethers can often withstand the low temperature, Lewis acid-mediated, coupling conditions needed to execute the Schmidt glycosidation. Myers therefore felt reasonably confident that these groups could withstand his proposed coupling of 1 with 2. Myers was also aware that TES ethers can often be detached under mild conditions that do not interfere with other sensitive functionality that might be present within a target molecule. The example

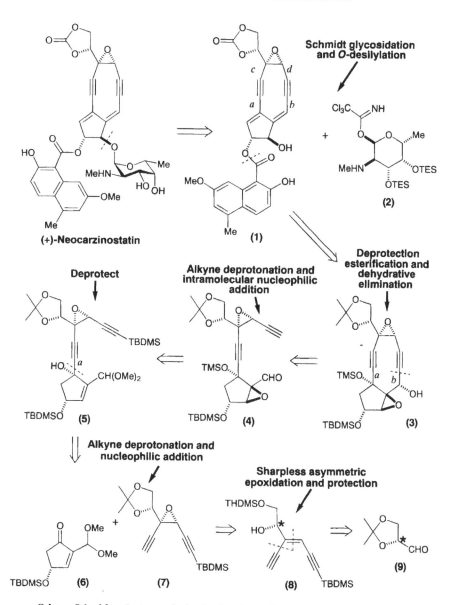

Scheme 8.1 Myers' retrosynthetic plan for the synthesis of (+)-neocarzinostatin.

given in Scheme 8.4 saliently illustrates this point. Thus, Myers' adoption of TES ethers for the problem at hand seemed ideal.

Myers next task was to disassemble aglycone **1** (see Scheme 8.1). This offers numerous opportunities for retrosynthetic simplification, since it

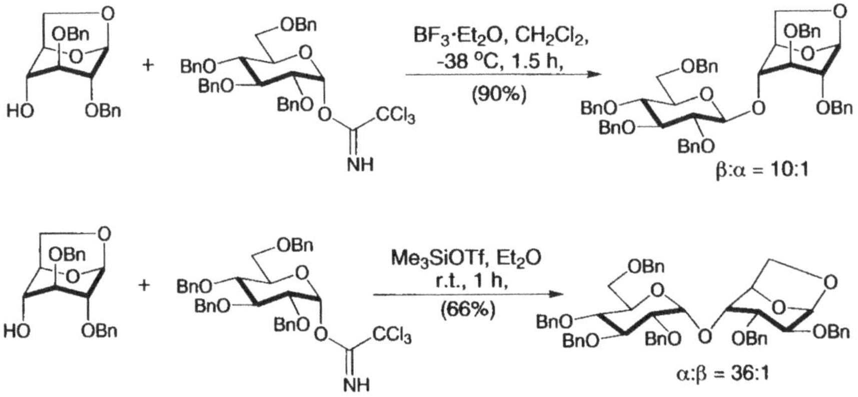

Scheme 8.2 Control of anomeric selectivity in glycosidations via Schmidt's trichloroacetimidate method.

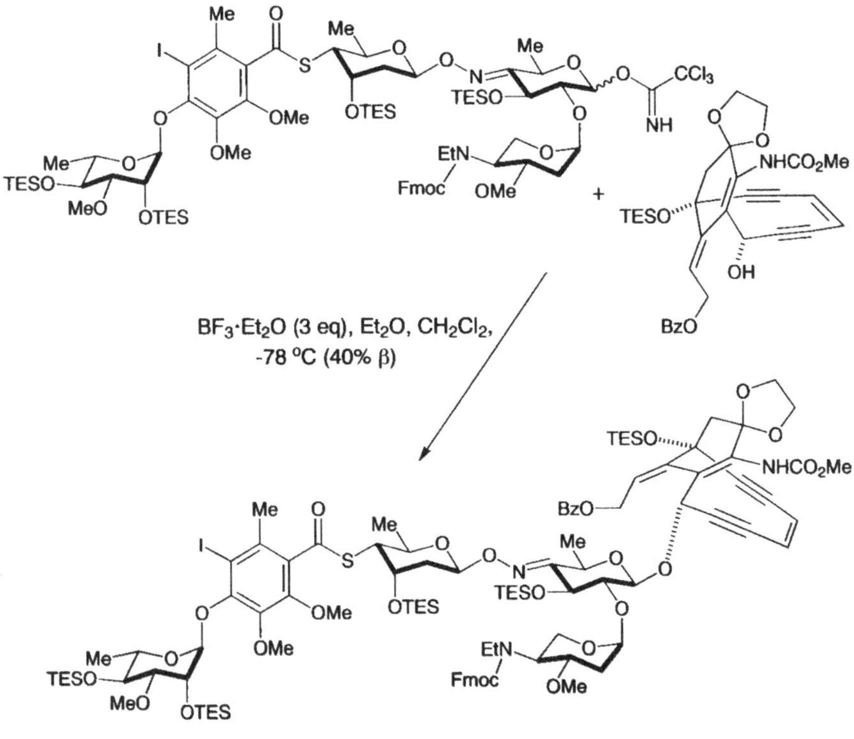

Scheme 8.3 Nicolaou's use of the Schmidt O-trichloroacemidate glycosidation protocol during his synthesis of calicheamycin γ^{1}.

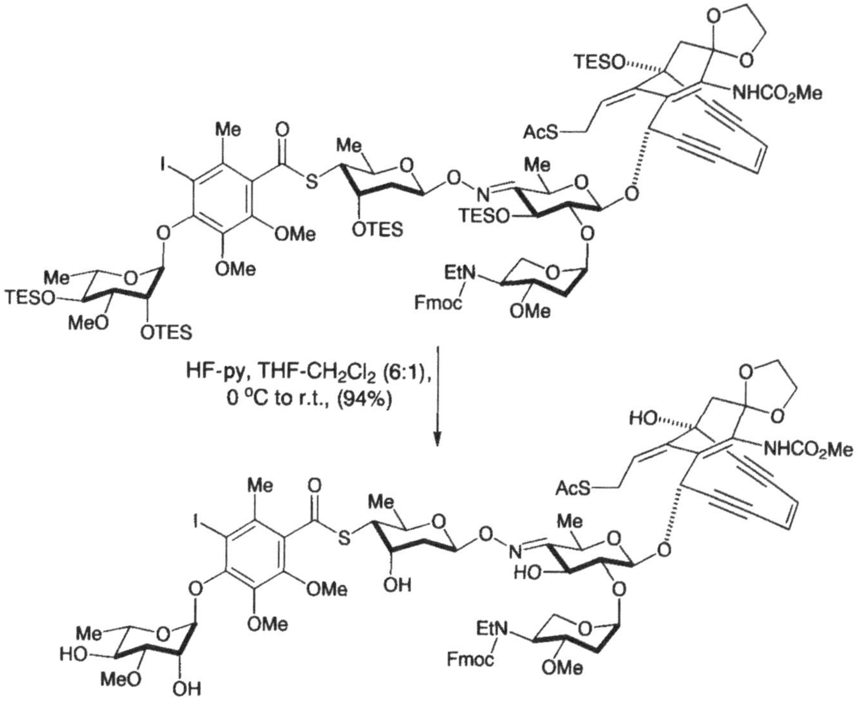

Scheme 8.4 Nicolaou's deprotection of TES ethers during his calicheamicin γ^1 synthesis.

contains four readily disconnectible alkynyl σ bonds (*a,b,c* and *d*). It also possesses two alkenes that are amenable to retrosynthetic functionalisation followed by bond cleavage. As mentioned previously, it would be strategically advantageous to attempt a partial retrosynthetic clearance of the epoxydienediyne unit quite early on in proceedings, since this would reduce the reactivity of many intermediates in the forward synthesis. It would also relieve the considerable ring-strain present within the nine-membered carbocycle, which again would greatly facilitate its construction. Accordingly, Myers hydrated diene 1 in a 1,3-manner, to prime the alkynyl σ bonds *a* and *b* for retrosynthetic cleavage. Although, conceptually, both these bonds can be disconnected concurrently to give a stabilised dianion and an α-aldehydocyclopentenone as precursors of **3**, the corresponding forward reaction would almost certainly produce a mixture of regioisomers if successful, which clearly would not constitute good synthetic planning if efficiency was the goal. Myers therefore staggered his order of bond cleavage. He initially disconnected bond *b* in **3** to obtain alkynyl aldehyde **4**, and he proposed closure of the medium-ring by intramolecular nucleophilic addition of an acetylenic anion onto the tethered aldehyde group. One might now ponder about why Myers would wish to preserve two potentially reactive epoxides in compound **4** during such a ring-closure, since surely these could potentially engage in undesired side-reactions with the acetylenic anion. Fortunately, a survey of the literature reveals that acetylenic anions do not react readily with di- or trisubstituted epoxides at low temperatures (e.g. $-78°$C), particularly in the absence of strong Lewis acids, chelating ligands, or dipolar aprotic solvents. Aldehydes, on the other hand, usually react rapidly at these temperatures, without promoters. The example shown in Scheme 8.5 from Danishefsky's synthesis of calicheamicinone,[6] nicely illustrates the

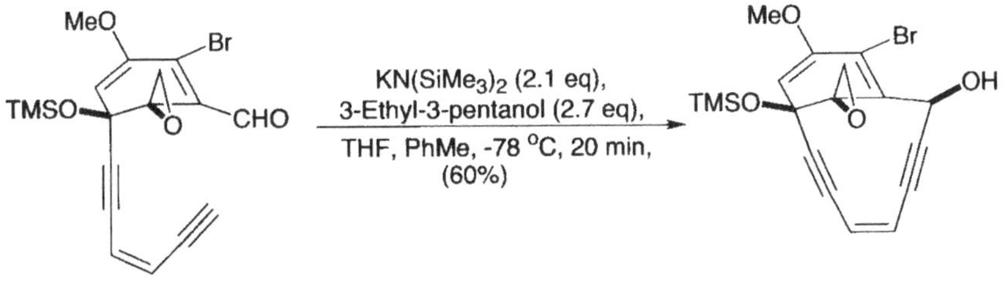

Scheme 8.5 Acetylenic anion ring-closure in Danishefsky's calicheamicinone synthesis.

striking difference in electrophilicity between these two functional groups. Thus, the chemoselectivity issue facing Myers was not quite as serious as one might initially think. Let us now return to the question of why Myers would want to perform the aforementioned cyclisation with the diynyl-epoxide present. Probably his main reason for retaining the epoxide was to facilitate ring-closure. The presence of a *cis*-epoxide would assist in bringing the two reactive termini of **4** into reasonably close proximity to one another. Restricting the rotational degrees of freedom of a medium-ring precursor is often a good ploy for facilitating cyclisation, so long as the restraining element correctly positions the two termini in an orientation where they can intramolecularly react. Sometimes, a con-straining group can be employed for the opposite purpose, of keeping two reactive groups apart, thereby preventing cyclisation. However, this was clearly not the intention here for compound **4**!

Let us now give some thought to Myers' retrosynthetic analysis of **4** (Scheme 8.1). After a partial clearance of reactive epoxide functionality from within **4**, Myers disconnected bond *a* in compound **5** to yield the partially protected diyne **7** and ketone **6** as possible building blocks. Note how one of the terminal alkyne units in **7** has been retrosynthetically protected with a *t*-butyldimethylsilyl group to permit regiospecific alkynyl anion generation in the impending forward synthesis. It will also be appreciated that the reactive aldehydo group in **4** has been temporarily protected as a dimethyl acetal, to allow the prospective nucleophilic addition to ketone **6** to proceed successfully. Retrosynthetic clearance of the *O*-isopropylidene and epoxide groups from **7** followed next. This led to the enediyne alcohol **8**. It is now that the possibility of introducing the epoxide unit by Sharpless asymmetric epoxidation[7] starts to become apparent. It will also be seen that the absolute stereochemistry of **8** matches that of (*R*)-glyceraldehyde acetonide **9**. It should not be a surprise, therefore, to learn that Myers selected this as the chiral starting material for his eventual total synthesis.

8.3   Myers' total synthesis of (+)-neocarzinostatin

The first issue confronted by Myers was preparation of homochiral epoxide **7**, the key intermediate needed for his intended nucleophilic addition reaction to enone **6**. Its synthesis began with the addition of lithium trimethylsilylacetylide to (*R*)-glyceraldehyde acetonide (Scheme 8.6).[8] This afforded a mixture of propargylic alcohols that underwent oxidation to alkynone **10** with pyridinium dichromate (PDC). A Wittig reaction next ensued to complete installation of the enediyne unit within **11**. A 3:1 level of selectivity was observed in favour of the desired olefin isomer. After selective desilylation of the more labile trimethylsilyl group from the product mixture, deacetalation with 1N HCl in tetrahydrofuran (THF) enabled both alkene components to be separated, and compound **12** isolated pure.

In order to prevent competing homoallylic asymmetric epoxidation (AE, which, it will be recalled, preferentially delivers the opposite enantiomer to that of the allylic alcohol AE), the primary alcohol in **12** was selectively blocked as a thexyldimethylsilyl ether. Conventional Sharpless AE[7] with the oxidant derived from (−)-diethyl tartrate, titanium tetraisopropoxide, and *t*-butyl hydroperoxide next furnished the anticipated α,β-epoxy alcohol **13** with excellent stereocontrol (for a more detailed discussion of the Sharpless AE see section 8.4). Selective *O*-desilylation was then effected with HF–triethylamine complex. The resulting diol was protected as a base-stable *O*-isopropylidene acetal using 2-methoxypropene and a catalytic quantity of *p*-toluenesulfonic acid in dimethylformamide (DMF). Note how this blocking protocol was fully compatible with the acid-labile epoxide.

Deprotonation of the terminal acetylene in epoxydiyne **7** was next attempted with lithium hexamethyldisilazide in toluene at low temperature. The resulting lithium acetylide added readily to cyclopentenone **6**[9] at $-78°C$ to provide the 1,2-addition product **5** with $> 20:1$ selectivity. The nucleophile preferentially approached the β face of ketone **6** because of the bulky OTBDMS group hindering attack from the α side. After global *C*- and *O*-desilylation, the less hindered secondary alcohol was selectively reprotected as an OTBDMS ether to obtain **14**. A transketalisation was then implemented with *p*-toluenesulfonic acid and acetone to unmask the dimethoxyacetal. These were conditions that left the *O*-isopropylidene and the OTBDMS groups undamaged. The tertiary alcohol was now protected as a trimethylsilyl ether. The sole product of silylation was aldehyde **15**, obtained in 88% yield. Diisobutylaluminium hydride (DIBAL) was the reagent of choice for the selective reduction of enal **15** to the primary allylic alcohol. Sharpless AE[7] with (+)-diethyl tartrate [(+)-DET] (see Section 8.4.1) and Dess–Martin oxidation[10] subsequently afforded the α,β-epoxy aldehyde **4** in good yield. This was the key cyclisation precursor needed for constructing the nine-membered ring.

It transpired that epoxy aldehyde **4** cyclised efficiently when alkyne deprotonation was attempted with lithium diphenyltetramethyldisilazide[11] at low temperature; LiCl was added to the reaction mixture as a mild Lewis acid promoter. The end result was compound **3**, formed as a single entity in 79% yield. After protection of the newly created alcohol as an *O*-chloroacetate ester, chemoselective *O*-desilylation with HF–triethylamine complex furnished diol **16**. Again, the capacity of the two acid-sensitive epoxides to withstand these deprotection conditions is worthy of comment. By attaching the *O*-chloroacetate group to the nine-membered carbocycle, it now became possible selectively to acylate the cyclopentanol with napthoic acid **17**. An amine work-up sufficed for cleaving the more electrophilic *O*-chloroacetate group to provide compound **18**.

In yet another powerful display of highly chemoselective chemistry, the *O*-isopropylidene group of **18** was selectively detached with *p*-toluenesulfonic acid in methanol. Significantly, transesterification or epoxide ring-opening did not compete. The resulting terminal 1,2-diol was then selectively converted to a cyclic carbonate with carbonyldiimidazole, and the phenolic and secondary hydroxyls blocked as *O*-triethylsilyl ethers. This paved the way for an efficient dehydrative elimination of the tertiary alcohol with Martin sulfurane[12] to obtain epoxyenediyne **19** (see section 8.4 for mechanistic details). After cleavage of the two *O*-triethylsilyl groups from **19**, a dehydrative epoxide ring-opening was performed with triphenylphosphine and iodine[13] to integrate the 1,3-diene unit into the nine-membered ring system (see Section 8.4). It is ironic that relief of angular strain within the three-membered epoxide ring serves as the driving force for an increase in strain in the nine-membered ring!

However, it is the high P=O bond energy in Ph₃P=O which clearly helps offset this dubious energy trade-off.

With the neocarzinostatin aglycone **1** now successfully assembled, the time had arrived for connecting the sugar unit to the cyclopentene ring.[2b] Somewhat surprisingly, a wide range of Lewis acid promoters proved

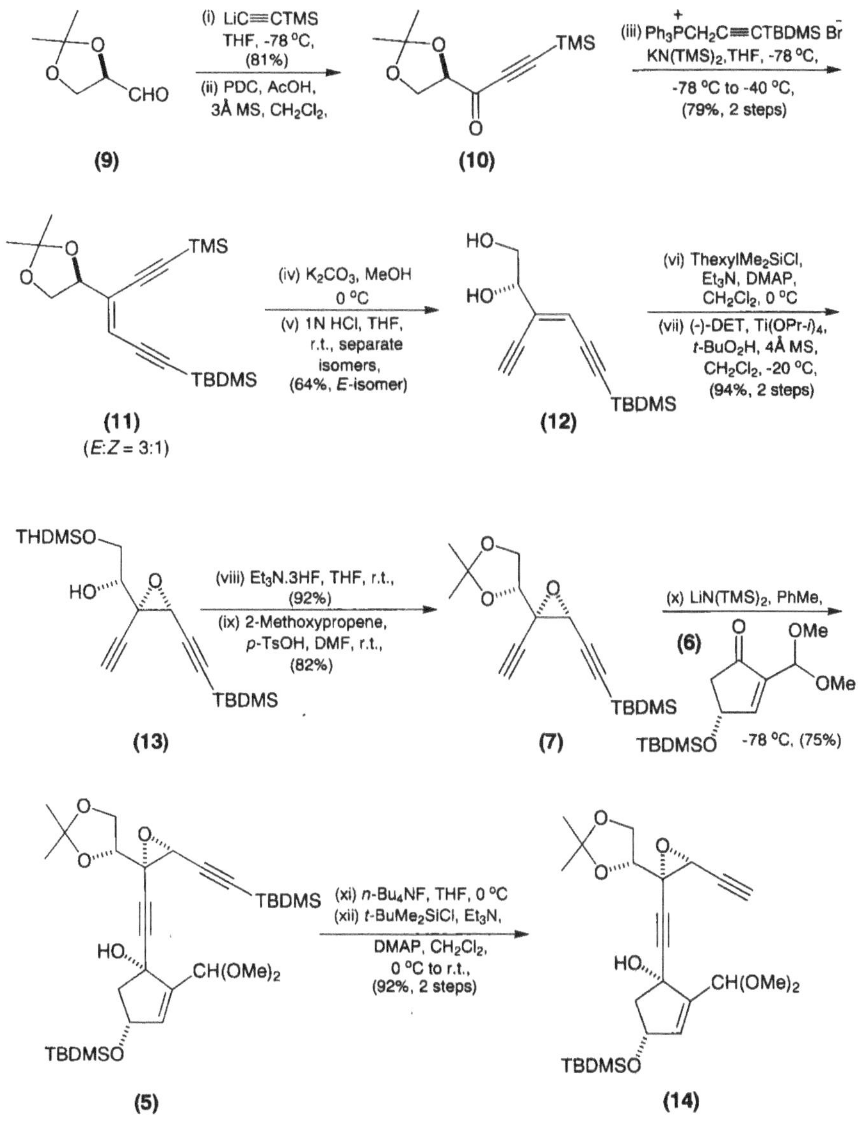

Scheme 8.6 Myers' total synthesis of the (+)-neocarzinostatin aglycone.

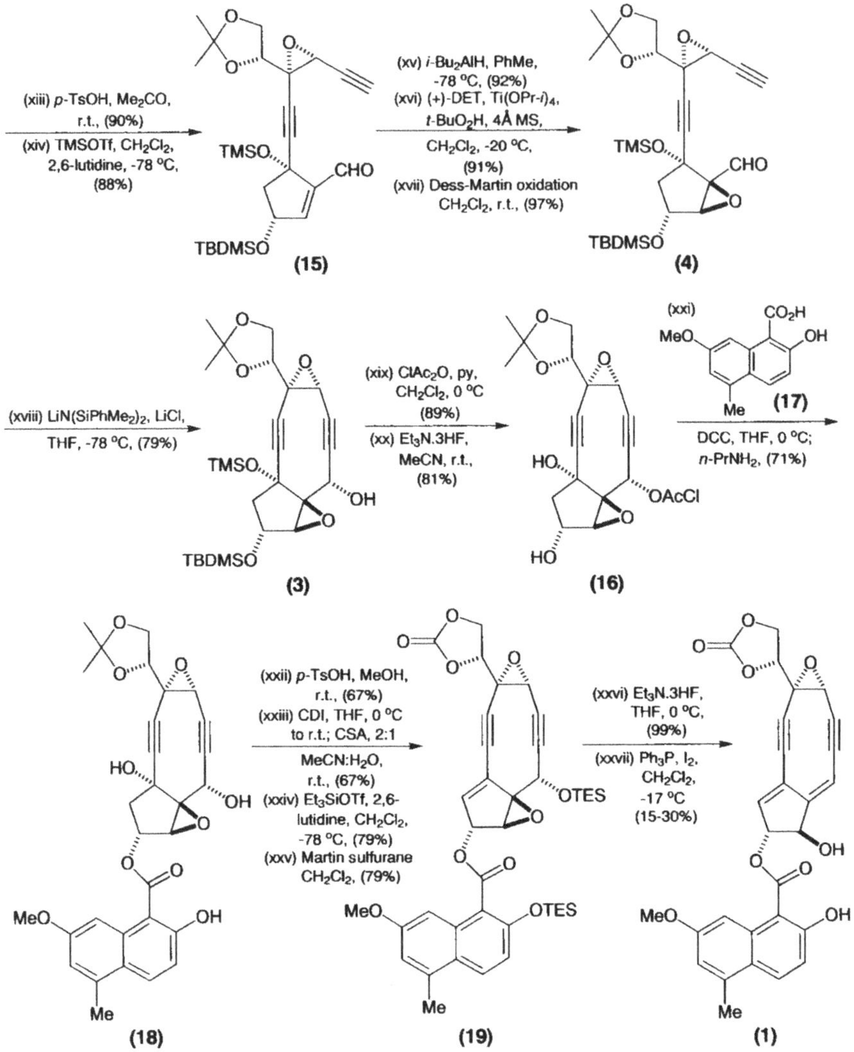

Scheme 8.6 (Continued).

successful at instigating the desired α-glycosidation between **1** and **2**. The optimal coupling conditions exposed **1** and **2** to BF$_3$ · Et$_2$O in toluene at −30°C for 1 h. This led exclusively to the α-linked product **20** (Scheme 8.7).

It is likely that the phenolic hydroxyl of **1** is much less nucleophilic than the allylic OH because of its involvement in strong intramolecular hydrogen bonding with the *ortho* carbonyl of the ester. This would

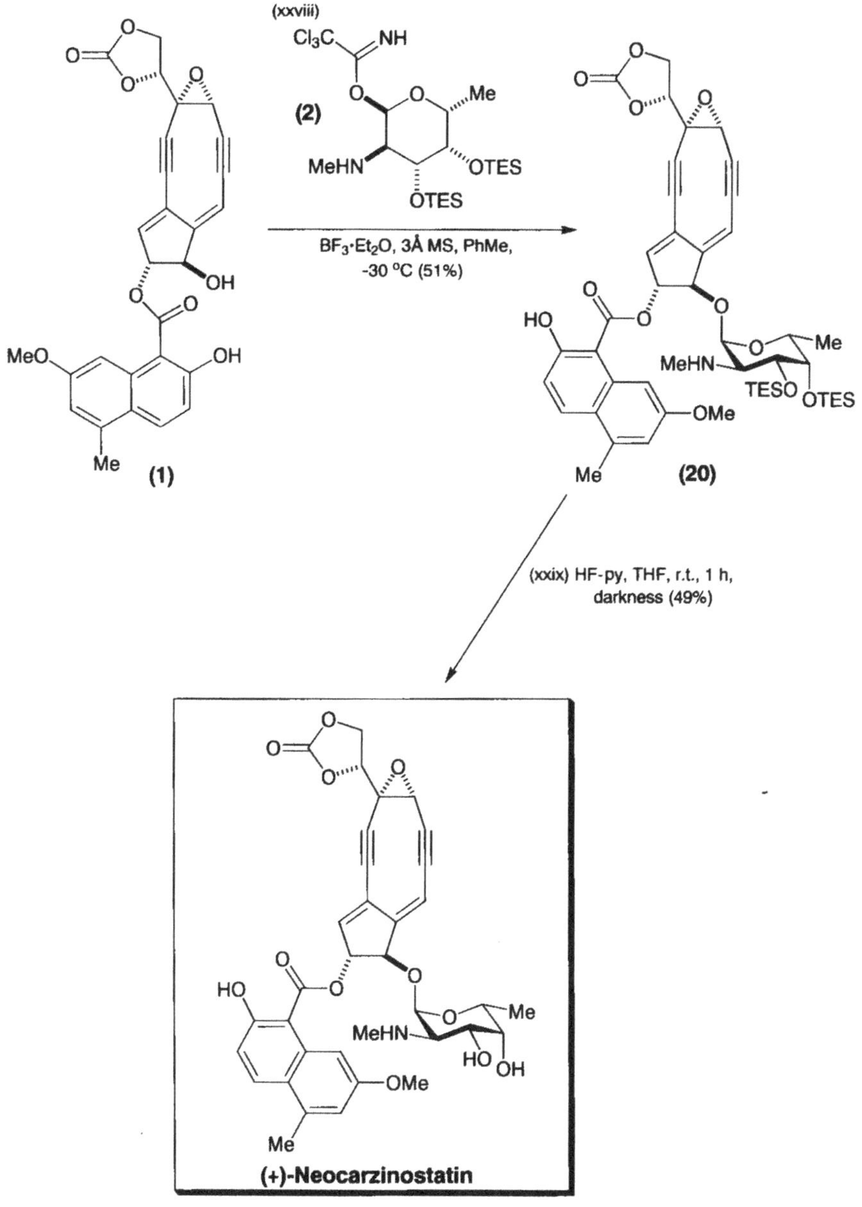

Scheme 8.7 Myers' total synthesis of (+)-neocarzinostatin.

explain the high regiospecificity observed in this glycosidation, which is most striking. The last step of the total synthesis was deprotection of the OTES groups, a process that was complete within one hour.

(+)-Neocarzinostatin was isolated in 49% yield. When one considers the high chemical instability of this molecule, this really was a superlative recovery of the final product.

Myers' total synthesis of (+)-neocarzinostatin[2] is outstanding not only for its bold use of chemoselective and regioselective reactions, but also for its powerful expansion of the scope of the Schmidt trichloroacetimidate glycosylation protocol.[3,5] The unearthing of this highly stereocontrolled method for forming 1,2-*cis*-linked 2-alkylamino-glycosides represents a major advance in glycosidation chemistry, and provides a powerful illustration of how 'real-life' complex molecule synthesis can drive important methodological breakthroughs. There can be no doubt that the Myers synthesis of (+)-neocarzinostatin is a *tour de force* of the modern synthetic age.

8.4 Mechanistic analysis of the key steps in Myers' (+)-neocarzinostatin synthesis

The conversion of 21 to 22

This is an example of the Sharpless asymmetric epoxidation reaction of allylic alcohols;[7] one of the most versatile, reliable, and synthetically

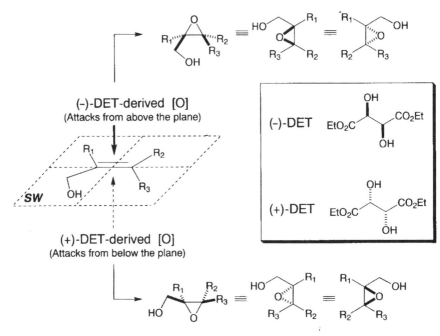

Scheme 8.8 Sharpless rule for predicting product stereochemistry.

useful reactions ever discovered. As with the asymmetric dehydroxylation reaction, Sharpless has formulated a very simple rule for predicting the stereochemical outcome of the AE, which is shown in Scheme 8.8. Basically, the chiral oxidant obtained from (−)-diethyl tartrate (DET), titanium tetraisopropoxide and *t*-butyl hydroperoxide preferentially epoxidises allylic alcohols from the top side when the alcohol group is positioned in the SW quadrant. The enantiomeric oxidant preferentially epoxidises the same substrate from its underside when drawn in the same arrangement. The enantiomeric excesses (ees) attained in the AE are always very high (> 90%), which has made this one of the most popular reactions currently used in modern-day asymmetric synthesis.

The mechanism of the AE has been studied in considerable detail by Sharpless and co-workers.[14] For compound **21**, it is outlined in Scheme 8.9.

Scheme 8.9. E = CO$_2$Et

The conversion of 22 to 4

Mechanistic studies on the Dess–Martin oxidation have shown that the alcohol exchanges with one of the acetate groups in the periodinane to create a new iodinane which then undergoes α C—H bond cleavage in the manner shown in Scheme 8.10.

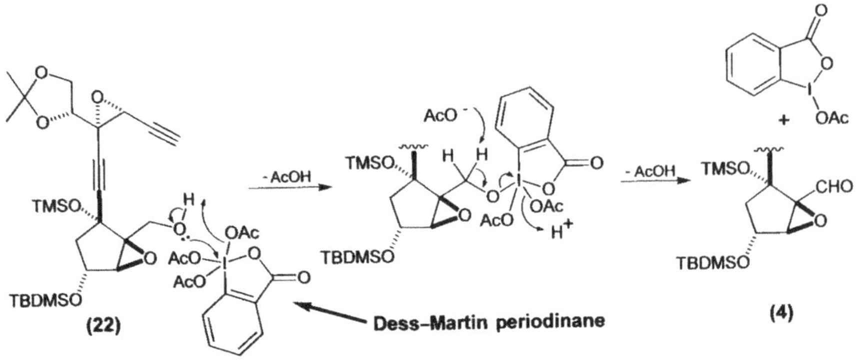

Scheme 8.10.

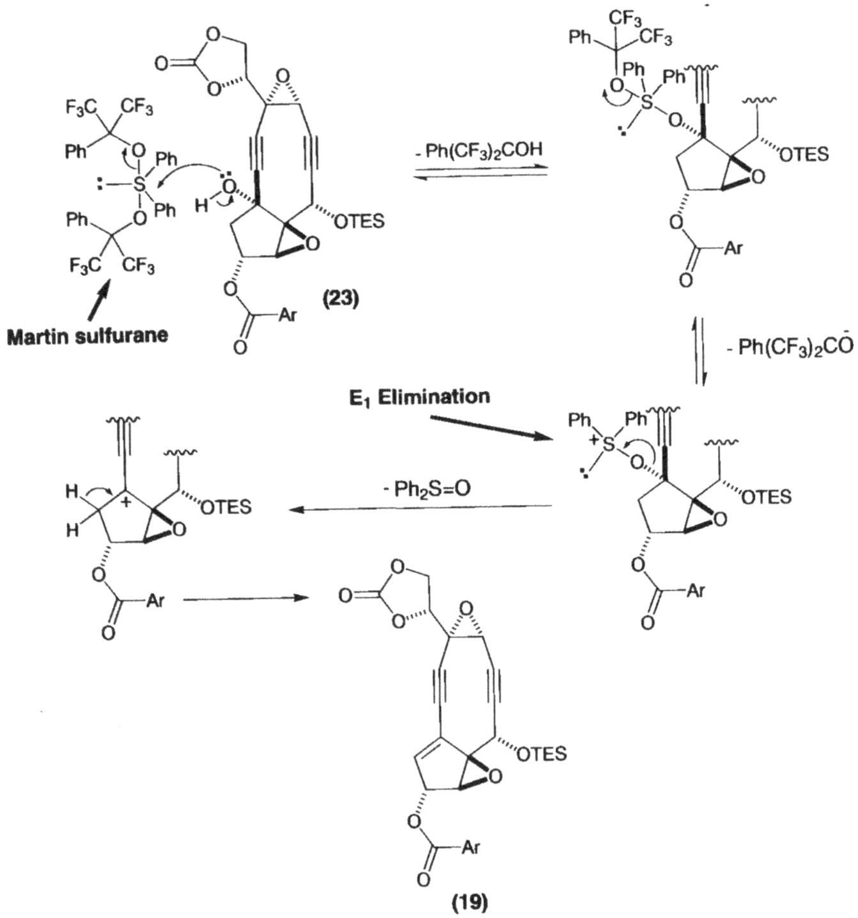

Scheme 8.11.

The conversion of 23 to 19
The Martin sulfurane is an excellent reagent for the rapid dehydration of tertiary alcohols.[12] Mechanistically, the dehydration involves ligand exchange around sulfur, diphenylsulfonium ion formation, and E1 elimination of diphenylsulfoxide to give the alkene (Scheme 8.11).

The conversion of 24 to 1
It seems likely that this dehydrative elimination proceeds via the inverted iodide **26** (Scheme 8.12),[13] although this intermediate was never detected in the reaction mixture.

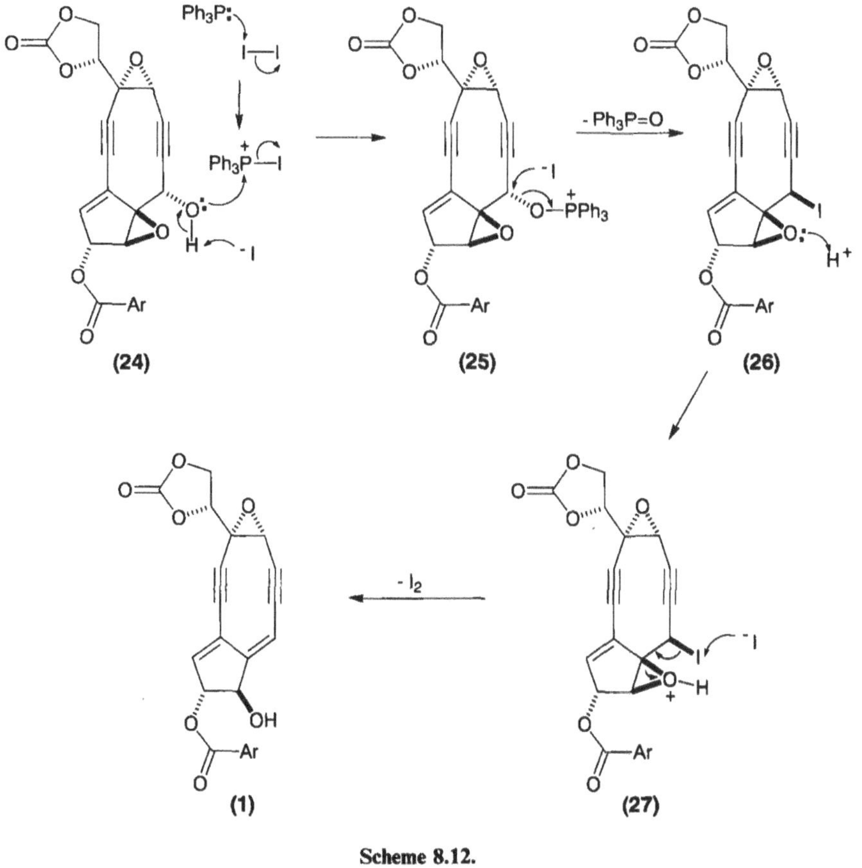

Scheme 8.12.

The conversion of 1 into 20

Myers attributes[2b] the α selectivity of this coupling to hydrogen bonding between the incoming alcohol and the 2-methylamino group of the gly-cosyl cation. This guides in the alcohol *cis* to the methylamino group (Scheme 8.13).

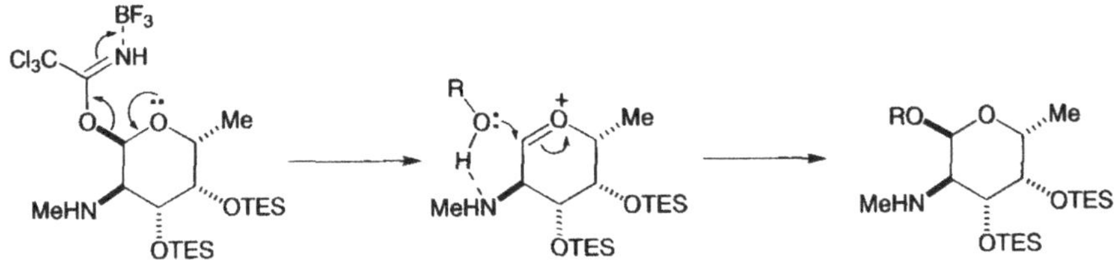

Scheme 8.13.

References

1. (a) A.L. Smith and K.C. Nicolaou, 1996, *J. Med. Chem.*, 39, 2103; (b) H. Lhermitte and D.S. Grierson, 1996, *Contemporary Organic Synthesis*, 41; (c) H. Lhermitte and D.S. Grierson, 1996, *Contemporary Organic Synthesis*, 93; (d) S.J. Danishefsky and M.D. Shair, 1996, *J. Org. Chem.*, 61, 16; (e) K.C. Nicolaou and W.-M. Dai, 1991, *Angew. Chem. Int. Ed. Engl.*, 30, 1387.
2. (a) A.G. Myers, M. Hammond, Y. Wu, J.-N. Xiang, P.M. Harrington and E.Y. Kuo, 1996, *J. Amer. Chem. Soc.*, 118, 10006; (b) A.G. Myers, J. Liang, M. Hammond, P.M. Harrington, Y. Wu and E.Y. Kuo, 1998, *J. Amer. Chem. Soc.*, 120, 5319.
3. (a) R.R. Schmidt, 1986, *Angew. Chem. Int. Ed. Engl.*, 25, 212; (b) R.R. Schmidt and W. Kinzy, 1994, *Adv. Carbohydr. Chem. Biochem.*, 50, 21; (c) R.R. Schmidt and K.-H. Jung, 1997, *Preparative Carbohydrate Chemistry* (S. Hanessian, ed.), Chapter 12.
4. Total synthesis of calicheamycin γ¹: (a) K.C. Nicolaou, 1993, *Angew. Chem. Int. Ed. Engl.*, 32, 1377; (b) K.C. Nicolaou, C.W. Hummel, M. Nakada, K. Shibayama, E.N. Pitsinos, H. Saimoto, Y. Mizuno, K.-U. Bladenius and A.L. Smith, 1993, *J. Amer. Chem. Soc.*, 115, 7625; (c) A.L. Smith, E.N. Pitsinos, C.-K. Hwang, Y. Mizuno, H. Saimoto, G.R. Scarlato, T. Suzuki and K.C. Nicolaou, 1993, *J. Amer. Chem. Soc.*, 115, 7612; (d) R.D. Groneberg, T. Miyazaki, N.A. Stylianides, T.J. Schulze, W. Stahl, E.P. Schreiner, T. Suzuki, Y. Iwabuchi, A.L. Smith and K.C. Nicolaou, 1993, *J. Amer. Chem. Soc.*, 115, 7593.
5. (a) W. Kinzy and R.R. Schmidt, 1989, *Carbohydr. Res.*, 193, 33; (b) G. Grundler and R.R. Schmidt, 1984, *Liebigs Ann. Chem.*, 1826; (c) R.R. Schmidt and G. Grundler, 1982, *Angew. Chem. Int. Ed. Engl.*, 21, 781; (d) G.J.P.H. Boons, M. Overhand, G.A. van der Marel and J.-H. van Boom, 1989, *Angew. Chem. Int. Ed. Engl.*, 28, 1504; (e) T. Kaneko, K. Takahashi and M. Hirama, 1998, *Heterocycles*, 47, 91.
6. (a) M. Paz Cabal, R.S. Coleman and S.J. Danishefsky, 1990, *J. Amer. Chem. Soc.*, 112, 3253; (b) J.N. Haseltine, M. Paz Cabal, N.B. Mantlo, N. Iwasawa, D.S. Yamashita, R.S. Coleman, S.J. Danishefsky and G.K. Schulte, 1991, *J. Amer. Chem. Soc.*, 113, 3850.

7. (a) T. Katsuki and K.B. Sharpless, 1980, *J. Amer. Chem. Soc.*, 102, 5974; (b) B.E. Rossiter, T. Katsuki and K.B. Sharpless, *J. Amer. Chem. Soc.*, 1981, 103, 464; (c) Y. Gao, R.M. Hanson, J.M. Klunder, S.Y. Ko, H. Masamune and K.B. Sharpless, 1987, *J. Amer. Chem. Soc.*, 109, 5765.

8. (a) D-Glyceraldehyde acetonide: C.R. Schmid, J. D. Bryant, M.J. McKennon and A.I. Meyers, 1993, *Org. Synth.* (D.L. Coffen, ed.), 72, 6; (b) L-glyceraldehyde acetonide: C. Hubschwerlen, J.-L. Specklin, J. Higelin, T.M. Heidelbaugh and L.A. Paquette, 1993, *Org. Synth.* (D.L. Coffen, ed.), 72, 1.

9. (a) A.G. Myers, M. Hammond and Y. Wu, 1996, *Tetrahedron Lett.*, 37, 3083; (b) L.A. Paquette, M.J. Earle, G.F. Smith, T. Kirrane and A.I. Meyers, 1995, *Org. Synth.* (R.K. Boeckman, ed.), 73, 36.

10. (a) D.B. Dess and J.C. Martin, 1983, *J. Org. Chem.*, 48, 4156; (b) R.E. Ireland and L. Lui, 1993, *J. Org. Chem.*, 58, 2899.

11. D.Y. Zhinkin, G.N. Mal'nova and Zh.V. Gorislavskaya, 1968, *Zh. Obshch. Khim.*, 38, 2800; (b) S. Masamune, J.W. Ellingboe and W. Choy, 1982, *J. Amer. Chem. Soc.*, 104, 5526.

12. R.J. Arhart and J.C. Martin, 1972, *J. Amer. Chem. Soc.*, 94, 5003.

13. (a) P.J. Garegg and B. Samuelsson, 1979, *J. Chem. Soc., Chem. Commun.*, 978; (b) P.J. Garegg and B. Samuelsson, 1979, *Synthesis*, 469; (c) P.J. Garegg and B. Samuelsson, 1979, *Synthesis*, 813.

14. B.H. McKee, T.H. Kalantar and K.B. Sharpless, 1991, *J. Org. Chem.*, 56, 6966.

9 (+)-Castanospermine

K.J. Hale

9.1 Introduction

(+)-Castanospermine is a polyhydroxylated alkaloid found in the plant *Castanospermum australe.*[1] Its ability to function as a selective inhibitor of α and β glycosidases has made it the focal point of much synthetic activity in recent years.[2] One particularly elegant synthesis of (+)-castanospermine is that of Pandit and Overkleeft.[3] It features a remarkable intramolecular olefin metathesis reaction for indolizidine ring assembly. We now analyse this interesting route, which showcases many important reagents and reactions used in contemporary organic synthesis.

9.2 The Pandit retrosynthetic analysis of (+)-castanospermine

Examination of the structure of (+)-castanospermine reveals that it is a close cousin of D-glucose, in which the ring-oxygen has been replaced by a ring-nitrogen, and a two-carbon bridge has been used to tether C(6) to this nitrogen. When considered in this way, most seasoned carbohydrate chemists will immediately start thinking about using D-glucose as a chiral starting material for its synthesis. So what, in principle, would one have to do to manipulate D-glucose into (+)-castanospermine? Pandit's retrosynthetic analysis of this problem (Scheme 9.1) had him selectively protecting the three piperidine ring hydroxyls as *O*-benzyl ethers, and oxidatively functionalising adjacent to the ring nitrogen. These combined retrosynthetic manoeuvres not only facilitated disconnection of the target, but also more clearly defined its structural relationship with the six-carbon framework of this starting sugar.

Pandit's next retrosynthetic move was to *syn*-hydroxylate at the carbon next to the pyrrolidine hydroxyl to create diol **1**. Establishing a *syn*-relationship between these two hydroxyls was tactically advantageous for it allowed these stereogenic centres to be installed by OsO$_4$ dihydroxylation. The latter reaction would almost certainly proceed from the β face of **2** to give **1**, owing to the C(4)-*O*-benzyl group sterically shielding the α side of the cyclopentene ring, a fact that can be readily confirmed by examination of molecular models.

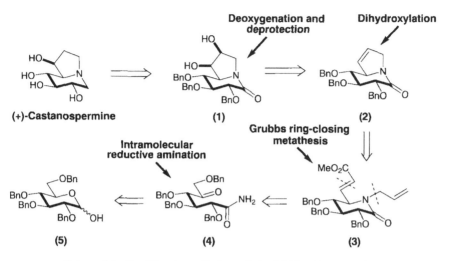

Scheme 9.1 Pandit's retrosynthetic analysis of (+)-castanospermine.

Whenever a target molecule contains a cyclic alkene, it is often worthwhile exploring the possibilities for disconnecting and reforming the alkene ring-system by ring-closing metathesis (RCM).[4] Pandit's application of an RCM transform to **2** yielded diene **3** as a subtarget. The *N*-allyl group and enoate double bond of **3** were the next features he selected for clearance. Further structural simplifications then led to keto amide **4**. In a forward synthesis, iminium ion formation from **4**, followed by reduction from the α side (as shown in Scheme 9.3), could be predicted to deliver the desired trihydroxylated piperidine ring system. Tracing back the lineage of **4** to commercially available 2,3,4,6-tetra-*O*-benzyl-D-glucose now becomes quite straightforward.

9.3   Pandit's total synthesis of (+)-castanospermine

The first step in Pandit's synthesis was a Moffatt oxidation of tetra-*O*-benzyl-D-glucose **5** with acetic anhydride and dimethylsulfoxide (DMSO) to obtain the 1,5-glucuronolactone (Scheme 9.2). The lactone ring was then fissured with concentrated ammonia in methanol to obtain **6**. Another Moffatt oxidation converted the C(5)-OH to the ketone needed for stereospecific intramolecular reductive amination[5] with sodium cyanoborohydride and formic acid. Having securely ensconced the ring-nitrogen within **7**, Pandit next performed an *N*-allylation under phase-transfer conditions, and chemoselectively cleaved the C(6) *O*-benzyl ether by acetolysis and base hydrolysis. Alcohol **8** was then

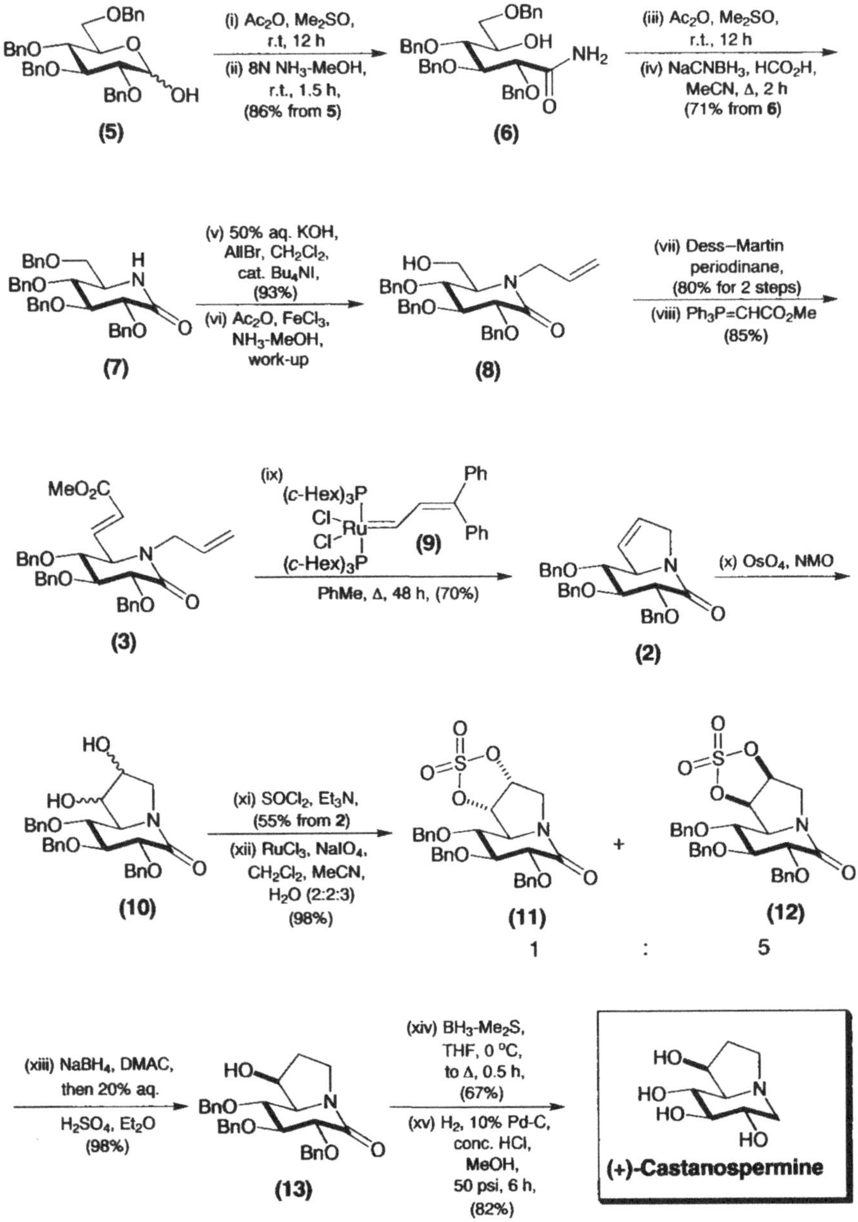

Scheme 9.2 Pandit's total synthesis of (+)-castanospermine.

oxidised with the Dess–Martin periodinane, and the C(6)-aldehyde condensed with the stabilised ylide methyl (triphenylphosphoranylidene) acetate to obtain **3** in 85% yield. Despite the widely-held belief that α,β-unsaturated esters are not generally good substrates for the ring-closing metathesis reaction, Pandit forged ahead and implemented the RCM process on **3**. Significantly, when performed in toluene at reflux in the presence of 5 mol% of the Grubbs ruthenium catalyst **9**,[5] the aforementioned RCM process worked extremely well, it delivering the cyclic alkene **2** in a sturdy 70% yield! So why did this reaction perform so admirably? In our view, a likely explanation comes from the acrylonitrile cross-metathesis work of Crowe and Goldberg.[6] These workers observed that when 1 mol% of the Schrock complex, Mo(CHCMe$_2$Ph)(NAr) [OCMe(CF$_3$)$_2$]$_2$,[7] is mixed with a pair of alkenes, one electron-rich, the other electron-poor, the cross-metathesis reaction between both alkenes generally works well. However, when a similar cross- or self-metathesis reaction is attempted between two electron-deficient alkenes, the reactions often do not proceed at all. It will be seen that with diene **3**, we now have a situation where one electron-rich and one electron-deficient alkene are both present. By analogy, therefore, one might expect this substrate to cross-metathesise successfully with **9**, and indeed it did!

Having assembled the five-membered cycloalkene ring system, attention was directed towards installing the pyrrolidine-hydroxyl of the target by *syn*-dihydroxylation. Steric shielding by the C(4) *O*-benzyl group helped ensure that the desired diol predominated in the 5:1 mixture of diastereomers **10** that emerged. Without separation, both constituents were converted to the cyclic sulfates **11** and **12** via the Sharpless method.[8] Sharpless views cyclic sulfates rather like reactive epoxide synthons.[8a, 9] They readily undergo ring-opening with a wide range of nucleophiles, which include carbanions. In the present case, sodium borohydride was the active nucleophile in dimethylacetamide. It regioselectively deoxygenated **12** at the less-hindered cyclic sulfate carbon.

Earlier work[2e] by Miller and Chamberlin had already shown that borane-dimethyl sulfide was the reagent of choice for deoxygenating the amide of **13**. The final step of the synthesis was *O*-debenzylation by catalytic hydrogenation. Note how the addition of acid helped prevent the catalyst from getting poisoned by the amine, ensuring that deprotection was rapid and efficient.

Pandit's synthesis of (+)-castanospermine stands out for its seminal demonstration of the applicability of the Grubbs RCM reaction[5] to dienes containing one electrophilic alkene and one nucleophilic alkene. When considered alongside Crowe and Goldberg's landmark work on the cross-metathesis of electron-deficient alkenes,[6] Pandit's synthesis has contributed enormously to improving our understanding of the types of diene that can perform well in the RCM process.

9.4 Mechanistic analysis of the Pandit synthesis of (+)-castanospermine

The conversion of 4 into 7
Treatment of amide **4** with formic acid triggers a nucleophilic addition of the amide NH_2 to the protonated ketone, and an elimination of water, to give iminium ion **14**. Reduction of **14** then occurs exclusively from the α face, probably via the half-chair transition state shown in Scheme 9.3.

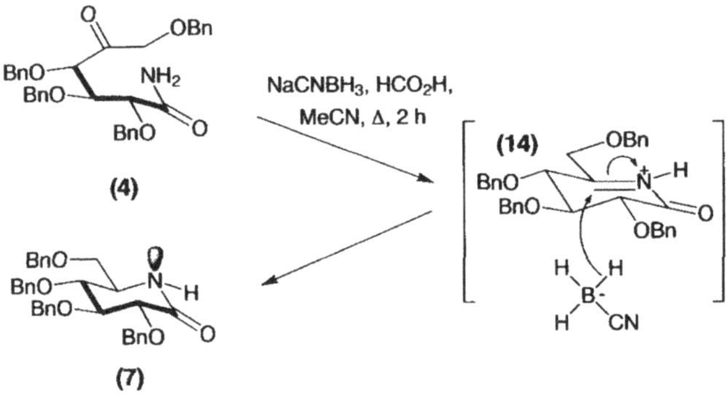

Scheme 9.3.

This maximises overlap of the developing nitrogen lone pair with the lactam carbonyl.[10]

The conversion of 3 into 2
Mechanistically, one can think of this RCM reaction proceeding as shown in Scheme 9.4.

The 'electrophilic' Ru alkylidene complex **9** initially engages in a [2 + 2]-cycloaddition[11] with the less hindered 'nucleophilic' terminal alkene of **3** to give the metallocyclobutane **15**. A cycloreversion then takes place to generate **16**, an intermediate in which the Ru alkylidene can be consid- ered 'nucleophilic'. As a result it willingly participates in a second, intramolecular, [2 + 2]-cycloaddition with the nearby electrophilic enoate to generate **17**. Finally, after yet another cycloreversion, and a loss of the 'electrophilic' Ru complex **18**, the catalytic cycle rekindles in the manner shown (Scheme 9.4).

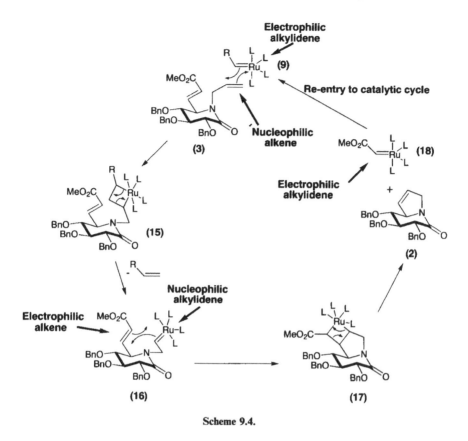

Scheme 9.4.

References

1. L.D. Hohenschutz, E.A. Bell, P.J. Jewess, D.P. Leworthy, R.J. Pryce, E. Arnold and J. Clardy, 1981, *Phytochemistry*, 20, 811.

2. (a) H. Ina and C. Kibayashi, 1993, *J. Org. Chem.*, 58, 52; (b) M. Gerspacher and H. Rapoport, 1992, *J. Org. Chem.*, 57, 3700; (c) H. Hamana, N. Ikota and B. Ganem, 1987, *J. Org. Chem.*, 52, 5492; (e) S.A. Miller and A.R. Chamberlin, 1990, *J. Amer. Chem. Soc.*, 112, 8100.

3. H.S. Overkleeft and U.K. Pandit, 1996, *Tetrahedron Lett.*, 37, 547.

4. (a) Reviews: R.H. Grubbs, S.J. Miller and G.C. Fu, 1995, *Acc. Chem. Res.*, 28, 446; (b) H.-G. Schmalz, 1995, *Angew. Chem.*, 107, 1981; (c) R.H. Grubbs and S.H. Pine, 1991, *Comprehensive Organic Synthesis* (B.M. Trost and I. Fleming, eds.), Pergamon, Oxford, Vol. 5, 1115; (d) R.H. Grubbs and S. Chang, 1998, *Tetrahedron*, 54, 4413.

5. (a) S.T. Nguyen, L.K. Jonsen and R.H. Grubbs, 1992, *J. Amer. Chem. Soc.*, 114, 3974; (b) S. T. Nguyen, R.H. Grubbs and J.W. Ziller, 1993, *J. Amer. Chem. Soc.*, 115, 9858; (c) Z. Wu, S.T. Nguyen, R.H. Grubbs and J.W. Ziller, 1995, *J. Amer. Chem. Soc.*, 117, 5503.

6. W.E. Crowe and D.R. Goldberg, 1995, *J. Amer. Chem. Soc.*, 117, 5162.

7. (a) J.S. Murdzek and R.R. Schrock, 1987, *Organometallics*, 6, 1373; (b) J.S. Murdzek and R.R. Schrock, G.C. Bazan, J. Robbins, M. DiMare and M. O'Regan, 1990, *J. Amer. Chem.*

Soc., 112, 3857; (c) H.H. Fox, K.B. Yao, J. Robbins, S. Cai and R.R. Schrock, 1992, *Inorg. Chem.*, 31, 2287.

8. (a) Y. Gao and K.B. Sharpless, 1988, *J. Amer. Chem. Soc.*, 110, 7538; (b) B.B. Lohray, Y. Gao and K.B. Sharpless, 1989, *Tetrahedron Lett.*, 30, 2623; (c) B.M. Kim and K.B. Sharpless, 1989, *Tetrahedron Lett.*, 30, 655; (d) R. Oi and K.B. Sharpless, 1991, *Tetrahedron Lett.*, 32, 999; (e) K.J. Hale, S. Manaviazar and V.M. Delisser, 1994, *Tetrahedron*, 50, 9181.

9. Review: B.B. Lohray, 1992, *Synthesis*, 1035.

10. H.S. Overkleeft, J. van Wittenburg and U.K. Pandit, 1994, *Tetrahedron*, 50, 4215.

11. (a) J. Kress, J.A. Osborn, R.M.E. Greene, K.J. Ivin and J.J. Rooney, 1985, *J. Chem. Soc., Chem. Commun.*, 874; (b) J. Kress, J.A. Osborn, R.M.E., Greene, K.J. Ivin and J.J. Rooney, 1987, *J. Amer. Chem. Soc.*, 109, 899; (c) E.L. Dias, S.T. Nguyen and R.H. Grubbs, 1997, *J. Amer. Chem. Soc.*, 119, 3887.

10 (−)-Silphiperfolene

K.J. Hale

10.1 Introduction

(−)-Silphiperfolene is an unusual sesquiterpenoid triquinane obtained from the roots of *Silphium perfolatium.*[1] Its structure was first deduced in 1980 by Bohlmann and co-workers, and its synthesis in enantiomerically pure form was first achieved by Paquette *et al.* in 1984.[2] It is probably fair to say that if most organic chemists were asked to devise a synthetic route to (−)-silphiperfolene, many would probably not consider utilising a monosaccharide as their chiral starting material, primarily because the target is devoid of heteroatom functionality, and it bears little real resemblance to a carbohydrate. However, when the eminent carbohydrate chemist Bert Fraser-Reid set his mind to this problem, he did see a potential link between (−)-silphiperfolene and D-mannose, a relationship that culminated in its eventual total synthesis in 1990.[3] In the coming paragraphs, we will retrace Bert Fraser-Reid's retrosynthetic footsteps on the (−)-silphiperfolene problem, and comment on the synthesis that was eventually developed. His route provides a beautiful illustration of how one can cleverly exploit the conformational bias and multiple chiral centres of a pyranoside to assemble multiring carbocyclic target structures.

10.2 The Fraser-Reid retrosynthetic analysis of (−)-silphiperfolene

The disconnection of enantiopure carbocycles is never an easy task, but this is especially so when the target molecule possesses multiple ring systems comprised entirely of carbon and hydrogen atoms linked together via σ bonds. When confronted with such structures it is often advisable (or even necessary) to retrosynthetically functionalise one or more of the ring systems present to permit a simplifying disconnection to be made. Once a hydroxyl, an amine, a carbonyl, or some other multiply bonded unit is positioned within a ring, this usually primes adjacent or nearby bonds for immediate disconnection, or for further functionalisation followed by disconnection.

The presence of a tetrasubstituted C=C bond in the A-ring of (−)-silphiperfolene provides a logical starting point from where to commence retrosynthetic disassembly. It was Fraser-Reid's view that the option of further functionalisation followed by disconnection was the more strategic in this particular instance. He performed an imaginary hydration reaction on the double bond in ring A to obtain chiral tertiary alcohol 1 (Scheme 10.1). He then disconnected the C(4)-exo-methyl group to access ketone 2. Then, rather than retrosynthetically introducing the C(5) methyl group of 2 by enolate chemistry, he resolved to retro-synthetically reduce and block the C(4) carbonyl group, in order to

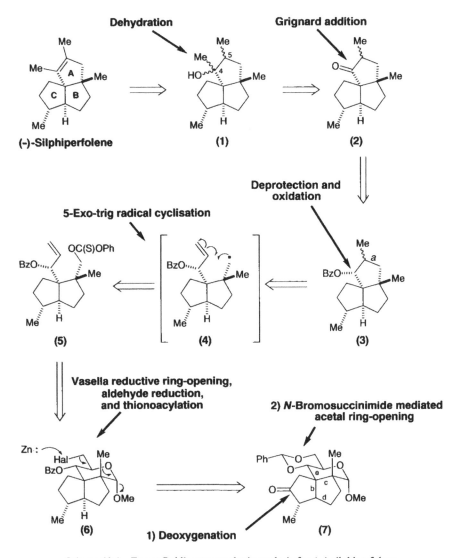

Scheme 10.1 Fraser-Reid's retrosynthetic analysis for (−)-silphiperfolene.

explore the effect of antithetically rupturing bond *a* in subtarget **3**. Aware that highly functionalised five-membered carbocycles can often be readily constructed through 5-exo-trig free radical ring-closure,[4] Fraser-Reid implemented this transform on compound **3** to obtain the alkenyl thionocarbonate **5** as a subtarget. It is now that the experienced carbohydrate chemist can start to see the six-carbon hexopyranose framework emerging from within **5**, particularly if they are aware of the Vasella reductive ring-opening reaction of 6-halo-hexopyranosides.[5] This is a novel reductive elimination that allows sugar hexopyranosides to be converted to acyclic olefinic aldehydes under very mild conditions. Fraser-Reid knew of this methodology, and retrosynthetically used it to very powerful effect on **5** to acquire **6**.

Having identified the 'hidden' carbohydrate framework within (−)-silphiperfolene, Fraser-Reid next attempted to dismantle the complex bicyclooctane ring system of **6**, which now is an array consisting entirely of C and H atoms linked together via σ bonds. Significantly, the presence of an exo-methyl group in ring-C of **6** helped guide a retrosynthetic oxidation of that ring to provide ketone **7** as a subtarget. In turn, this one operation opened up a plethora of disconnective possibilities for 5-exo-trig radical ring-cleavage of the bonds β to the newly introduced carbonyl that is, bonds *b*, *c*, *d* and *e* in ketone **7** (Scheme 10.1).

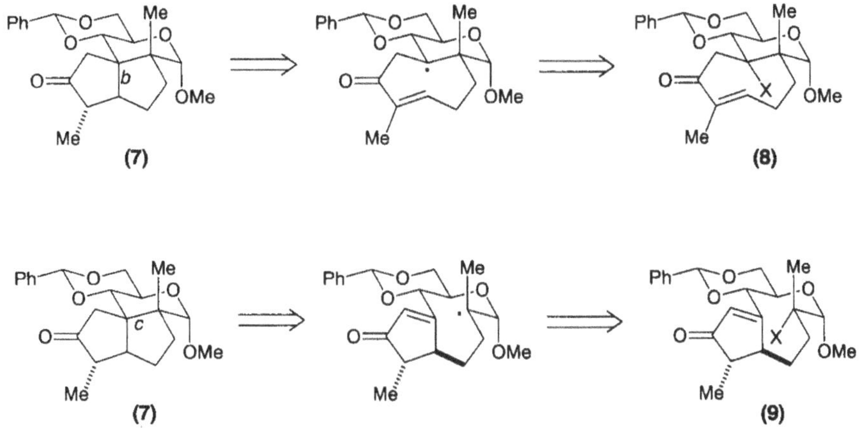

Scheme 10.2   Retrosynthetic radical cleavage of bonds *b* and *c* in **7**. Disadvantages: the cleavage of bonds *b* and *c* does not afford precursors that have been significantly simplified. It leads to highly functionalised medium ring precursors that will be very difficult to synthesise.

It will soon be appreciated that rupture of bonds *b*, *c* and *e* offers no real tactical advantages. For example, radical cleavage of bond *b* creates an eight-membered unsaturated carbocycle **8** that is even more challenging to synthesise than **7** (Scheme 10.2). Under no circumstances can this be viewed as a simplifying transform. Likewise, cleavage of bond *c* disturbs the pyranoside framework, and provides the equally demanding problem of having to create a remotely functionalised nine-membered ether ring fused onto a cyclopentenone (as in compound **9**). Disconnection of bond *c* also offers the daunting prospect of having to generate a bond between two highly hindered quaternary carbon centres, whilst simultaneously controlling their relative and absolute stereoche-mistries. Again, even for most highly seasoned synthetic campaigner, this would be a dreaded task! By retrosynthetically analysing subtarget **7**, in this methodical way, Fraser-Reid eventually fractured bond *d* to obtain **10** (Scheme 10.3). It will soon be appreciated that a cyclisation of this sort would generate a stabilised α-keto radical that would prefer to abstract a

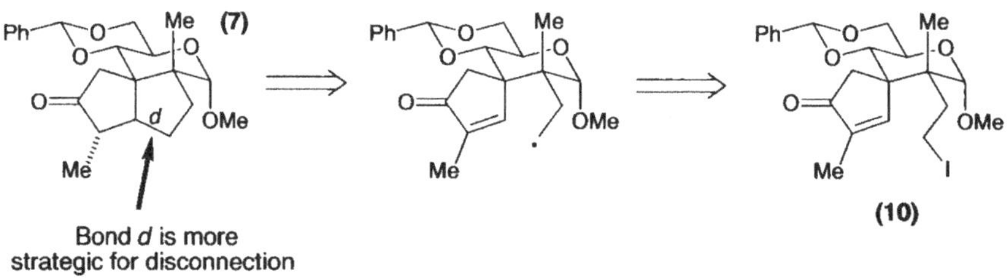

Bond *d* is more
strategic for disconnection

Scheme 10.3 Free radical retrosynthetic disassembly of bond *d* in **7**. Advantages: it does not necessitate the construction of a functionalised medium-sized ring. It forms the bicyclooctane ring system by a favourable 5-exo-trig pathway. Conformational and steric bias will allow the stereochemistry of the methyl α to the ketone to be controlled.

hydrogen atom from Bu_3SnH, from the exo-side of the newly formed tricyclic array, for steric reasons. The correct ring junction stereochem-istry would emerge as a consequence of the stereochemical arrangement between the substituents at C(2) and C(3) in the pyran framework. As an added bonus, retrosynthetic olefination of ring C to obtain **10** would facilitate disassembly of this ring system by regioselective double-bond hydration and retro-aldol reaction (Scheme 10.4). In a forward synthesis, the prospects for achieving an aldol reaction/dehydration in a single step from **11** would be excellent, which added to the attractiveness of this simplifying operation.

Thus, by using logical retrosynthetic thinking, Fraser-Reid was able to reduce the (−)-silphiperfolene problem to the issue of setting two quaternary carbon stereocentres at adjacent positions within a hexopy-ranoside framework. Fortunately, he had already provided a partial solution to this particular methodological challenge in the mid-1980s,

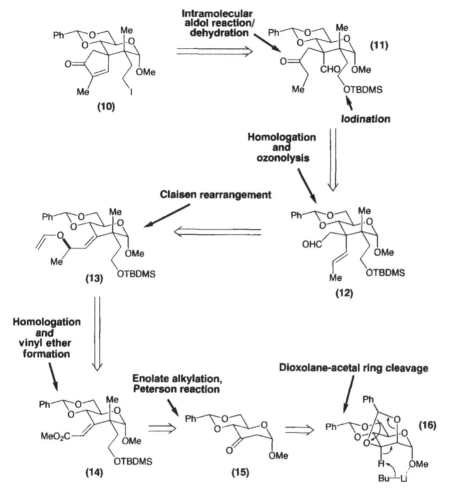

Scheme 10.4 Fraser-Reid's retrosynthetic analysis for (−)-silphiperfolene.

after he showed that C(3)-allylic vinyl ethers, such as **13**, readily engage
in stereospecific Claisen rearrangements[6] to form hexopyranosides with
quaternary stereocentres at C(3). Because compound **13** could now potentially be derived from enoate **14**, a Wittig disconnection now appeared
feasible, which consequently allowed ketone **15** to be selected as a subtarget. It is now that the possibilities for performing a C(2)-enolate
double alkylation on **15** become apparent. Since ketone **15** was available
from the *O*-benzylidene acetal cleavage of **16** with *n*-BuLi, Fraser-Reid
opted to examine the viability of this strategy in more detail.

10.3   Fraser-Reid's total synthesis of (−)-silphiperfolene

Fraser-Reid's synthesis of (−)-silphiperfolene (Scheme 10.5) commenced with a regioselective and chemoselective cleavage of the dioxolane acetal in methyl 2,3;4,6-di-*O*-benzylidene-α-D-mannopyranoside **16** with *n*-BuLi.[7] This afforded an enolate that reacted with ethyl bromoacetate to produce **17**.[6b] It is likely that the axial alkylation product is initially formed in this reaction, and that this then isomerises to the more stable equatorial isomer, under the basic reaction conditions. A second enolisation was then effected at C(2) with potassium hydride in tetrahydrofuran (THF).[8] Again alkylation proceeded from the less-hindered axial direction, but now the lack of an acidic hydrogen at C(2) precluded loss of stereochemical integrity at this position. Because Wittig and Peterson technology both failed to olefinate this ketone, an alternative strategy had to be put in place to convert it to the primary allylic alcohol **20**. One pathway added vinyl magnesium bromide to the ketone, from the less-hindered β face to provide the tertiary allylic alcohol **18**. After ester reduction and alcohol protection, a thionyl-chloride-induced rearrangement on **19** yielded the primary allylic chloride. The latter was then subjected to a nucleophilic displacement with sodium acetate, and a transesterification reaction, to obtain the desired allylic alcohol **20** in good overall yield.

Alcohol oxidation and Grignard addition were next implemented to access **21**, along with its hydroxy epimer. The undesired alcohol was recycled by a two-step procedure, involving a Mitsunobu reaction with benzoic acid, and ester cleavage with methyllithium. Compound **21** was converted into enol ether **13**, and a Claisen rearrangement effected to obtain **12** as a single product. After addition of ethylmagnesium bromide to **12**, Swern oxidation and ozonolysis subsequently provided keto-aldehyde **11**. Intramolecular aldol condensation and dehydration, mediated by potassium *t*-butoxide, transformed **11** into the desired spirocyclopentenone **22**. *O*-Desilylation and iodination finally yielded the B-ring radical cyclisation precursor **10**. Treatment of this iodide with tributylstannane brought about the desired radical conjugate addition to produce ketone **7** with the correct methyl group stereochemistry at C(10).

Since the C(11) keto group in **7** had fulfilled its role in assembly of the diquinane unit, it was converted to the enol triflate **23**,[9] and deoxygenation effected by palladium(0)-catalysed hydrogenolysis.[10] The benzylidene acetal of **24** was then cleaved by a Hanessian–Huller reaction,[11] and bromide **25** converted to iodide **26** by a Finkelstein reaction with sodium iodide. Reductive acetal ring-cleavage with zinc amalgam[6] next provided aldehyde **27**, which was reduced to the alcohol and thionoacylated[12] to obtain **5**. Radical-mediated ring-closure of **5** furnished a mixture of methyl epimers at the newly-introduced stereocentre. This loss of stereo-chemical control was of little real consequence, however, since this carbon was later to become sp^2-hybridised in the target.

In order to arrive at ketone **2**, benzoate **3** was reduced with lithium aluminium hydride, and the resulting alcohol oxidised with pyridinium

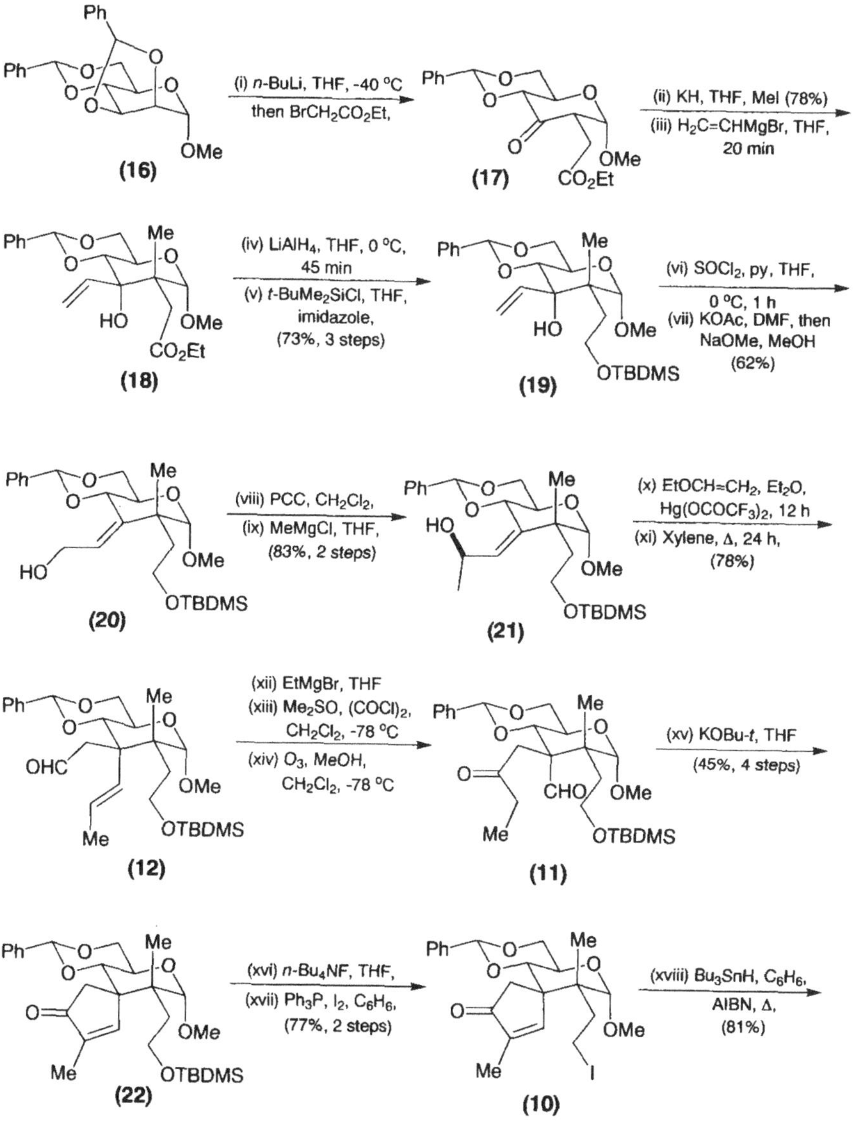

Scheme 10.5 Fraser-Reid's synthesis of (−)-silphiperfolene.

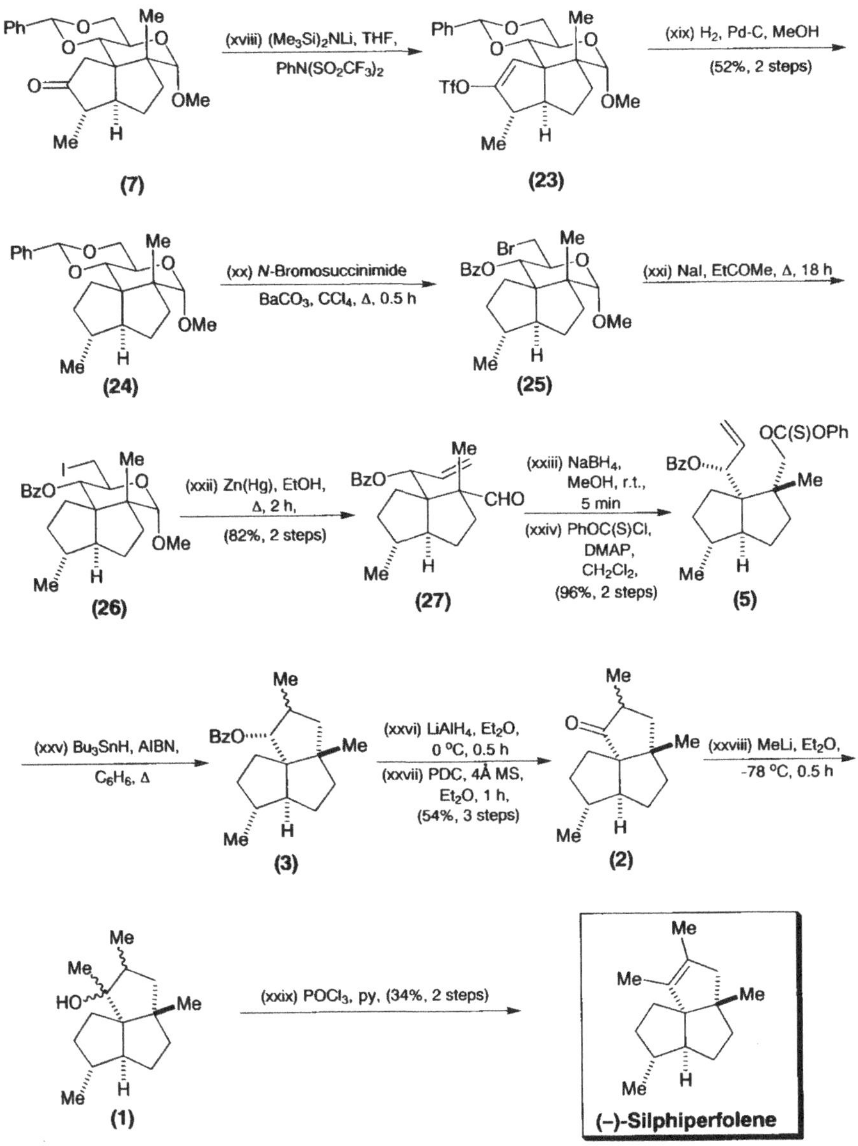

Scheme 10.5 (Continued).

dichromate. Addition of methyllithium to **2** delivered a mixture of tertiary alcohols **1**, that underwent smooth dehydration to the target, (−)-silphiperfolene, in good yield.

10.4 Mechanistic analysis of the Fraser-Reid (−)-silphiperfolene synthesis

The conversion of 19 to 29, and 29 to 30
The first step is an allylic chloride displacement on a chlorosulfite ester generated from the tertiary vinyl carbinol and thionyl chloride. Attack of chloride on an allylic carbocation intermediate can equally well be envisaged. The subsequent conversion is simply an S_N2 displacement by acetate ion (Scheme 10.6).

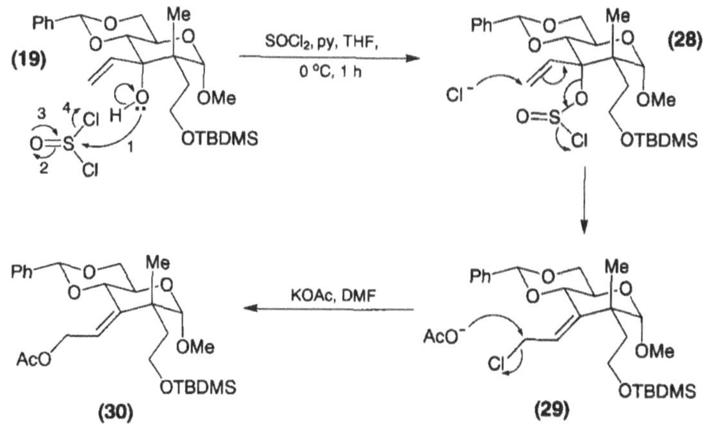

Scheme 10.6.

The conversion of 13 into 12
Here a thermally-induced [3,3]-sigmatropic rearrangement occur (Scheme 10.7).[13] It probably proceeds via the chair-like transition state shown.

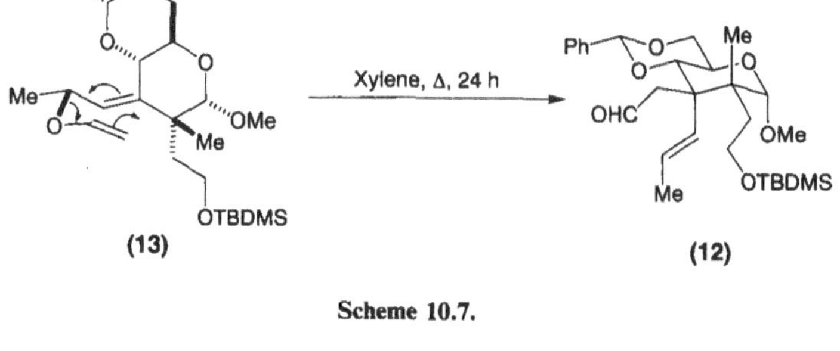

Scheme 10.7.

References

1. F. Bohlmann and J. Jakupoviv, 1980, *Phytochemistry*, 19, 259.
2. L.A. Paquette, R.A. Roberts and G.J. Drtina, 1984, *J. Amer. Chem. Soc.*, 106, 6690.

3. J.K. Dickson and B. Fraser-Reid, 1990, *J. Chem. Soc., Chem. Commun.*, 141.

4. For a detailed discussion of the rules for radical ring-closure, see: A.L.J. Beckwith, C.J. Easton and A.K. Serelis, 1980, *J. Chem. Soc., Chem. Commun.*, 482. For Baldwin's earlier rules for ring-closure, see: J.E. Baldwin, 1976, *J. Chem. Soc., Chem. Commun.*, 734.

5. B. Bernet and A. Vasella, 1979, *Helv. Chim. Acta*, 62, 1990 and 2400.

6. (a) D.B. Tulshian, R. Tsang and B. Fraser-Reid, 1984, *J. Org. Chem.*, 49, 2347; (b) J.K. Dickson, R. Tsang, J.M. Llera and B. Fraser Reid, 1989, *J. Org. Chem.*, 54, 5380; (c) H. Pak, J.K. Dickson and B. Fraser-Reid, 1989, *J. Org. Chem.*, 54, 5357.

7. (a) A. Klemer and G. Rodemeyer, 1974, *Chem. Ber.*, 107, 2612; (b) D. Horton and W. Weckerle, 1975, *Carbohydr. Res.*, 44, 227; (c) G. Rodemeyer and A. Klemer, 1976, *Chem. Ber.*, 109, 1708.

8. Y. Chapleur, 1983, *J. Chem. Soc. Chem. Comm.*, 141.

9. J.E. McMurry and W.J. Scott, 1982, *Tetrahedron Lett.*, 24, 979.

10. V.B. Jigajinni and R.H. Wightman, 1983, *Tetrahedron Lett.*, 23, 117.

11. S. Hanessian and N.R. Plessas, 1969, *J. Org. Chem.*, 34, 1035 and 1045.

12. (a) M.J. Robins and J.S. Wilson, 1981, *J. Amer. Chem. Soc.*, 103, 932; (b) M.J. Robins, J.S. Wilson and F. Hansske, 1983, *J. Amer. Chem. Soc.*, 105, 4059.

13. (a) W.S. Johnson, L. Werthermann, W.R. Bartlett, T.J. Brocksom, T. Li, D.J. Faulkner and M.R. Petersen, 1970, *J. Amer. Chem. Soc.*, 92, 741; (b) Mechanism: P. Vittorelli, T. Winkler, H.-J. Hansen and H. Schmid, 1968, *Helv. Chim. Acta*, 51, 1457.

11   (−)-Allosamizoline

K.J. Hale

11.1   Introduction

In 1986 the novel chitinase inhibitor (−)-allosamidin was discovered in fermentation broths of *Streptomyces* sp. 1713 by Sakuda and co-workers.[1] (−)-Allosamidin is unusual inasmuch as it contains two *N*-acetyl-D-allosamine residues glycosidically linked β-1,4 to one another. In turn, this disaccharide is connected to a cyclopentanoid known as (−)-allosamizoline in the manner shown in Figure 11.1.

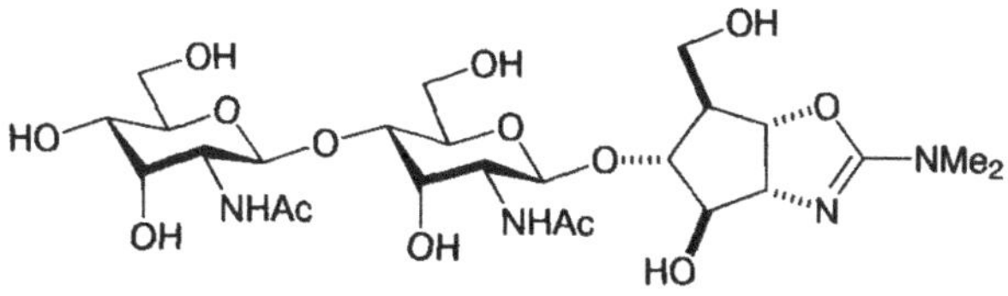

Figure 11.1   (−)-Allosamidin.

Synthetic interest in (−)-alloasamizoline has risen steadily ever since its structure was first announced in the mid-1980s. Trost and Van Vranken[2] were the first to report a racemic synthesis, in 1990, and there later followed two enantiospecific syntheses by Kuzuhara *et al.*[3] and Simpkins and Stokes[4] in 1991 and 1992, respectively.

A careful comparison of the (−)-allosamizoline structure with that of D-glucosamine reveals a perfect heteroatom chirality match at carbons 2, 3, and 4 (Scheme 11.1). Conceptually, one can arrive at (−)-allosamizoline if one establishes a C—C bond between C(1) and C(5) in D-glucosamine, and one then appends an amino-oxazoline ring onto the

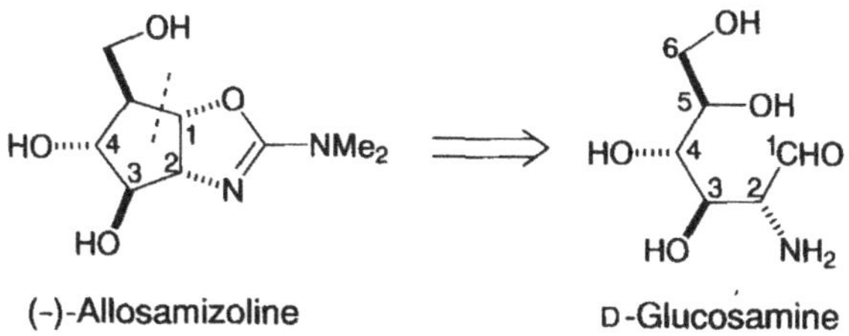

Scheme 11.1   The conceptual conversion of (−)-allosamizoline to D-glucosamine.

newly formed cyclopentane. Although this assessment of the problem can make it seem ever so straightforward, the stereocontrolled construction of highly functionalised cyclopentane rings is anything but a trivial task, with special tactics often being required before such systems succumb to total synthesis. In the coming section, we will examine the different synthetic strategies used by Kuzuhara[3] and Simpkins[4] and their co-workers in their respective total syntheses of (−)-allosamizoline.

11.2 The Kuzuhara retrosynthetic analysis of (−)-allosamizoline

One very useful stratagem for assembling heavily substituted rings is to construct a more readily accessible system of larger ring size (often with more functionality), and then to attempt its ring contraction. Whilst some target molecules can readily yield to this line of retrosynthetic thinking, others, such as (−)-allosamizoline, often require detailed analysis before such a transform becomes apparent. Clearly, planning of this type benefits enormously from a thorough working knowledge of the reactions available for ring contraction. In Scheme 11.2 we list some ring-contraction processes that have previously been used to build up smaller ring systems from larger ring precursors.[5] These include the diazotisation reaction of cyclic-1,2-aminoalcohols,[6] the reductive 1,2-hydroxytosylate rearrangement,[7] and the Taguchi dibutylzirconocene-mediated ring-contraction[8] of monosaccharide cyclic allylic acetals.

Let us return now to the (−)-allosamizoline question. If Kuzuhara's intention was to derive three of the target's chiral centres from

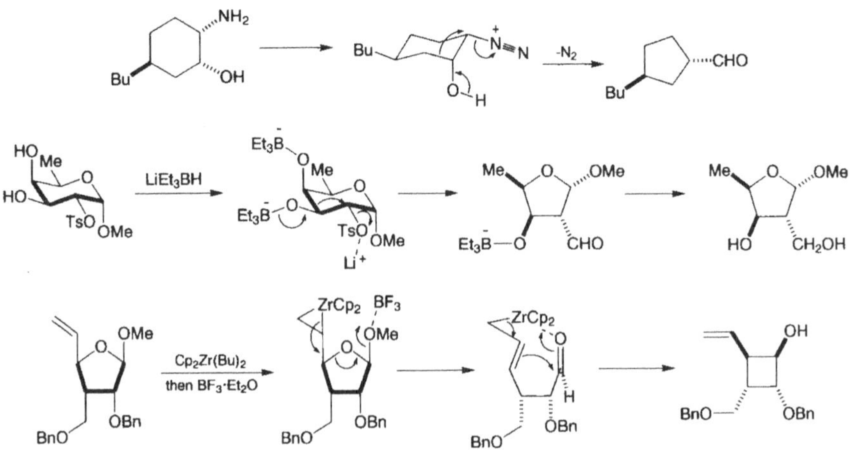

Scheme 11.2   Examples of ring-contraction processes.

D-glucosamine, then obviously he would have to avoid disrupting these carbons during his retrosynthetic analysis. Bearing this in mind, his application of a reductive 1,2-hydroxy-tosylate rearrangement tactic[6] to (−)-allosamizoline demanded that this rearrangement be confined to C(1), C(5) and C(6) of a potential six-membered carbocycle precursor. After careful consideration of the various options, Kuzuhara concluded that hydroxy-tosylate 2 would be the ideal candidate for such a ring-contraction. He then retrosynthetically cleaved the tosyloxy group from 2 to obtain diol 3, and retrosynthetically installed its cis-diol unit by a syn-dihydroxylation on alkene 4 (Scheme 11.3). Because the latter had functionality on either side of the double bond, retro-S_N2' type disconnection did appear feasible at either allylic position. However, owing to the necessity for preserving stereochemical integrity at O(4), throughout the retrosynthetic analysis, Kuzuhara opted to implement this transform in a manner that allowed simultaneous disjunction of the dimethylaminooxazoline ring. This enabled the allylic mesylate 5 to be selected as a subtarget. Kuzuhara's positioning of an allylic O-mesylate at C(5) was highly strategic since it permitted oxygenation of the cyclitol at that position, which allowed enone 6 to be selected as the progenitor of 5. In turn, enone 6 was considered accessible from the modified D-glucosamine derivative 7 by Ferrier rearrangement.[9] With these tactical considerations in mind, we now outline the Kuzuhara synthesis[3] of (−)-allosamizoline from D-glucosamine.

11.3 The Kuzuhara total synthesis of (−)-allosamizoline

Kuzuhara's journey towards (−)-allosamizoline[3] began with the conversion of D-glucosamine hydrochloride 8 into known 9 in five steps. Compound 9 was then N-acylated with dimethylcarbamyl chloride and O-debenzylidenation performed under mildly acidic conditions to obtain diol 10 (Scheme 11.4). Selective iodination of the less hindered primary hydroxyl in 10 was accomplished by treatment with N-iodosuccinimide and triphenylphosphine.[10] The resulting iodide was O-silylated with tert-butyldimethyl silyl (TBDMS) triflate[11] to give 11 in 71% yield for the last two steps. E2-elimination of the iodo group from 11 was then performed with potassium t-butoxide in tetrahydrofuran (THF). Glycal 7 was then subjected to Ferrier rearrangement with mercuric sulfate in acidic acetone.[9] The mixture of β-hydroxy ketones that emerged was next subjected to O-mesylation and in situ E1cb elimination to provide enone 6. After stereoselective reduction of the keto group in 6 with $NaBH_4$-$CeCl_3 \cdot 7H_2O$, the resulting allylic alcohol underwent an

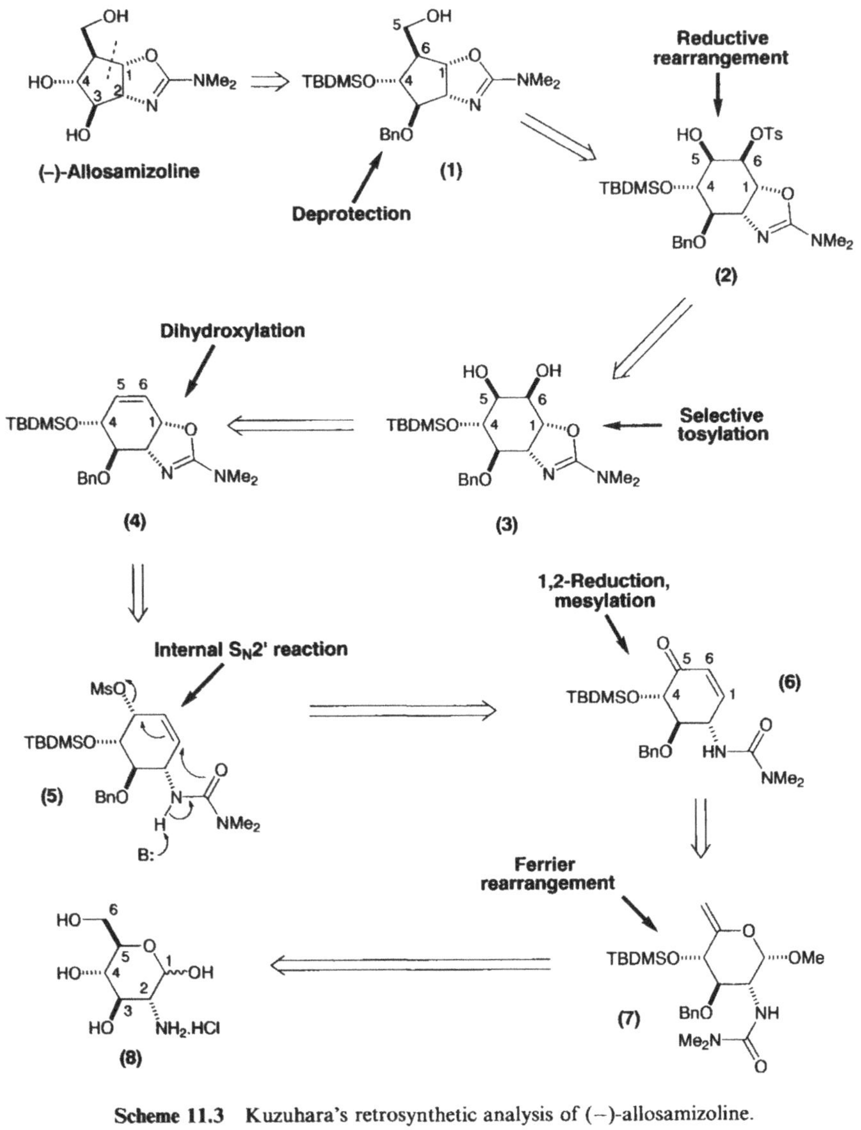

Scheme 11.3 Kuzuhara's retrosynthetic analysis of (−)-allosamizoline.

internal S_N2' reaction when mesylated with methanesulfonic anhydride and triethylamine. Although the stereochemical outcome of this reduction was not specified, topological considerations suggest that it probably proceeded from the less hindered β side of ketone **6**. Given also the preference for S_N2' reactions to proceed via transition states which

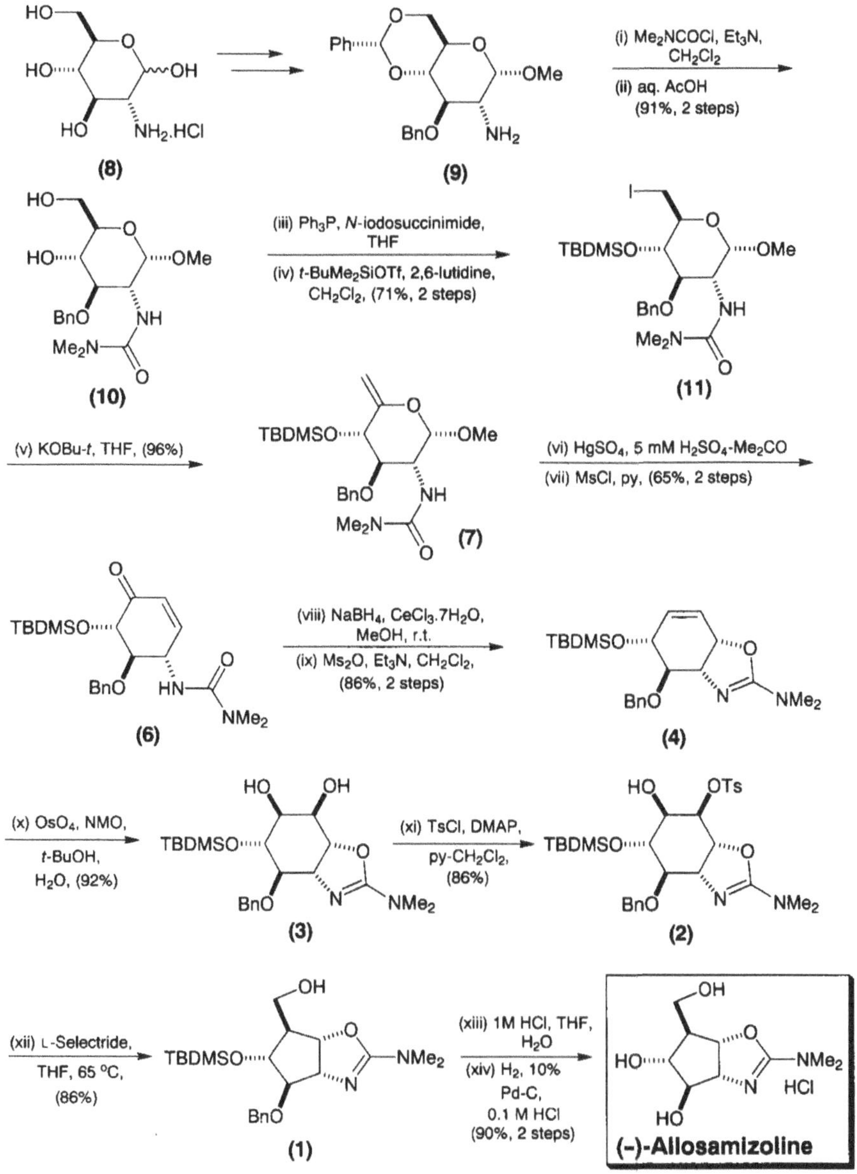

Scheme 11.4   The Kuzuhara synthetic route to (−)-allosamizoline.

position the incoming nucleophile *syn* to the leaving group,[12] it also seems likely that **4** arises from **5**.

Osmylation of **4** proceeded stereospecifically from the β face of the double bond to give diol **3** as the sole of the product reaction. The stereochemical outcome of this reaction can again be attributed to steric hindrance in the approach of the reagent to the α face of the alkene. Diol 3 was selectively sulfonylated with *p*-toluenesulfonyl chloride at its less hindered OH. Although equatorial hydroxyls on six-membered cyclitols are generally more susceptible to *O*-sulfonation than their axial counterparts, the presence of a bulky OTBDMS group next to the equatorial hydroxyl of 3 helped hinder sulfonylation of this alcohol. In turn, the correct positioning of the leaving group at C(6) in 2 now permitted reductive rearrangement[6] with L-selectride. The advantage of using L-selectride to mediate this reaction lay in the fact that the initially-formed aldehyde was reduced immediately to the primary alcohol 1. As a result, (−)-allosamizoline was obtainable in only two more steps, by acid-promoted *O*-desilylation and hydrogenolytic cleavage of the *O*-benzyl group at C(3).

Clearly, in the Kuzuhara synthesis of (−)-allosamizoline, we have seen yet another powerful strategy for ring synthesis, namely ring-contraction, being used to excellent effect.

11.4   Mechanistic analysis of the Kuzuhara (−)-allosamizoline synthesis

The conversion of 7 into 6
The first step is a Ferrier mercuration[9] reaction on the enol ether double bond, which initiates ring-opening of the pyranoside to form 13 (Scheme 11.5). An intramolecular aldol addition reaction then ensues to give 14. After *O*-mesylation, 15 undergoes an E1*cb* elimination reaction via enolate 16.

The conversion of 2 into 1
This is an example of a reductive hydroxy-tosylate 1,2-rearrangement. For such a rearrangement to proceed, the migrating bond must be antiperiplanar to the C—O bond of the leaving group. Initially, treatment of 2 with L-selectride[6] leads to aldehyde 18, which is then reduced *in situ* to 1 (Scheme 11.6).

11.5   Simpkins' retrosynthetic strategy for (−)-allosamizoline

The most outstanding feature of Simpkins' retrosynthetic analysis[4] of (−)-allosamizoline was its proposed useage of a Bartlett monosaccharide-oxime radical cyclisation reaction[13] to assemble the amino-cyclopentane ring system of the target. Simpkins recognised that this methodology was tailor-made for solving the C(1)–C(5) bonding issue. His remaining

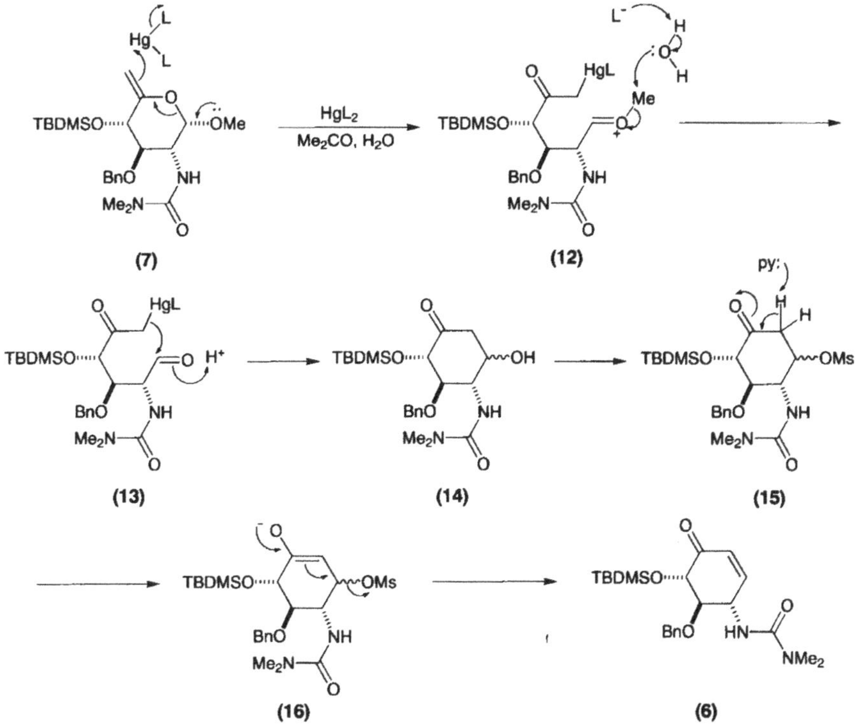

Scheme 11.5.

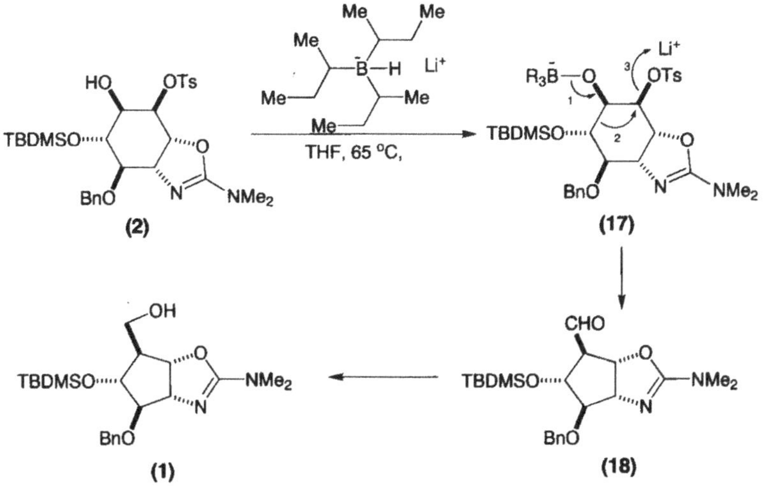

Scheme 11.6.

analysis therefore focused on the design of a viable pathway for the selective deamination and stereospecific oxygenation of C(1), which again would be far from trivial tasks (Scheme 11.7).

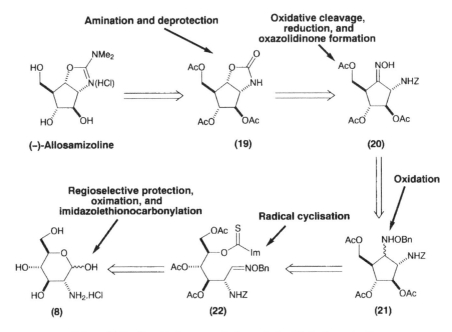

Scheme 11.7 Simpkins' retrosynthetic analysis of (−)-allosamizoline.

11.6 Simpkins' total synthesis of (−)-allosamizoline

After converting D-glucosamine hydrochloride **8** into the Z-protected peracetate **23** in two steps, Simpkins attempted a chemoselective transesterification of the more reactive O(1)-acetate to access the hemiacetal (Scheme 11.8). This was then reacted with O-benzylhydroxylamine hydrochloride in pyridine and dichloromethane to bring about ringopening and formation of the oxime. Compound **24** was then reacted with thiocarbonyldiimidazole to obtain **22**, and a Bartlett 5-exo-trig free-radical cyclisation[13] was implemented with tributylstannane and azobisisobutyronitrile (AIBN) to form the cyclopentane **21** in moderate yield. Simpkins' next task was oxygenation at C(1), a transformation that required two more steps. Initially **21** was oxidised to the oxime **20** with m-chloroperoxybenzoic acid (m-CPBA).[14] A second oxidation was then

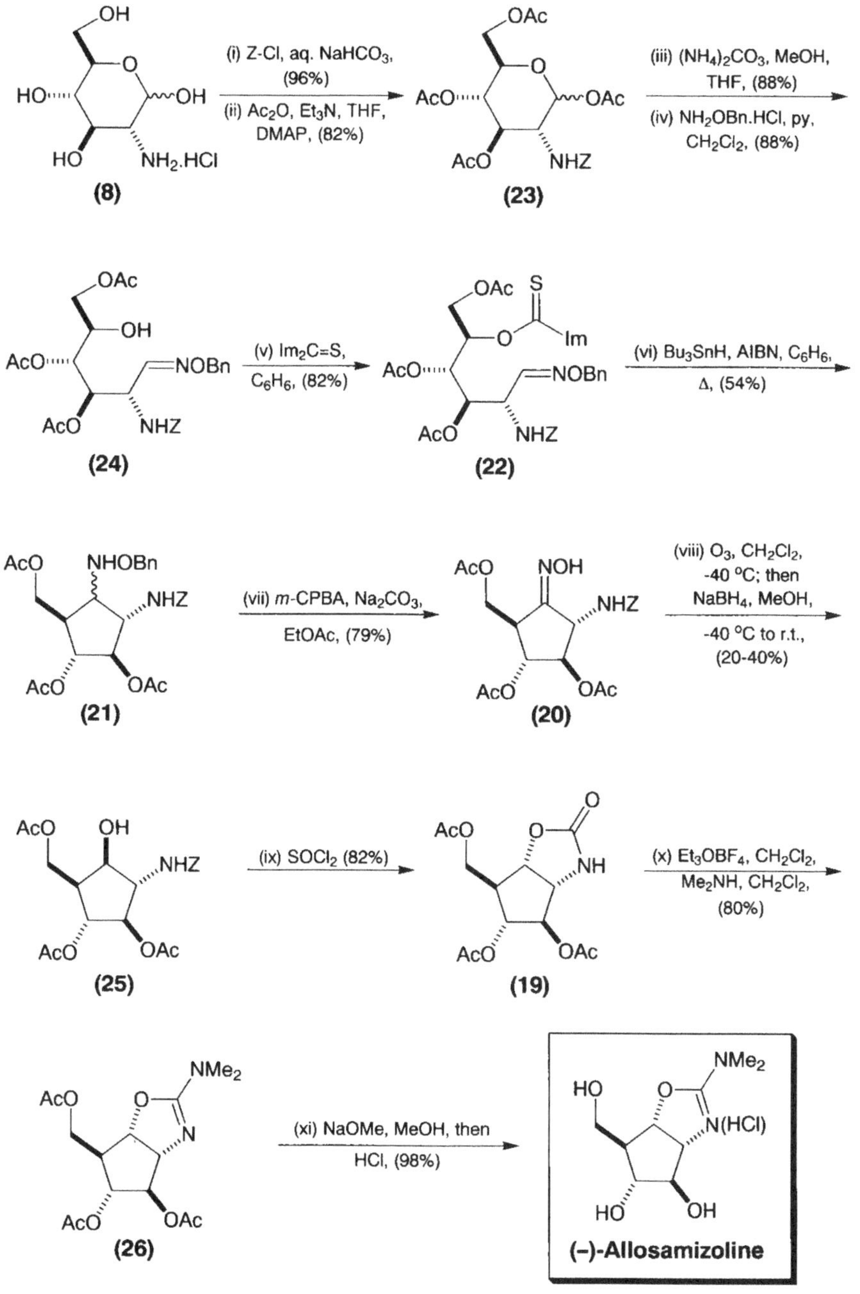

Scheme 11.8 Simpkins' enantiospecific route to (−)-allosamizoline.

performed with ozone[15] to create a secondary ozonide which underwent reduction to alcohol **25**. The latter was isolated in 20–40% overall yield from **20**.

The next phase of the synthesis was installation of the dimethylamino-oxazoline ring system. This was constructed from the oxazolidinone precursor **19**. Oxazolidinone formation occurred when **25** was reacted with thionyl chloride. The more nucleophilic carbonyl of **19** was then *O*-alkylated with the Meerwein reagent to give an iminium ion that readily participated in a nucleophilic addition/elimination reaction with dimethylamine to give **26**. The final step of the synthesis was *O*-deacetylation of **26** with sodium methoxide to provide (−)-allosamizoline hydrochloride in 98% yield after acidification.

11.7 Mechanistic analysis of some key steps in the Simpkins (−)-allosamizoline synthesis

The transformation of 22 into 21
In the Bartlett monosaccharide-oxime free-radical cyclisation reaction,[13] the thionocarbonylimidazolide grouping is initially attacked by a tri-*n*-butyltin radical at its sulfur terminus (Scheme 11.9). This leads to a highly

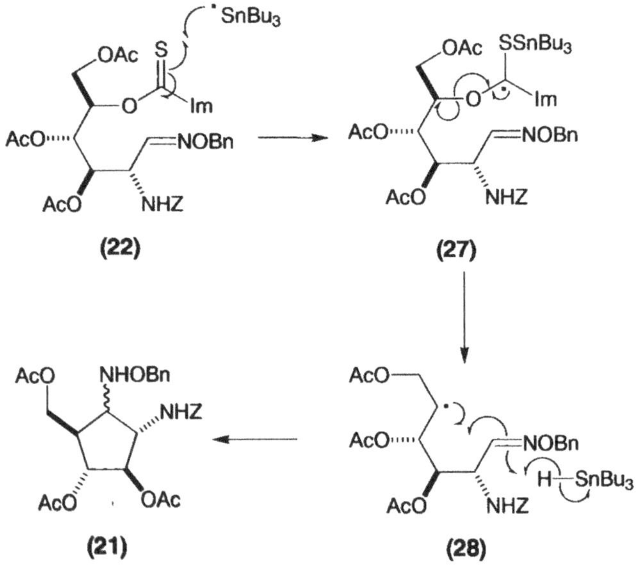

Scheme 11.9.

stabilised tertiary free radical which fragments to a carbon-centred secondary free radical. The latter then cyclises to give a highly reactive nitrogen-centred free radical which then goes on to abstract a hydrogen atom from tri-*n*-butyltin hydride to produce **21**.

The conversion of 21 into 20
One possible mechanism for this reaction is shown in Scheme 11.10. It commences with a nucleophilic attack of the nitrogen grouping on the

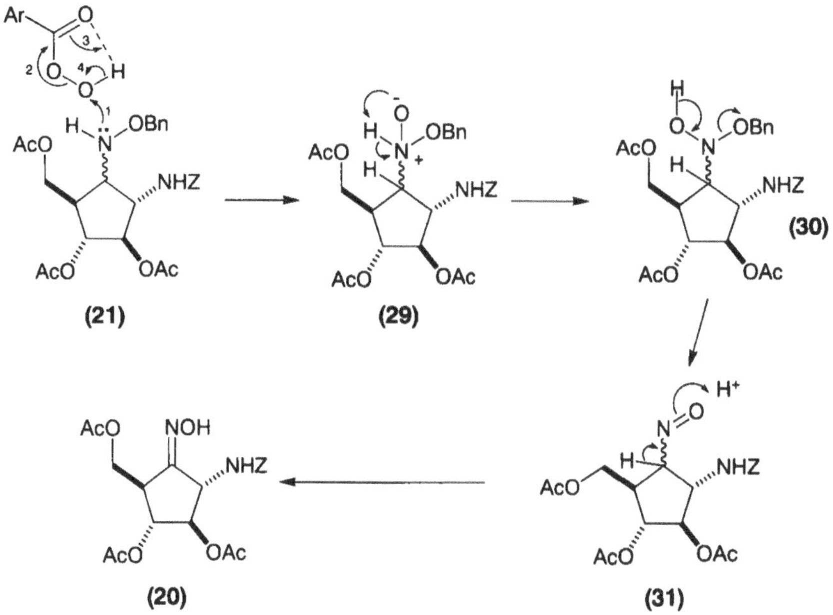

Scheme 11.10.

highly electrophilic terminal OH of the peroxyacid. This causes fission of the very weak O—O bond to give *m*-chlorobenzoic acid and intermediate **29**. After proton transfer and loss of benzyl alcohol from **30**, the nitroso intermediate **31** tautomerises to produce **20**.

The formation of oxazolidinone 19 from 25
The reaction of **25** with thionyl chloride almost certainly leads to the chlorosulfite ester **32** (Scheme 11.11). An internal nucleophilic displacement then occurs by the adjacent benzyloxycarbonyl group to form the

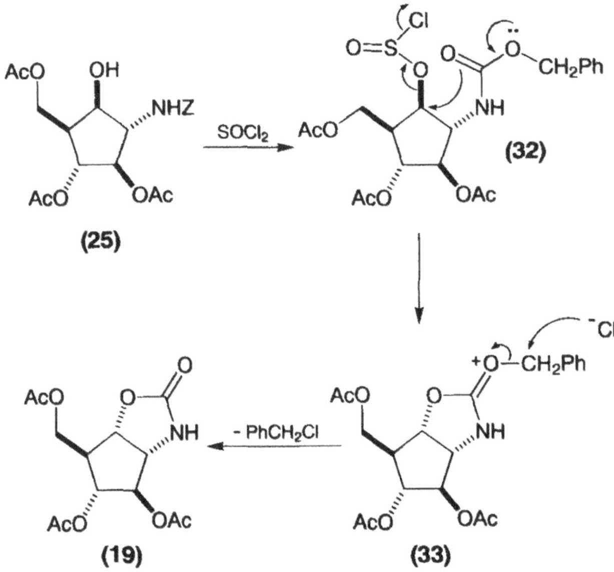

Scheme 11.11.

O-alkylated oxazolidinone **33**, SO_2 and chloride ion. The liberated chloride then attacks the activated benzyl group to form benzyl chloride and **19**.

11.8 Epilogue

The most striking difference between the Kuzuhara and Simpkins syntheses lies in their respective lengths, the Simpkins synthesis being much shorter, because of its more rapid assembly of the cyclopentane ring system by free-radical chemistry. However, both routes are elegant in their own right, each marvellously showcasing the utility of monosaccharide starting materials for the construction of complex natural products.

References

1. (a) S. Sakuda, A. Isogai, S. Matsumoto, A. Suzuki and K. Koseki, 1986, *Tetrahedron Lett.*, 27, 2475; (b) S. Sakuda, A. Isogai, T. Makita, S. Matsumoto, K. Koseki, H. Kodama and A. Suzuki, 1987, *Agric. Biol. Chem.*, 51, 3251; (c) S. Sakuda, A. Isogai, S. Matsumoto, A. Suzuki, K. Koseki, H. Kodama and Y. Yamada, 1988, *Agric. Biol. Chem.*, 52, 1615; (d) Y. Nishimoto, S. Sakuda, S. Takayama and Y. Yamada, 1991, *J. Antibiot.*, 44, 716.
2. B.M. Trost and D.L. Van Vranken, 1990, *J. Amer. Chem. Soc.*, 112, 1261.
3. S. Takahashi, H. Terayama and H. Kuzuhara, 1991, *Tetrahedron Lett.*, 32, 5123.

4. N.S. Simpkins and S. Stokes, 1992, *Tetrahedron Lett.*, 33, 793.
5. Reviews: H. Redlich, 1994, *Angew. Chem. Int. Ed. Engl.*, 33, 1345; (b) T.S. Stevens and W.E. Watts, 1973, *Selected Molecular Rearrangements*, Van Nostrand Reinhold, London.
6. Review: J.M. Williams, 1975, *Adv. Carbohydr. Chem. Biochem.*, 31, 9.
7. (a) H.H. Baer, D.J. Astles, H.-C. Chin and L. Siemsen, 1985, *Can. J. Chem.*, 63, 432; (b) H.H. Baer, 1989, *Pure and Appl. Chem.*, 61, 1217.
8. H. Ito, Y. Motoki, T. Taguchi and Y. Hanzawa, 1993, *J. Amer. Chem. Soc.*, 115, 8835.
9. (a) R.J. Ferrier, 1979, *J. Chem. Soc. Perkin Trans. 1*, 1455; (b) R. Blattner, R.J. Ferrier and S. Hains, 1985, *J. Chem. Soc. Perkin Trans. 1*, 2413; (c) Review: R.J. Ferrier and S. Middleton, 1993, *Chem. Rev.*, 93, 2779; (d) For examples of the catalytic Ferrier rearrangement with 1 mol % $Hg(OCOCF_3)_2$ in 33% aq. Me_2CO, see: N. Chida, M. Ohtsuka and S. Ogawa, 1991, *Tetrahedron Lett.*, 32, 4525, and N. Chida, M. Ohtsuka, K. Ogura and S. Ogawa, 1991, *Bull. Chem. Soc. Jpn*, 64, 2118.
10. S. Hanessian, M.M. Ponpipom and P. Lavalee, 1972, *Carbohydr. Res.*, 24, 45.
11. E.J. Corey, H. Cho, C. Rucker and D.H. Hua, 1981, *Tetrahedron Lett.*, 22, 3455.
12. (a) G. Stork and W.N. White, 1953, *J. Amer. Chem. Soc.*, 75, 4119; (b) G. Stork and W.N. White, 1956, *J. Amer. Chem. Soc.*, 78, 4609; (c) Review: R.M. Magid, 1980, *Tetrahedron*, 36, 1901.
13. P.A. Bartlett, K.L. McLaren and P.C. Ting, 1988, *J. Amer. Chem. Soc.*, 110, 1633.
14. E.J. Corey and S.G. Pyne, 1983, *Tetrahedron Lett.*, 24, 2821.
15. R.E. Erickson, P.J. Andrulis, J.C. Collins, M.L. Lungle and G.D. Mercer, 1969, *J. Org. Chem.*, 34, 2961.

12 (−)-Reiswigin A

K.J. Hale

12.1 Introduction

(−)-Reiswigin A is a naturally-occurring diterpene collected from the marine organism *Epipolasis reiswigi* at depths of 330 m (Figure 12.1).[1] It has a potent antiviral profile, with specific activity against Herpes simplex virus type I and murine A59 hepatitis virus. Kim and co-workers recently reported[2] a fully stereocontrolled route to (−)-reiswigin A which again highlights the utility of the Johnson orthoester Claisen rearrangement for fashioning stereocentres and stereodefined olefin systems. Below, we discuss their synthesis.

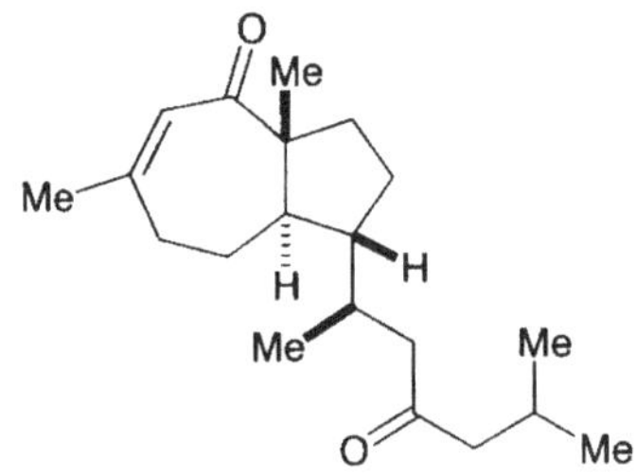

Figure 12.1   (−)-Reiswigin A.

12.2 The Kim retrosynthetic analysis of (−)-reiswigin A

The presence of an alkene in the seven-membered ring of (−)-reiswigin A provides an immediate disconnection point for this target. Kim and co-workers capitalised on this fact and retrosynthetically cleaved this ring to obtain keto-phosphonate **1** (Scheme 12.1).[3] A further clearance and protection of the reactive functionality within **1** then led to cyclopentane **2**.

While scission of the three bonds α to the ester in **2** provides a very rich bounty of stabilised anion/electrophile partnerships, a detailed evaluation of each respective forward reaction reveals that only one is tactically viable. For example, the cleavage of bond *a* in **2** would not be strategic for it would require enolate **11** to undergo methylation from its more hindered top-face (Scheme 12.2), which would be most unlikely. Such a plan would almost certainly lead to a product with incorrect stereochemistry at the newly introduced quaternary carbon centre. In similar

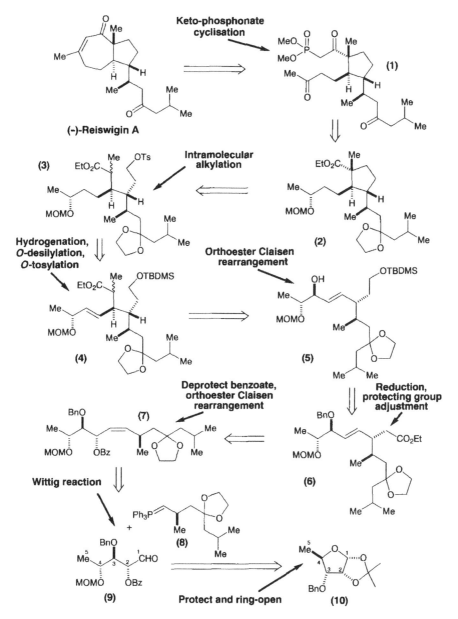

Scheme 12.1   The Kim retrosynthetic analysis of (−)-reiswigin A.

vein, heterolytic fission of bond *b* in compound **2** would produce the enolates **12** and **13** (Scheme 12.3). Careful scrutiny of these intramolecular alkylations again reveals tactical flaws. The cyclisation of **12** would

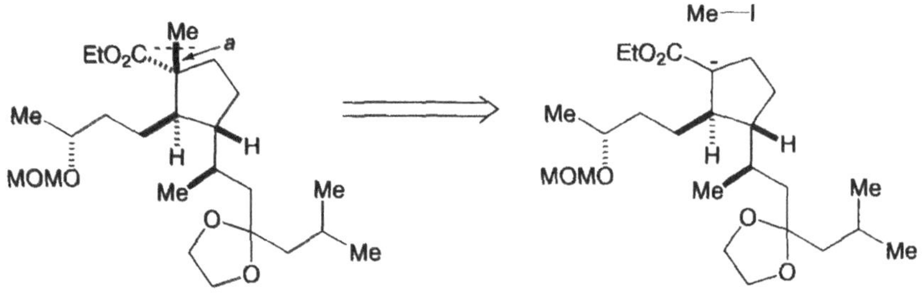

(2) (11)

Scheme 12.2.

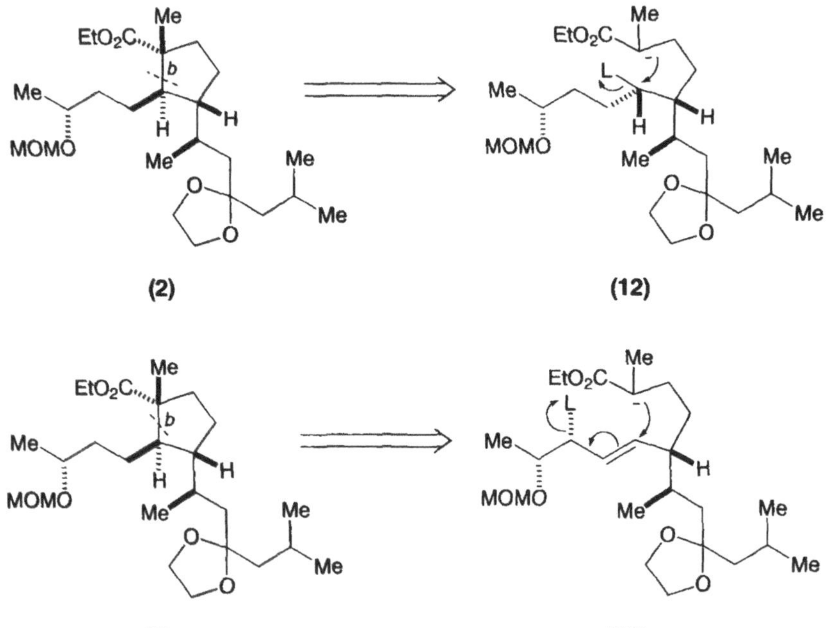

(2) (12)

(2) (13)

Scheme 12.3.

require a leaving group (L) to be displaced from a secondary carbon atom that is branched. Geometric constraints would also deter cyclisation. The internal S_N2' alkylation/hydrogenation sequence needed to convert **13** into **2** would likewise be potentially problematic. Not only would elimination of the leaving group be a possible side reaction, the cyclisation itself would afford a product with incorrect stereochemistry at the newly established ring junction. In light of these considerations, both disconnective possibilities can be discounted.

Cleavage of bond *c*, however, looks most expedient. The formation of enolate **14** seems perfectly feasible (Scheme 12.4), as does its cyclisation to **2**. The latter reaction would be predicted to be facile as the carbon now

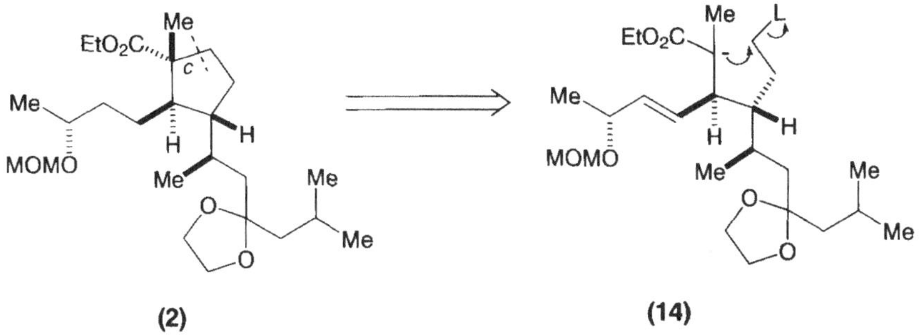

Scheme 12.4.

undergoing displacement is primary and unhindered. This disconnection has added value in as much as it is highly simplifying, it dismantling the cyclopentane ring system and one of the quaternary carbon stereocentres simultaneously. This was the stratagem eventually selected by Kim and co-workers for their synthesis (see Scheme 12.6).

The three contiguous alkyl stereocentres present within **3** (Scheme 12.1) offer a special challenge for retrosynthetic simplification. As discussed already, the successful dismantling of a target can often require the functionalisation of a previously unfunctionalised C—C bond within a carbon skeleton. With this in mind, Kim retrosynthetically positioned a C=C bond adjacent to the OMOM group in **3** and converted its *O*-tosyl ester to a more robust OTBDMS group to obtain **4**. Although, in principle, Kim had the option of retrosynthetically breaking bond *d* in **4** to obtain enolate **15** and halide **16** (Scheme 12.5), a more circumspect evaluation of this operation reveals that the forward reaction is tactically compromised by a potentially competitive S$_N$2 alkylation. Given this weakness, Kim decided to evaluate other methods for breaching the allylic C—C bond of **4**. One artifice that is always worth considering, when dismantling allylic side-chains containing two or more carbon atoms, is the reverse Johnson orthoester Claisen rearrangement.[4] This transform invariably leads to a chiral allylic alcohol of defined stereochemistry, when made at an allylic asymmetric centre. Kim's use of this tactic on **4** led to the stereodefined alcohol **5** (Scheme 12.1). After

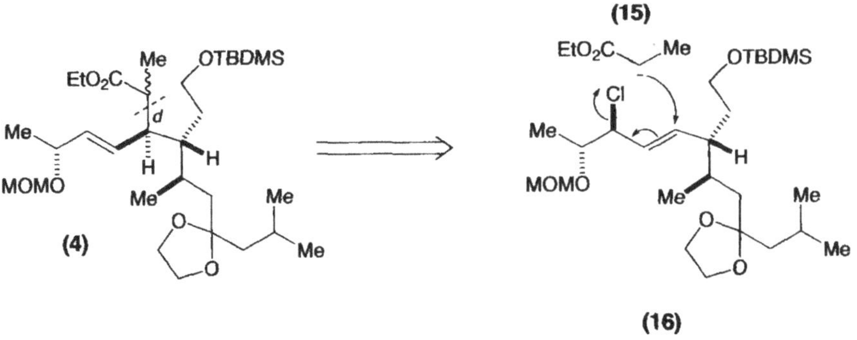

Scheme 12.5.

retrosynthetic oxidation of the CH$_2$OTBDMS group in **5** to an ester, and protection of the secondary allylic alcohol as an *O*-benzyl ether, a further opportunity for applying a Johnson orthoester transform manifested itself, which allowed compound **7** to be derived from **6**.

Despite the obvious possibilities for disengaging the allylic methyl group of **7** by retro-S$_N$2' reaction, Kim preferred to disjoin this molecule by a (*Z*)-selective Wittig reaction to obtain aldehyde **9** and ylide **8**. Since further analysis of **9** also indicated that it could potentially be reached from the readily available 5-deoxy-D-xylose derivative **10**, this was the strategy eventually settled upon.

With this as background, we now give an account of Kim's synthetic route to (−)-reiswigin A.[2]

12.3 The Kim total synthesis of (−)-reiswigin A

Compound **10** had previously been prepared from D-xylose by Kiss and co-workers.[5] It was transformed into **9** by the following sequence (Scheme 12.6). Initially, compound **10** was reacted with iodine in methanol[6] to bring about transacetalation of the 1,2-*O*-isopropylidene group. The resulting α,β mixture of methyl furanosides was then *O*-benzoylated to give **17**. Ring-opening and thioacetalation with ethanethiol, followed by protection of the remaining hydroxyl with methoxymethyl chloride next provided **18**. Finally, the aldehyde group was unmasked from **18** by dethioacetalation with iodine in aqueous acetone at low temperature. The base-mediated Wittig reaction[7] of **9** with phosphonium salt **19** (prepared according to Scheme 12.7) produced a single geometrical isomer when performed in the presence of hexamethylphosphoric triamide (HMPA). Saponification of the benzoate ester from **7** subsequently provided the key alcohol precursor needed for the first Johnson orthoester Claisen rearrangement.[4] The latter reaction was performed under mildly acidic conditions with triethyl orthoacetate, and initially produced a mixed ketene acetal that [3,3]-sigmatropically rearranged to the (*E*)-γ,δ-unsaturated ester **6** in high yield.

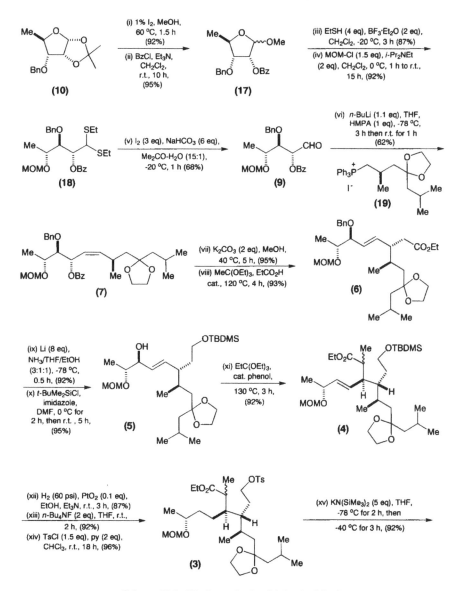

Scheme 12.6   Kim's synthesis of (−)-reiswigin A.

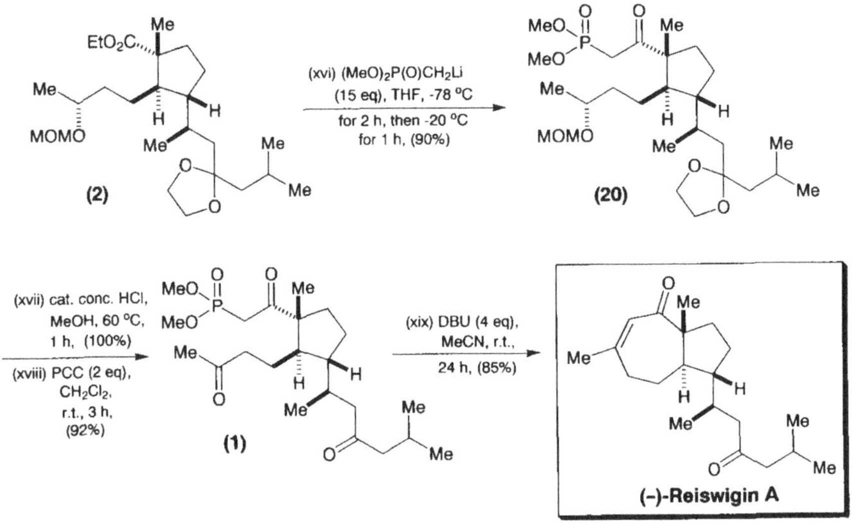

Scheme 12.6 (Continued).

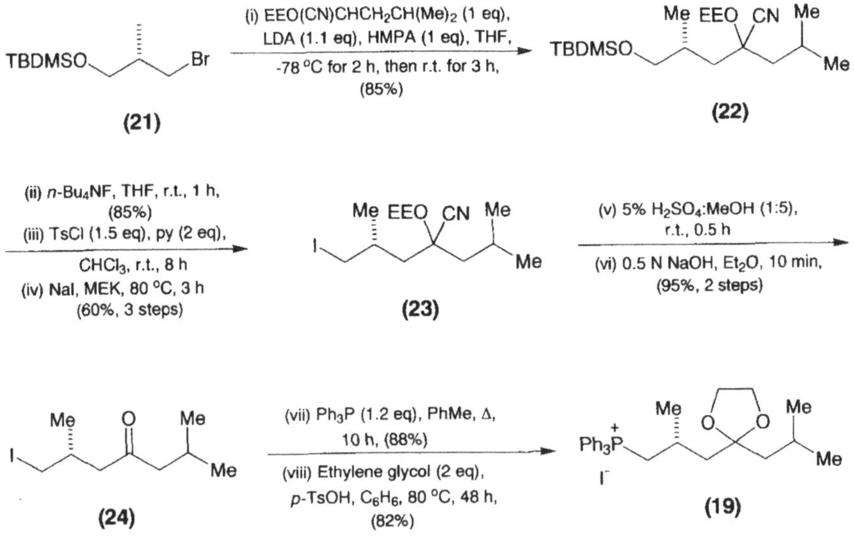

Scheme 12.7 Kim's route to the phosphonium salt **19** used in the synthesis of (−)-reiswigin A.

The next step was reduction of the ester and *O*-benzyl ether groups in **6** with lithium in liquid ammonia. Note how the allylic alcohol survived these reduction conditions. Sometimes, allylic deoxygenation can be problematic in dissolving metal reductions. It now proved necessary to block the primary hydroxyl in the product selectively as a TBDMS ether to obtain **5**. This set the stage for a second Johnson orthoester Claisen rearrangement,[4] on this occasion with triethyl orthopropionate. This reaction proceeded smoothly at 130°C over 3 h to deliver alkene **4** in 92% yield.

Hydrogenation of olefin **4**, followed by *O*-desilylation and *O*-tosylation next procured tosylate **3**, which cyclised readily when exposed to excess potassium hexamethyldisilazide. Elaboration of the ketophosphonate side chain of **20** was accomplished by condensing ester **2** with the lithiated anion of dimethyl methylphosphonate. After concurrent removal of the 1,3-dioxolane acetal and the MOM ether from **20**, the resulting secondary alcohol was oxidised with pyridinium chlorochromate (PCC) to produce methyl ketone **1**.

The last step of the synthesis was a regioselective Wadsworth–Horner–Emmons cyclisation[3] to give (−)-reiswigin A in 85% yield. Although, in principle, both a seven- and an eight-membered ring could have resulted from this cyclisation, only the product with smaller ring size was observed in practice for entropic (less loss in rotational freedom) and enthalpic (lower strain energy) reasons.

12.4   Mechanistic points of interest in the (−)-reiswigin A synthesis

The conversion of 10 into 27
The cleavage of acetals with catalytic iodine in methanol is a reaction first reported by Walter Szarek and co-workers[6] in the 1980s. Although a mechanism has never been proposed in the literature, it is conceivable that the C(1)-exo-acetal oxygen of **10** could attack iodine to form **25** and iodide ion (Scheme 12.8). Loss of the C(1)-*O*-substituent could then occur to create **26**, which could capture iodide ion and methanol to form **27** and acetone. Such a mechanism would regenerate the catalytic quantity of iodine needed to propagate the catalytic cycle.

The formation of 31 from 17
One mechanistic possibility for this thioacetalation reaction invokes furanoside ring-opening of **17**, initiated by complexation with boron trifluoride etherate (Scheme 12.9). Oxonium ion **28** could then get intercepted by the ethanethiol to produce **29**. After further complexation with BF$_3$, in the manner shown, thionium ion formation can again occur to give **30**, which can then engage in yet another nucleophilic addition with the ethanethiol to produce **31** after protonation.

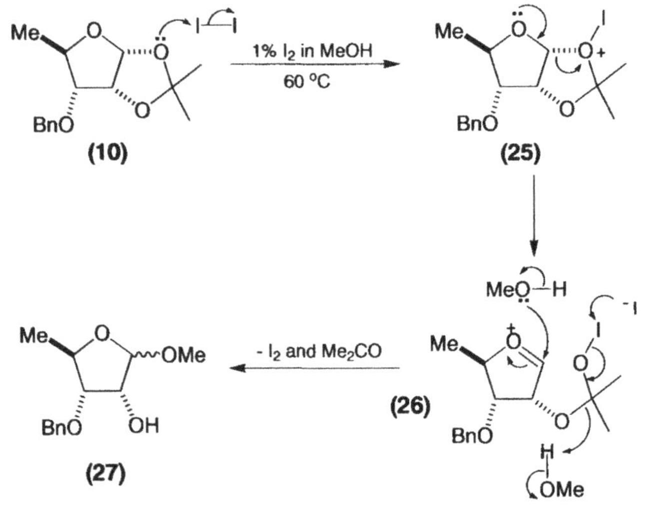

Scheme 12.8.

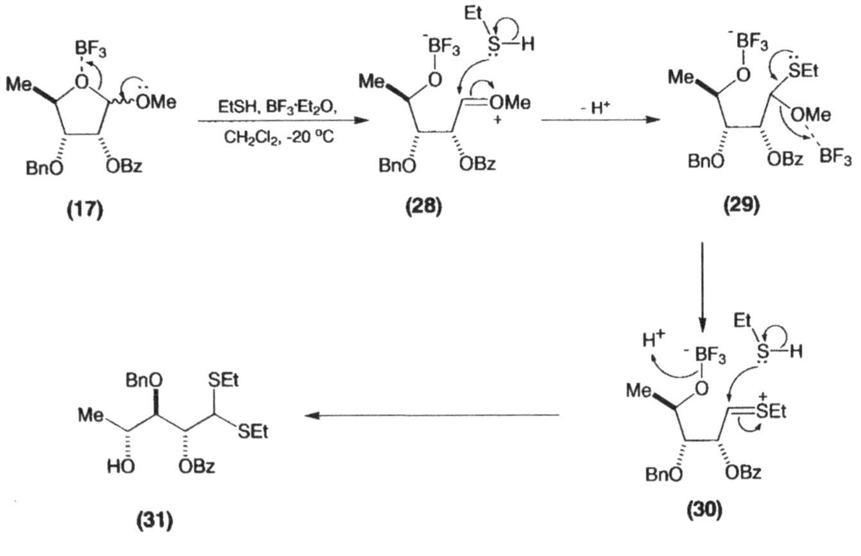

Scheme 12.9.

The conversion of 6 into 32

Two reductions occur concurrently in this dissolving-metal reaction
(Scheme 12.10). The first is a Birch reduction[8] of the benzyl ether to give a

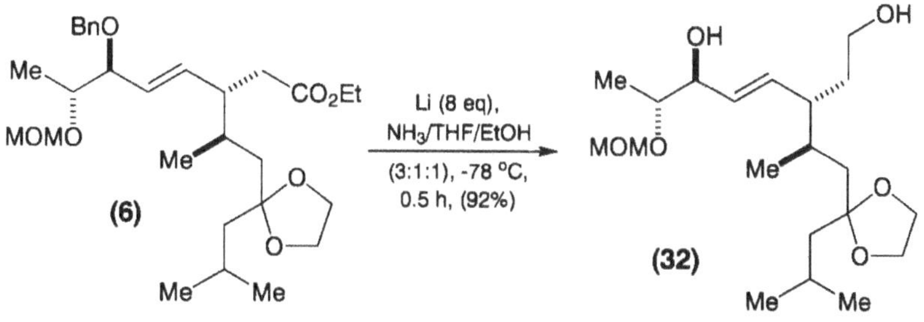

Scheme 12.10.

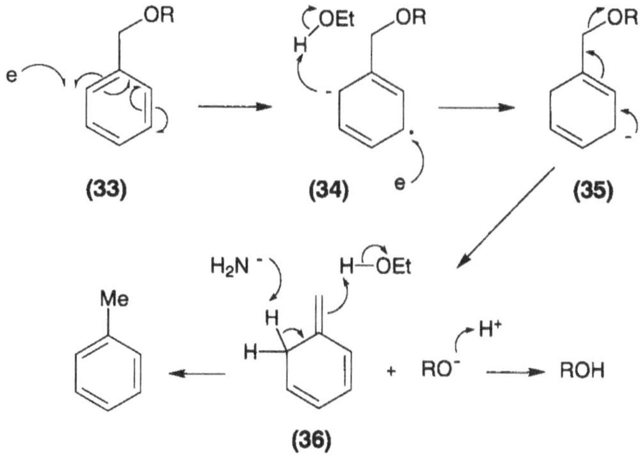

Scheme 12.11.

radical anion that protonates and captures an electron to create anion
35 (Scheme 12.11). This then eliminates to form the requisite alcohol and
36. The latter species is probably only generated transiently, it being
immediately converted to toluene.

In dissolving-metal ester reduction, the ester carbonyl is believed to ac-
cept an electron to form a radical oxyanion **37** (Scheme 12.12). Chelation
with a lithium counterion then ensues to produce a tertiary radical **38**
which then captures a second electron to become a carbanion. Protonation of **39** next yields **40**, whose fate is to collapse to aldehyde **41**.
Another multiple electron transfer/protonation sequence subsequently
yields the product alcohol **46**.

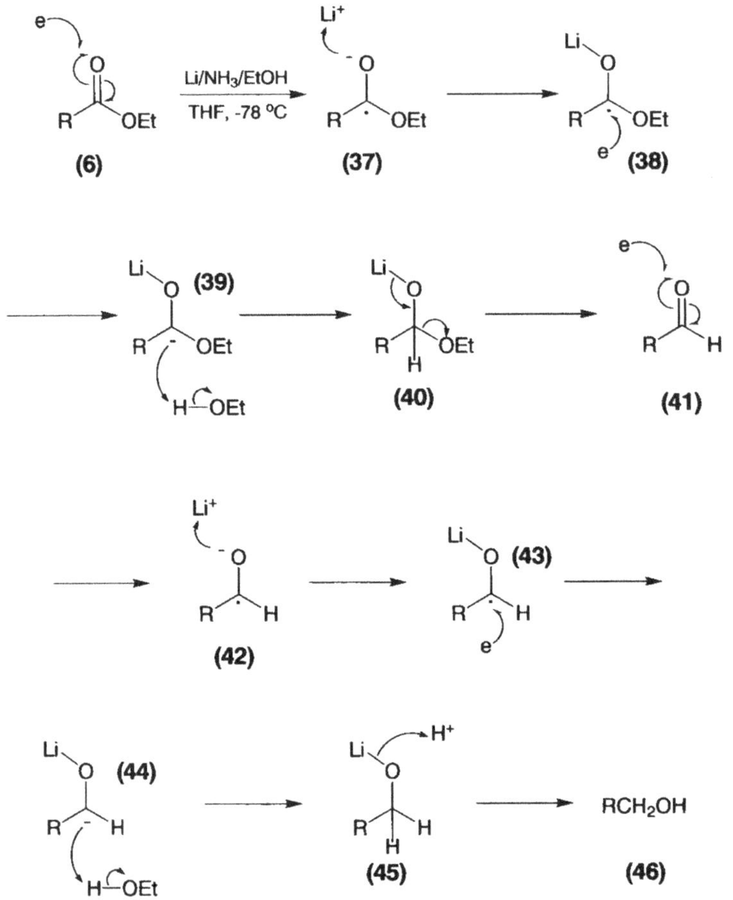

Scheme 12.12.

References

1. (a) Y. Kashman, S. Hirsh, F. Koehn and S. Cross, 1987, *Tetrahedron Lett.*, 28, 5461; (b) For the first total synthesis of (−)-reiswigin A, which established its relative and absolute stereochemistry, see: B.B. Snider and K. Yang, 1989, *Tetrahedron Lett.*, 30, 2465 and B.B. Snider and K. Yang, 1990, *J. Org. Chem.*, 55, 4392.
2. D. Kim, K.J. Shin, I.Y. Kim and S.W. Park, 1994, *Tetrahedron Lett.*, 35, 7957.
3. (a) Review: K. Becker, 1980, *Tetrahedron*, 36, 1717; (b) K.C. Nicolaou, R.A. Daines, T.K. Chakraborty and Y. Ogawa, 1988, *J. Amer. Chem. Soc.*, 110, 4685; (c) K.C. Nicolaou, S.P. Seitz and M.R. Pavia, 1982, *J. Amer. Chem. Soc.*, 104, 2030; (d) A.B. Smith, III, T.A. Rano, N. Chida, G.A. Sulikowski and J.L. Wood, 1992, *J. Amer. Chem. Soc.*, 114, 8008.
4. For another elegant example of an acyclic Claisen condensation being applied in an enantiospecific synthesis of prostaglandin $F_{2\alpha}$, see: G. Stork, T. Takahashi, I. Kawamoto and T. Suzuki, 1978, *J. Amer. Chem. Soc.*, 100, 8272.

5. J. Kiss, R. D'Souza, J. van Koeveringe and W. Arnold, 1982, *Helv. Chim. Acta*, 65, 1522.
6. W.A. Szarek, A. Zamojski, K.N. Tiwari and E.R. Isoni, 1986, *Tetrahedron Lett.*, 27, 3827.
7. Review: B.E. Maryanoff and A.B. Reitz, 1989, *Chem. Rev.*, 89, 863.
8. E.J. Reist, V.J. Bartuska and L. Goodman, 1964, *J. Org. Chem.*, 29, 3725.

13 (−)-Octalactin A

K.J. Hale

13.1 Introduction

(−)-Octalactin A is an unusual medium ring lactone of marine origin that displays powerful cytotoxic effects against murine and human tumour cell lines (Figure 13.1).[1] Its remotely functionalised eight-membered ring is significantly strained due to it adopting a boat-chair conformation with unfavourable *cis*-lactone geometry. The latter feature causes significant intramolecular dipolar repulsion. Given this demanding topography, and the potential sensitivity of this molecule to acids, bases, and strong nucleophiles, it poses quite a significant challenge for total synthesis. In this chapter we will examine how Buszek addressed the issues of remote stereocontrol and medium-ring lactone construction in his synthetic route to (−)-octalactin A.[2]

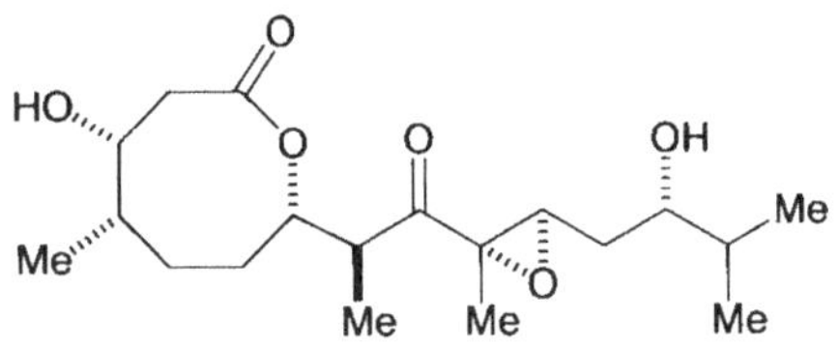

Figure 13.1   (−)-Octalactin A.

13.2   The Buszek retrosynthetic analysis of (−)-octalactin A

When attempting to devise a retrosynthetic plan for (−)-octalactin A, it is important to remember that its significantly strained lactone ring will readily fracture if the molecule is exposed to powerful nucleophiles or strong bases. Two features that confer particular lability upon it are the *cis*-ester unit, which can split apart by nucleophilic addition–elimination, and the β-hydroxy ester motif, which can cleave through base-mediated retro-aldol reaction. Given these synthetic snares, it would probably be prudent to defer lactonisation until late in a total synthesis, for, once installed, the lactone ring will clearly jeopardise all further synthetic manoeuvring, and will thus compromise the entire success of the venture.

As a countermeasure to the possibility of retro-aldol cleavage, Buszek took the precaution of retrosynthetically protecting the C(3)-hydroxyl group as a *p*-methoxybenzyl (PMB) ether[3] (Scheme 13.1). This would allow the C(3)-OH to be unmasked at the final step of the synthesis with dichlorodicyanoquinone (DDQ).[3] Although in principle, the C(13)-hydroxyl could also be protected as a *p*-methoxybenzyl (PMB) ether, *tert*-butyldiphenyl silyl (TBDPS) protection was eventually selected, primarily because further retrosynthetic analysis indicated that the C(10)–C(16) sector could be derived from a known starting material which contained this protecting group. Furthermore, this mode of protection was equally well suited to the task at hand. The next stage of the analysis was retrosynthetic reduction of the C(9)-ketone to obtain a 2,3-epoxy alcohol, and clearance of the epoxide moiety. A hydroxyl-directed epoxidation[4] on 1 was envisaged for stereospecifically installing the latter motif. Nucleophilic epoxidation of an enone was not deemed appropriate here owing to the presence of the base-sensitive lactone.

Because a ketone resides at C(9) in (−)-octalactin A, the issue of controlling alcohol stereochemistry at this site in 1 only takes on relevance from the perspective of determining the stereochemical outcome of the hydroxyl-directed epoxidation. Buszek's plan for introducing the allylic hydroxyl of 1 was to perform a Nozaki–Kishi–Takai Ni(II)–Cr(II) coupling[5] between vinyl iodide 3 and aldehyde 2. Although it is difficult to see why any stereocontrol should be observed for this reaction, if poor stereoselectivity was encountered, Buszek had the option of oxidising the allylic alcohol mixture, and performing a stereocontrolled reduction of the enone with one of the Corey–Bakshi–Shibata (CBS) reagents.[6] Some examples of successful CBS enone reductions performed in the presence of ester and lactone functionality are shown in Scheme 13.2.

Thus, Buszek's plan for unifying 2 with 3 (Scheme 13.1) had a potential fall-back position, which is always desirable when planning a synthetic route.

After retrosynthetic aldehyde reduction and protection to obtain 4, Buszek next dismantled the eight-membered ring system. Recognising that formation of a saturated medium ring is often accompanied by a significant loss in rotational freedom and entropy, Buszek decided to reduce significantly the degrees of freedom in his cyclisation-precursor, during his first generation synthesis of 4 (Scheme 13.3). This was accomplished by positioning a *cis*-olefin unit within the chain. It was also expected that this device would assist in bringing the reactive hydroxy and thiopyridyl ester termini into close proximity. Whilst this tactic did eventually prove very successful at closing the lactone ring of 11, it later transpired that the double bond of this molecule could not be reduced

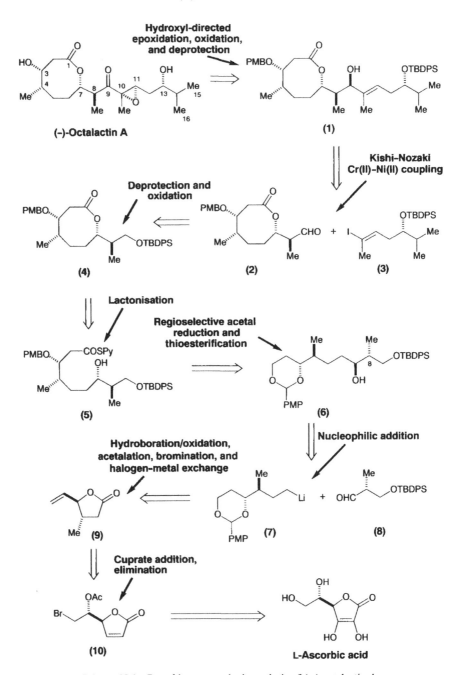

Scheme 13.1 Buszek's retrosynthetic analysis of (−)-octalactin A.

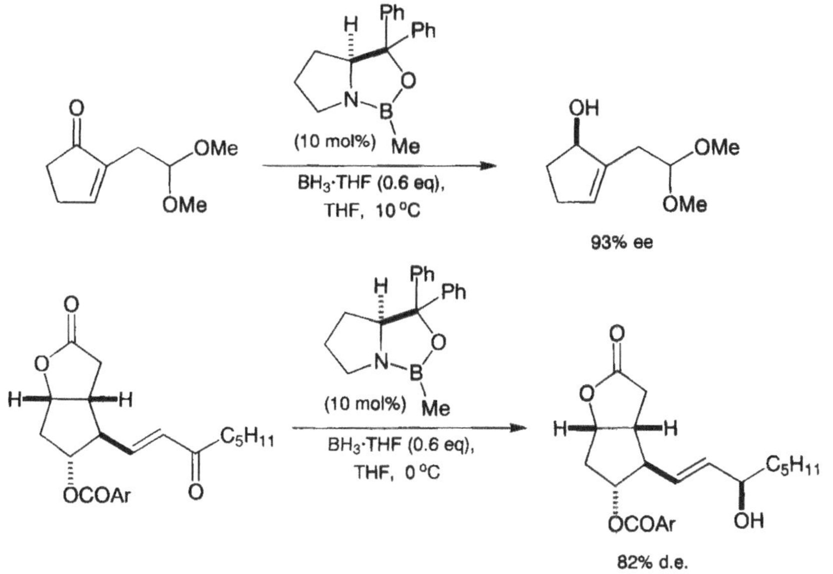

Scheme 13.2   Examples of successful Corey–Bakshi–Shibata enone reductions. $Ar = p\text{-}PhC_6H_4^-$.

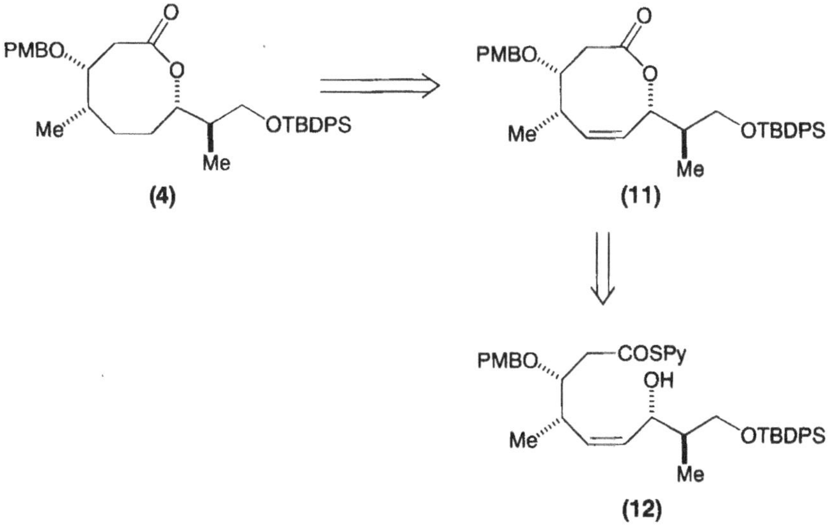

Scheme 13.3   Buzsek's first generation approach to lactone **4**.

under any of the conditions studied, which caused Buszek to rethink his approach to **4**.

In his second generation analysis of **4** (which is presented in full in Scheme 13.1), Buszek decided to attempt the much more difficult ring-closure of hydroxy-thiopyridyl ester **5**. This is a particularly challenging reaction to try, not only because of severe product ring strain, but also because adverse entropy effects conspire seriously to disfavour cyclisation. Notwithstanding these adverse considerations, however, there were some design features present within **5** which could potentially override these obstacles and favour ring-closure. The foremost of these was the Corey–Nicolaou thiopyridyl ester[7] that was incorporated to activate the C(1)-carbonyl. The main reason why these esters are so successful at closing many medium- and large-ring lactones stems from their ability to engage often in intramolecular hydrogen bonding with the terminal hydroxyl that is present within the chain. Not only does this greatly assist in internally bringing the two key termini into one another's vicinity, it also enhances the leaving-group potential of the thiopyridyl heterocycle, whilst significantly increasing the nucleophilicity of the attacking hydroxyl. Thus, when all three effects act synergistically, they can combine to favour markedly a lactonisation process that would ordinarily be predicted to be disfavoured.

Buszek also intended to add a 'soft acid' silver(I) salt[8] to the reaction mixture. Its role would be to complex with the 'softly basic' sulfur atom of the thioester moiety, and thereby assist in this group's departure. Thus, it will be recognised that every possible artifice was employed by Buszek to maximise his chances of forming **4** from **5**.

Having settled upon the intermediacy of ring-precursor **5** for his synthesis, Buszek next retrosynthetically converted it to **6**, and incised the latter at its midpoint to obtain **7** and **8**. Although high stereoselectivity would not be expected for this coupling, this strategy did guarantee the C(8)-methyl group stereochemistry for **6**. Furthermore, even if contra-selectivity did arise, it would not be an insurmountable problem to rectify, since recourse could be made to a Mitsunobu inversion tactic. Buszek therefore directed his attention at establishing the *anti*-relationship between the two stereocentres present within **7**. For this, he envisaged performing a stereoselective conjugate addition reaction on **10** with Me$_2$CuLi. In this, the oligomeric cuprate would preferentially add to the less-hindered underside of the butenolide to furnish **9** after reductive elimination. The latter could then be transformed into **7**. Significantly, compound **10** showed a stereochemical match with C(4) and C(5) in L-ascorbic acid. Buszek therefore deemed it an ideal starting material from where to commence a total synthesis of (−)-octalactin A.

13.3 Buszek's total synthesis of (−)-octalactin A

The key butenolide needed by Buszek, for his synthesis of (−)-octalactin A, had already been prepared by Godefroi and Chittenden and co-workers some years earlier (Scheme 13.4).[9] Their pathway to **10** provides it in excellent overall yield, in three straightforward steps from L-ascorbic acid. The first step entails stereospecific hydrogenation of the double bond to obtain L-gulono-1,4-lactone **13**. Reduction occurs exclusively from the sterically less-encumbered α face of the alkene in this reaction. Tetraol **13** was then converted to the 2,6-dibromide **14** with HBr and acetic anhydride in acetic acid. Selective dehalogenation of **14** with sodium bisulfite finally procured **10**. It is likely that the electron-withdrawing effect of the carbonyl in **14** preferentially weakens the adjacent C—Br bond, making this halide more susceptible to reductive elimination under these reaction conditions.

Although lower-order cuprate reagents will often engage in displacement reactions with alkyl halides, such reactions are usually slow. They are generally much less facile than 1,4-addition reactions to α,β-unsaturated enones or enoates. The latter processes are particularly facile when trimethylsilyl chloride is employed as an additive. It was Corey and Boaz[10] who first recognised the accelerating effect of trimethylsilyl chloride on cuprate addition reactions to α,β-unsaturated carbonyls. Buszek therefore capitalised on Corey's earlier observations in his reaction of **10** with lithium dimethylcuprate to obtain **15**.

In order to deoxygenate **15** at C(5), and subsequently reoxygenate at C(6), a reductive elimination was attempted with zinc dust in aqueous acetic acid to obtain lactone **9**. The lactone was then reduced with 9-borobicyclo[3.3.1]nonane (9-BBN), and the double bond simultaneously hydroborated and oxidised, to obtain triol **16**. *p*-Methoxybenzylidenation and bromination then furnished **17**. The latter underwent a rapid halogen–metal exchange reaction when treated with *t*-BuLi in ether and pentane at −90°C to produce the organolithium reagent **7**. This reacted readily with aldehyde **8** to provide a 1:1 mixture of **18** and **6** which were separated by chromatography. The undesired alcohol **18** was recycled towards **6** by a Mitsunobu inversion tactic. This entailed reacting **18** with triphenylphosphine, diethylazodicarboxylate, and *p*-nitrobenzoic acid to obtain the inverted nitrobenzoate ester, and then converting this ester to alcohol **6** by transesterification with potassium methoxide. After *O*-acetylation, reductive acetal cleavage with sodium cyanoborohydride and trimethylsilyl chloride furnished **19**. A two-step protocol was used to oxidise **19** to acid **20**. The latter was then converted to the hydroxythiopyridyl ester **5**, and this heated with silver tetrafluoroborate in boiling toluene for 96 h, to instigate the desired lactonisation to **4**. Lactone **4** underwent smooth *O*-desilylation, when exposed to tetra-*n*-butylammonium fluoride in acid buffered tetrahydrofuran (THF). The buffering helped prevent lactone ring-opening with this slightly basic reagent. The resulting alcohol was oxidised to aldehyde **2** with the Dess–Martin periodinane.

A Kishi–Nozaki–Takai Ni(II)–Cr(II) coupling[5] was next implemented between vinyl iodide **3** and aldehyde **2**. This afforded a separable 1.5:1 mixture of the two diastereoisomers **1** and **21**. Of particular note here was the superb chemoselectivity of this coupling, the labile lactones in **2**, **1** and **21** all surviving intact from this critical C—C bond forming step.

Having successfully knitted together the entire carbon skeleton of (−)-octalactin A, all that now remained was to form the epoxy ketone and deprotect. Pleasingly, compound **1** underwent hydroxyl-directed epoxidation with complete stereocontrol to give **22**. Oxidation to the ketone, and protecting group cleavage under standard conditions, then completed this fine total synthesis of (−)-octalactin A.

It also proved possible to convert compound **21** into the target by the sequence shown in Scheme 13.4. This pathway was, however, much less satisfactory owing to its epoxidation step being nonstereoselective. The latter reaction afforded a 1:1 mixture of epoxides that required chromatographic separation. Nevertheless, further useful quantities of the natural product were obtained.

The Buszek (−)-octalactin A synthesis is notable for its useage of the Corey–Nicolaou thiopyridyl ester[7] protocol for *saturated* eight-membered lactone construction. Prior to this synthesis, no eight-membered lactone ring had ever been prepared in high yield from the cyclisation of a saturated hydroxy carboxylic acid precursor. This reaction therefore broke important new ground in the arena of complex natural product synthesis.

13.4 Items of mechanistic interest in the Buszek (−)-octalactin A synthesis

The conversion of 20 into 4
Thiopyridyl ester **5** was prepared using the methodology of Mukaiyama *et al.*[11] This entailed reacting hydroxyacid **20** with 2,2′-dipyridyldisulfide and triphenylphosphine in CH_2Cl_2. Mechanistically, this reaction commences with a nucleophilic attack of the phosphine on the disulfide (Scheme 13.5). This produces a reactive thiopyridylphosphonium salt that reacts with **20** to give the acylphosphonium ion **24**. The latter then gets

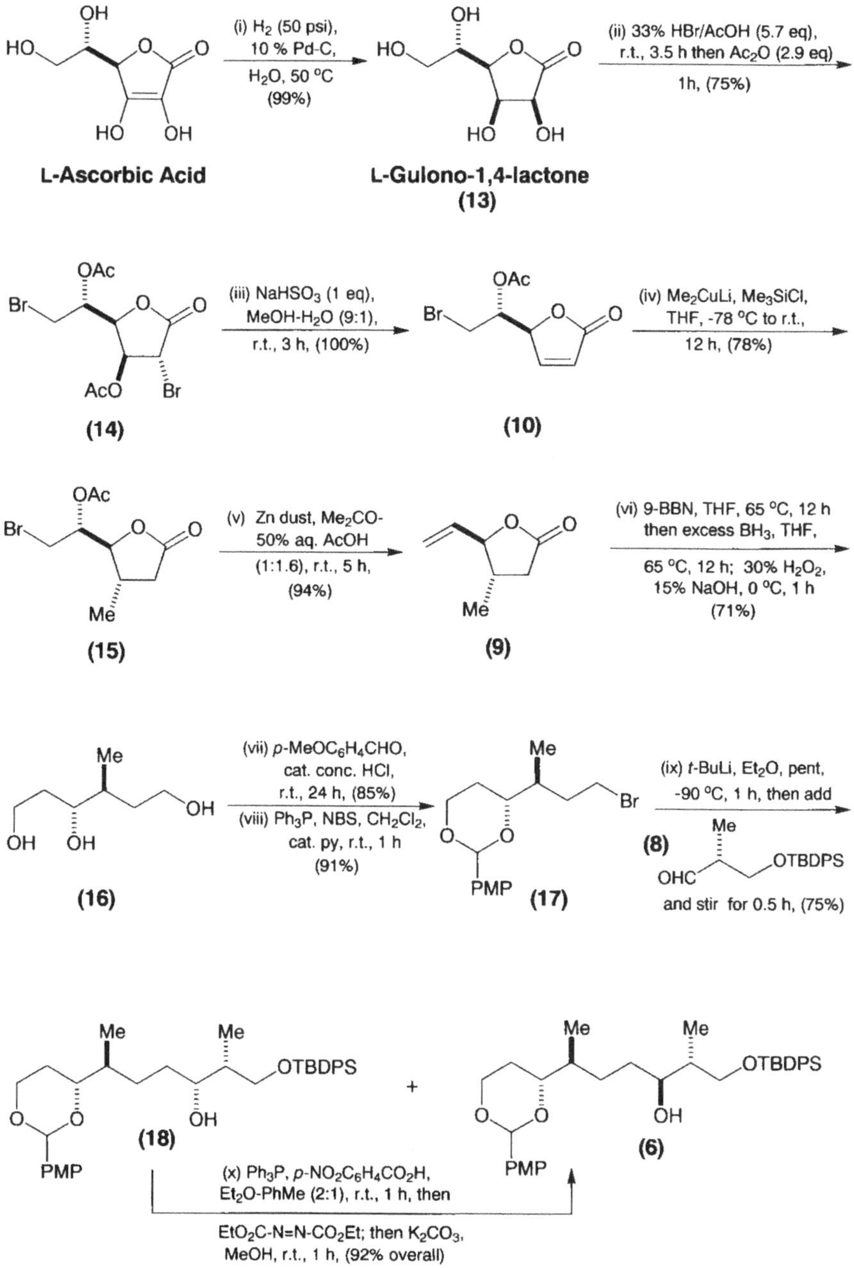

Scheme 13.4   Buszek's synthesis of (−)-octalactin A.

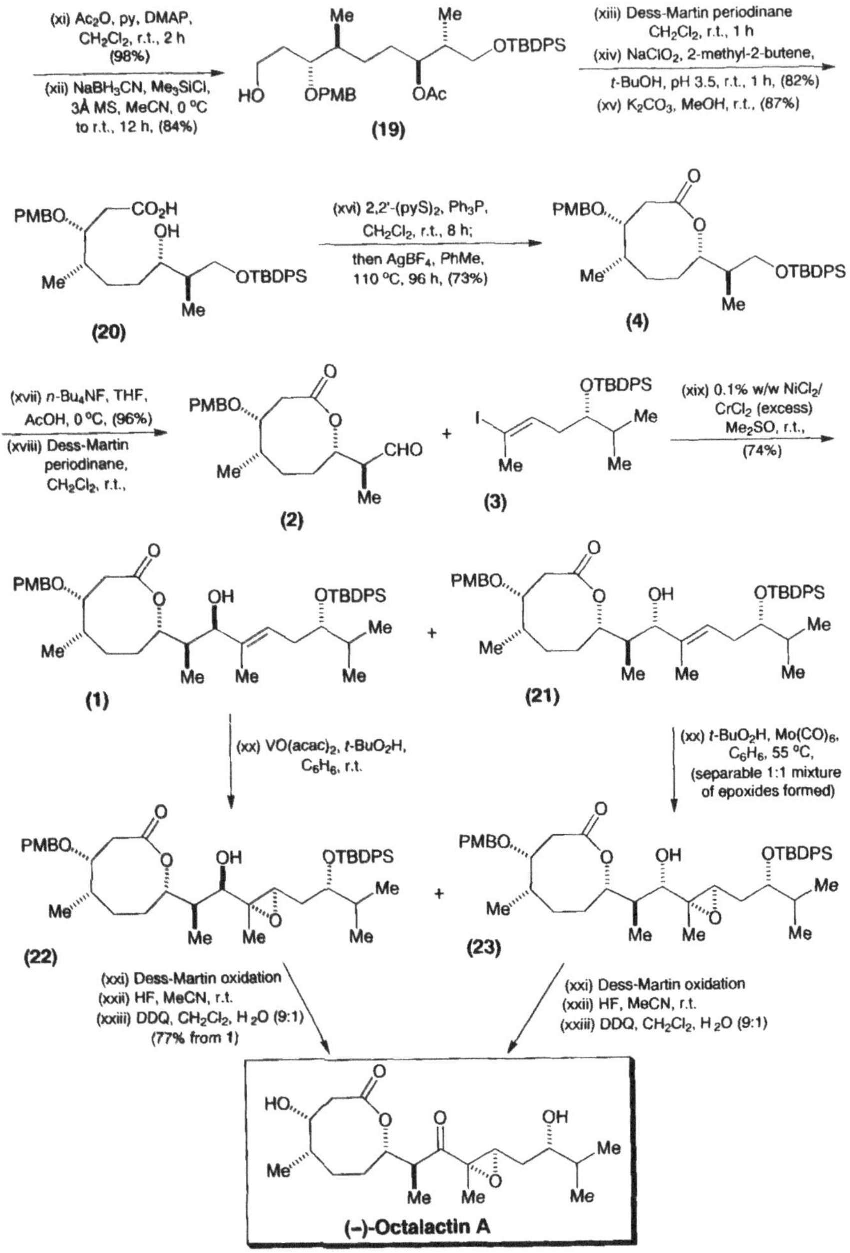

(xi) Ac₂O, py, DMAP, CH₂Cl₂, r.t., 2 h (98%)

(xii) NaBH₃CN, Me₃SiCl, 3Å MS, MeCN, 0 °C to r.t., 12 h, (84%)

(19)

(xiii) Dess-Martin periodinane CH₂Cl₂, r.t., 1 h
(xiv) NaClO₂, 2-methyl-2-butene, t-BuOH, pH 3.5, r.t., 1 h, (82%)
(xv) K₂CO₃, MeOH, r.t., (87%)

(20)

(xvi) 2,2'-(pyS)₂, Ph₃P, CH₂Cl₂, r.t., 8 h; then AgBF₄, PhMe, 110 °C, 96 h, (73%)

(4)

(xvii) n-Bu₄NF, THF, AcOH, 0 °C, (96%)
(xviii) Dess-Martin periodinane, CH₂Cl₂, r.t.,

(2)   +   **(3)**

(xix) 0.1% w/w NiCl₂/CrCl₂ (excess) Me₂SO, r.t., (74%)

(1)   +   **(21)**

(xx) VO(acac)₂, t-BuO₂H, C₆H₆, r.t.

(xx) t-BuO₂H, Mo(CO)₆, C₆H₆, 55 °C, (separable 1:1 mixture of epoxides formed)

(22)   +   **(23)**

(xxi) Dess-Martin oxidation
(xxii) HF, MeCN, r.t.
(xxiii) DDQ, CH₂Cl₂, H₂O (9:1) (77% from 1)

(xxi) Dess-Martin oxidation
(xxii) HF, MeCN, r.t.
(xxiii) DDQ, CH₂Cl₂, H₂O (9:1)

(−)-Octalactin A

Scheme 13.4 (Continued).

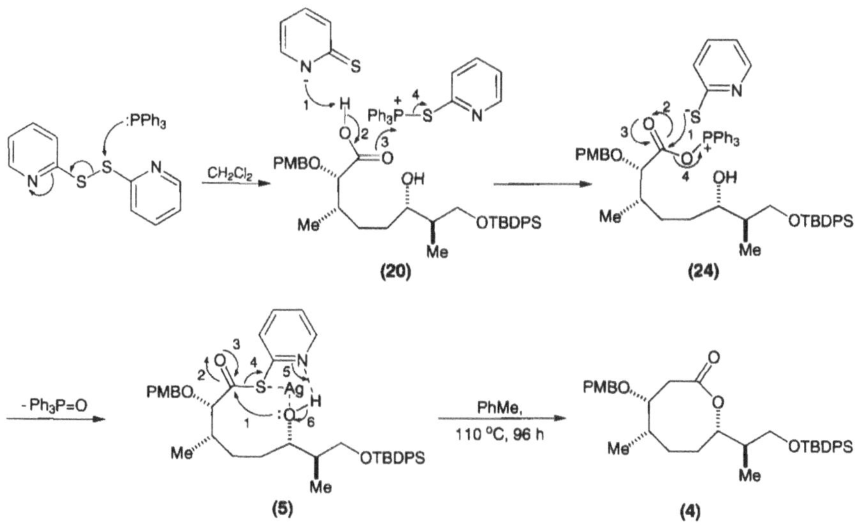

Scheme 13.5.

attacked by the liberated thiopyridone anion to form the thiopyridyl ester **5**. It is likely that complexation between the silver salt, the thioester group of **5**, and the tethered hydroxyl mutually assists in bringing these reactive groups close together, as well as in simultaneously activating the thioester unit for intramolecular nucleophilic attack. Intramolecular hydrogen-bonding between the thiopyridyl nitrogen and the hydroxyl would further serve to reinforce this metal-templating effect. The cyclisation itself almost certainly proceeds by a nucleophilic addition–elimination mechanism.

The conversion of 3 into 1/21

This is an example of the Kishi–Nozaki–Takai reaction. In this process a catalytic quantity of a nickel(II) salt is reduced by excess chromous(II) chloride to a nickel(0) species (Scheme 13.6). This then oxidatively adds to the vinyl iodide to form an alkenylnickel(II) intermediate **25**, with retention of olefin geometry. Transmetallation is then presumed to occur with a Cr(III) salt to give **26**, which engages in nucleophilic addition to the aldehyde carbonyl. Clearly, the transmetallation step regenerates the nickel(II) catalyst for re-entry into the catalytic cycle. The great beauty of these vinylchromium(III) reagents lies in their high reactivity towards aldehydes. Ketones, esters, nitriles, and epoxides are usually inert to such reagents.

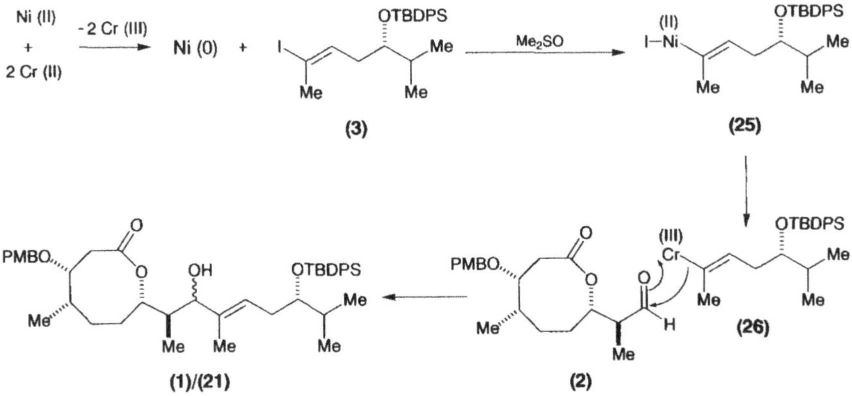

Scheme 13.6.

References

1. D.M. Tapiolas, M. Roman, W. Fenical, T.J. Stout and J. Clardy, 1991, *J. Amer. Chem. Soc.*, 113, 4682.

2. (a) K.R. Buszek, N. Sato and Y. Jeong, 1994, *J. Amer. Chem. Soc.*, 116, 5511; (b) K.R. Buszek and Y. Jeong, 1995, *Tetrahedron Lett.*, 36, 7189.

3. K. Horita, T. Yoshioka, T. Tanaka, Y. Oikawa and O. Yonemitsu, 1986, *Tetrahedron*, 42, 3021.

4. (a) K.B. Sharpless and R.C. Michaelson, 1973, *J. Amer. Chem. Soc.*, 95, 6136; (b) Review: K.B. Sharpless and T.R. Verhoeven, 1979, *Aldrichimica Acta*, 12, 63.

5. (a) H. Jin, J. Uenishi, W.J. Christ and Y. Kishi, 1986, *J. Amer. Chem. Soc.*, 108, 5644; (b) K. Takai, M. Tagashira, T. Kuroda, K. Oshima, K. Utimoto and H. Nozaki, 1986, *J. Amer. Chem. Soc.*, 108, 6048; (c) K. Takai, K. Kimura, T. Kuroda, T. Hiyama and H. Nozaki, 1983, *Tetrahedron Lett.*, 24, 5281; (d) Review: P. Cintas, 1992, *Synthesis*, 248.

6. (a) E.J. Corey, R.K. Bakshi and S. Shibata, 1987, *J. Amer. Chem. Soc.*, 109, 5551; (b) E.J. Corey, R.K. Bakshi, S. Shibata, C.-P. Chen and V.K. Singh, 1987, *J. Amer. Chem. Soc.*, 109, 7925; (c) E.J. Corey, C.-P. Chen and G.A. Reichard, 1989, *Tetrahedron Lett.*, 30, 5547; (d) E.J. Corey and J.O. Link, 1989, *Tetrahedron Lett.*, 30, 6275.

7. (a) E.J. Corey and K.C. Nicolaou, 1974, *J. Amer. Chem. Soc.*, 96, 5614; (b) K.C. Nicolaou, D.G. McGarry, P.K. Somers, B.H. Kim, W.W. Ogilvie, G. Yiannikouros, C.V.C. Prasad, C.A. Veale and R.R. Hark, 1990, *J. Amer. Chem. Soc.*, 112, 6263.

8. H. Gerlach and A. Thalman, 1974, *Helv. Chim. Acta*, 57, 2661.

9. J.A.J.M. Vekemans, G.A.M. Franken, C.W.M. Dapperens, E.F. Godefroi and G.J.F. Chittenden, 1988, *J. Org. Chem.*, 53, 627.

10. (a) E.J. Corey and N.W. Boaz, 1985, *Tetrahedron Lett.*, 26, 6015 and 6019; (b) K.J. Hale, S. Manaviazar and R. Tupprasoot, 1997, *Tetrahedron*, 53, 16365.

11. T. Mukaiyama, R. Matsueda and M. Suzuki, 1970, *Tetrahedron Lett.*, 1901.

14 (−)-ACRL toxin I

K.J. Hale

14.1 Introduction

A particularly harmful fungus of the genus *Alternaria citri* has recently been identified as producing a mixture of toxins that cause considerable damage to lemon and lime crops throughout the world.[1] The major component of this mixture is the (−)-ACRL toxin I. In 1991 Lichtenhaler and co-workers successfully achieved the first total synthesis of the (−)-ACRL toxin I from D-glucose[2] by a route we will now discuss.

14.2 The Lichtenhaler retrosynthetic analysis of (−)-ACRL toxin I

Lichtenhaler's protecting group strategy[2] for the (−)-ACRL toxin I masked the C(13)-hydroxyl as a *tert*-butyldimethyl silyl (TBDMS) ether and tied up the C(7)- and C(9)-hydroxyls as an *O*-isopropylidene acetal. These blocking groups were selected for 1 (Scheme 14.1), partly for their ease of removal at the final stages of the synthesis by acid hydrolysis, and partly for their known compatibility with carbanion chemistry. An imaginary lactone hydrolysis was next performed to obtain the corresponding β keto acid from 1. The latter was then retrosynthetically protected to obtain the 1,3-dioxinone 2. This operation actuated the C(4)—C(5) bond for scission, yielding the aldehyde 3 and extended dienolate 4 as subtargets. In the forward direction, a vinylogous aldol reaction was contemplated for connecting these two fragments together. Although enolate 4 could potentially add at either its α or γ positions, past literature precedent[3] suggested that such an addition would occur preferentially at the more sterically accessible γ site of the enolate. The prospects for controlling the stereochemical outcome of this reaction were far less certain, however. Based on Still and Schneider's earlier findings that organolithium reagents add to α-unsubstituted β-alkoxy-aldehydes with low levels of stereocontrol,[4] one could expect poor stereoselectivity to result in the addition of 4 to 3. However, one could reach an opposite conclusion if one was aware of the recent work of Vandewalle *et al.* on bryostatin 1.[5] They showed that a *single* diastereoisomer was formed in the aldol addition of lithium enolate 11 to aldehyde 12 (Scheme 14.2). From a predictive perspective, therefore, one could argue that this latter result shows that Lichtenhaler's proposed

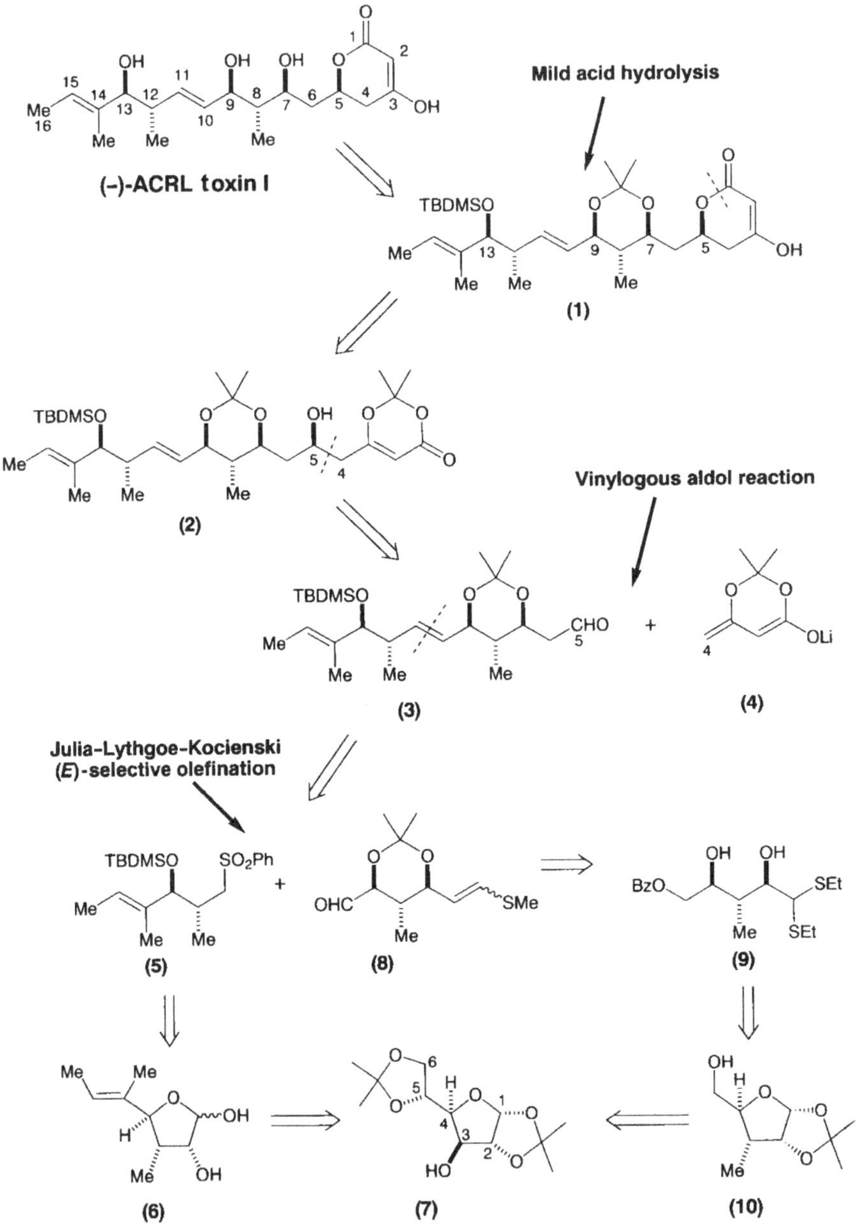

Scheme 14.1   The Lichtenhaler retrosynthetic analysis of (−)-ACRL toxin I.

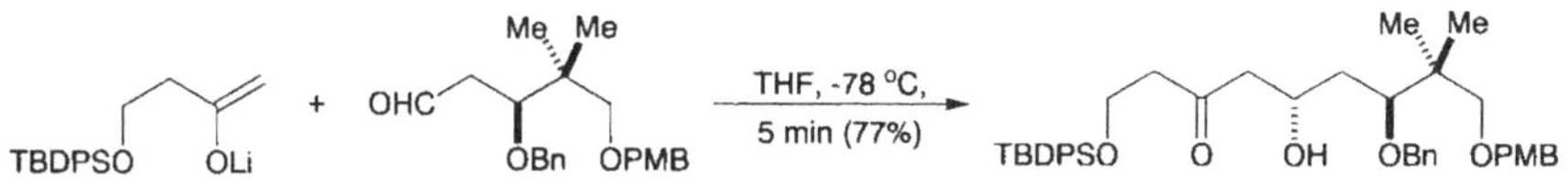

Scheme 14.2   Aldol addition of a lithium enolate to an aldehyde.

vinylogous aldol reaction was a perfectly sound and logical disconnection *a priori.* At this point, it is perhaps worthwhile reflecting on Professor Sir Jack Baldwin's sage words with regard to synthetic planning:

> Detailed mechanistic analysis can on occasion stifle the possibility of the unexpected, since frequently processes which appear mechanistically unsound or without precedent are discounted.

J.E. Baldwin[6]

To continue with Lichtenhaler's analysis of the (−)-ACRL toxin I, his next disconnection was made across the C(10)—C(11) olefinic bond of 3. Because this alkene is (*E*)-1,2-disubstituted, and it has a nearby alkyl substituent located at C(12), Lichtenhaler reasoned that a Julia–Lythgoe–Kocienski olefination[7] between 5 and 8 would be strategic for establishing this alkenic linkage. Such olefinations are usually very successful at providing (*E*)-1,2-disubstituted olefins from complex fragments.

Having settled upon the intermediacy of 5 and 8, Lichtenhaler next attempted to equate their chirality with members of the monosaccharide pool. Once again, D-glucose emerged as an ideal starting material for the preparation of both pieces. If one conceptually superimposes the structure of D-glucose onto aldehyde 8, a chirality congruence immediately becomes apparent between carbons 7 and 9 in 8, and carbons 2 and 4 in D-glucose. To convert D-glucose into 8 would require a methyl group being introduced at C(3) with inversion of configuration, and the C(6)-hydroxymethyl group being oxidatively excised. One very useful derivative of D-glucose, which allows C(3) to be readily modified, is 1,2;5,6-di-*O*-isopropylidene-D-glucose. Its 5,6-*O*-isopropylidene group is also more acid labile that its 1,2-acetal. This could therefore allow a periodate cleavage/reduction tactic to be used to remove the C(6)-hydroxymethyl as required. The use of 7 as a chiral starting material could therefore provide a potentially very rapid entry into 8.

In like fashion, if one overlays the structures of 5 and D-glucose, a chirality match immediately registers between the allylic hydroxyl of the subtarget and the C(4)-hydroxyl of D-glucose. As with 8, the C(3)-hydroxyl of D-glucose would again need to be converted to an inverted C(3)-methyl group. However, for 5, the controlled degradative removal of C(1) would also now be a requirement. A stereoselective two carbon-olefination at C(5) would additionally be needed. Yet again, 1,2;5,6-di-O-isopropylidene-D-glucose appears an excellent candidate for the preparation of this fragment.

In the coming section, we will detail how Lichtenhaler put such retro-synthetic thinking into practice in his synthesis of the (−)-ACRL toxin I.

14.3 Lichtenhaler's total synthesis of the (−)-ACRL toxin I

The nucleophilic displacement of 1,2;5,6-di-O-isopropylidene-3-O-tosyl-D-glucose with lithium dimethylcuprate would probably not be a viable reaction for introducing the methyl stereocentre of **15**. Such a displacement would probably suffer from severe dipolar repulsions in the S_N2 transition state, and would also be sterically impeded by the 1,2-O-isopropylidene group.[8] O-Desulfonylation would also be a likely side reaction. A less direct but more assured route to **15** was therefore sought. Previous work by Rosenthal and Sprinzl[9] had shown that **15** could be prepared from **7** in four relatively easy steps. The reactions needed are oxidation of the C(3)-alcohol to the corresponding ketone, Wittig ole-fination of the ketone with $Ph_3P{=}CH_2$, stereospecific hydrogenation of alkene **14**, and graded acid hydrolysis of the 5,6-O-isopropylidene group from the hydrogenation product (Scheme 14.3).

The oxidative removal of C(6) from **15** thereafter proved very straightforward, it being readily accomplished with aq. $NaIO_4$ in EtOH. Rather than isolating the product aldehyde, it was reduced *in situ* with sodium borohydride to obtain **10**. After protection of the primary alcohol in **10** as a benzoate ester, an acid-catalysed acetal ring-opening was performed with ethanethiol to obtain thioacetal **9** in good yield. Acetal exchange with 2,2-dimethoxypropane, and thioacetal hydrolysis with methyl iodide in aqueous acetone, next followed to provide aldehyde **16**. This was then subjected to one-carbon Wittig homologation and base-hydrolysis to access thioenol ether **17**. Swern oxidation completed the synthetic sequence to fragment **8**.

To gain access to methyl ketone **18** (Scheme 14.4), it was necessary to deoxygenate diol **15** at C(6) and oxidise the resulting deoxy sugar at C(5). Deoxygenation was achieved in two steps through selective tosylation of the C(6) hydroxyl, and sulfonate reduction with lithium aluminium hydride. The C(5) ketone was obtained by oxidation of the alcohol with pyridinium chlorochromate on alumina.[10] Wittig olefination of **18** with ethyl(triphenyl)phosphonium bromide and *n*-butyllithium next provided a 4.1:1 mixture of Z/E olefin isomers that readily isomerised when irradiated with ultraviolet (UV) light and phenyldisulfide. The double

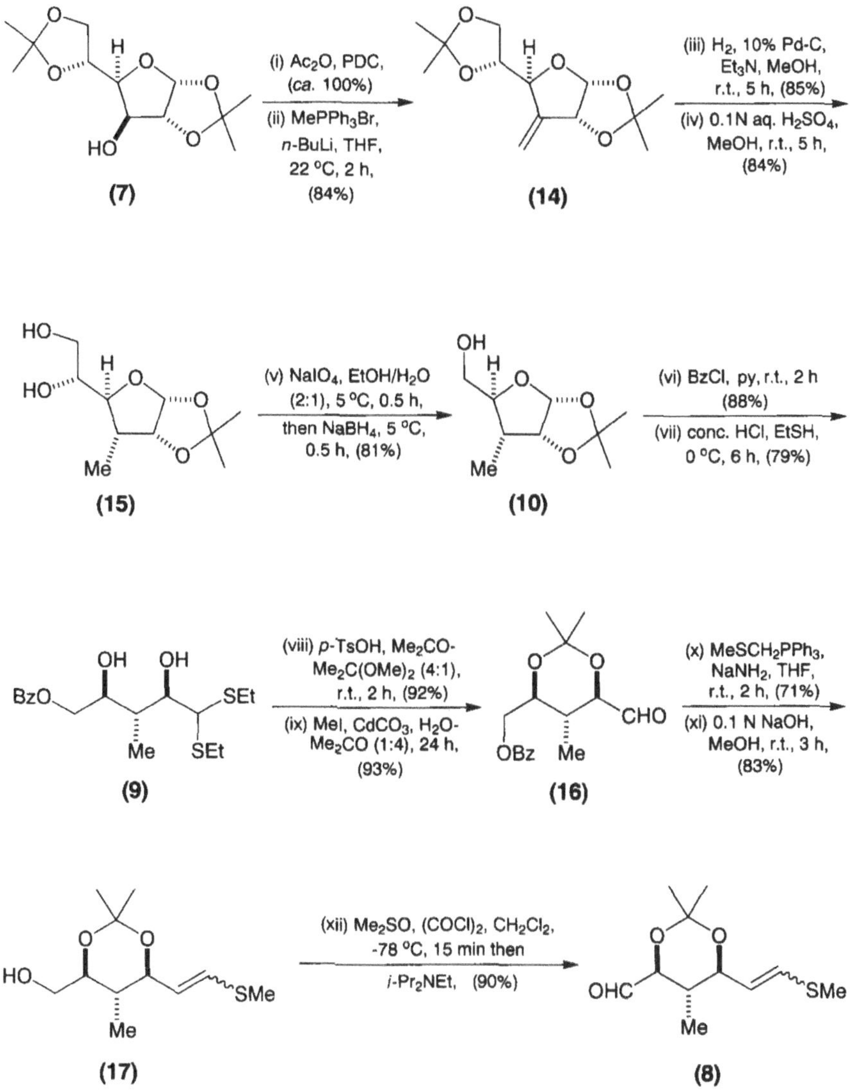

Scheme 14.3 Lichtenhaler's route to sub-target **8** in his (−)-ACRL toxin I synthesis.

bond isomerisation process afforded a 6.6:1 mixture of geometrical isomers, enriched in the desired (E)-olefin **19**. The latter was isolated pure in 67% yield. The 1,2-O-isopropylidene group was now detached from **19** by acid hydrolysis. Oxidative degradation of the 1,2-diol in **6** followed next. This furnished an aldehyde that was readily reduced by sodium borohydride. Selective tosylation and silylation of **20** delivered alkene **21**.

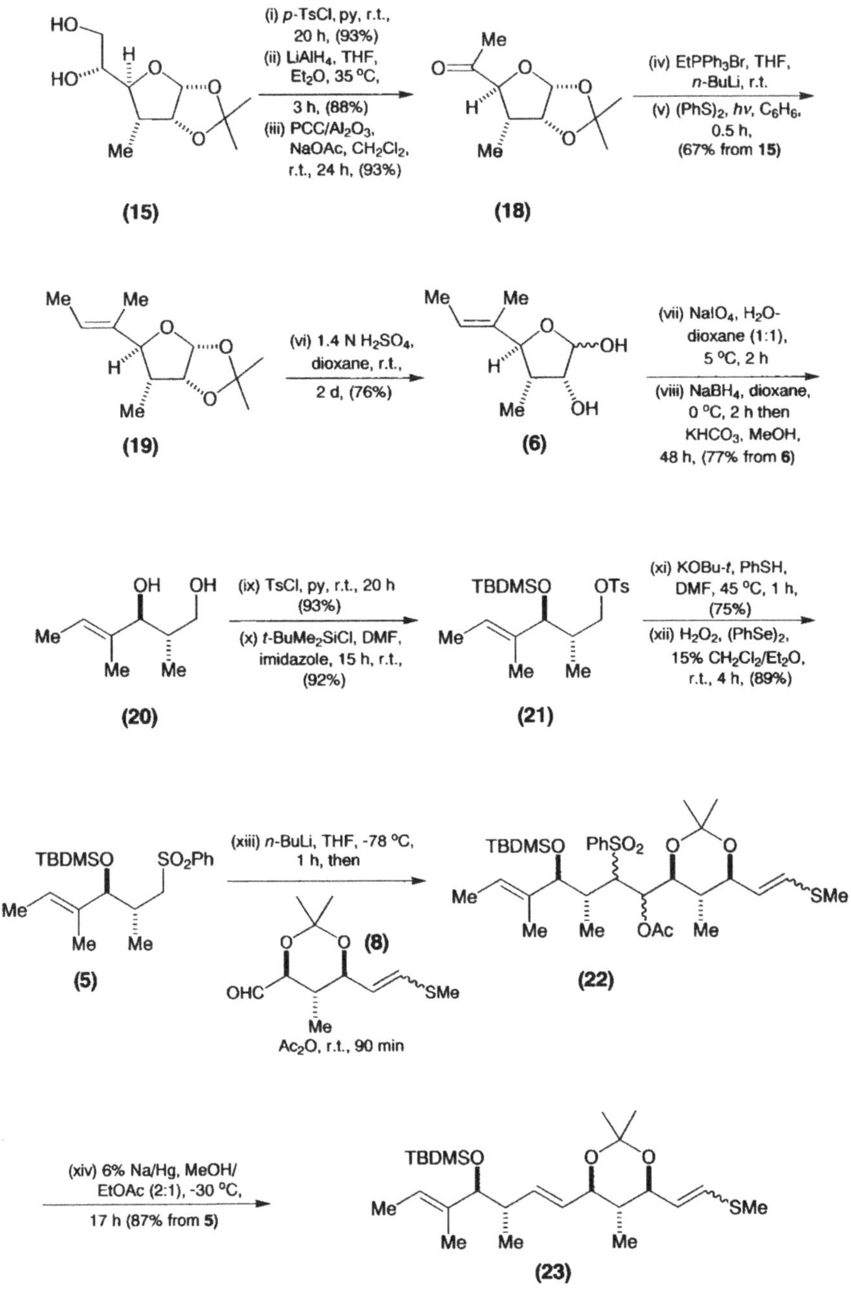

Scheme 14.4 Lichtenhaler's synthesis of (−)-ACRL toxin I.

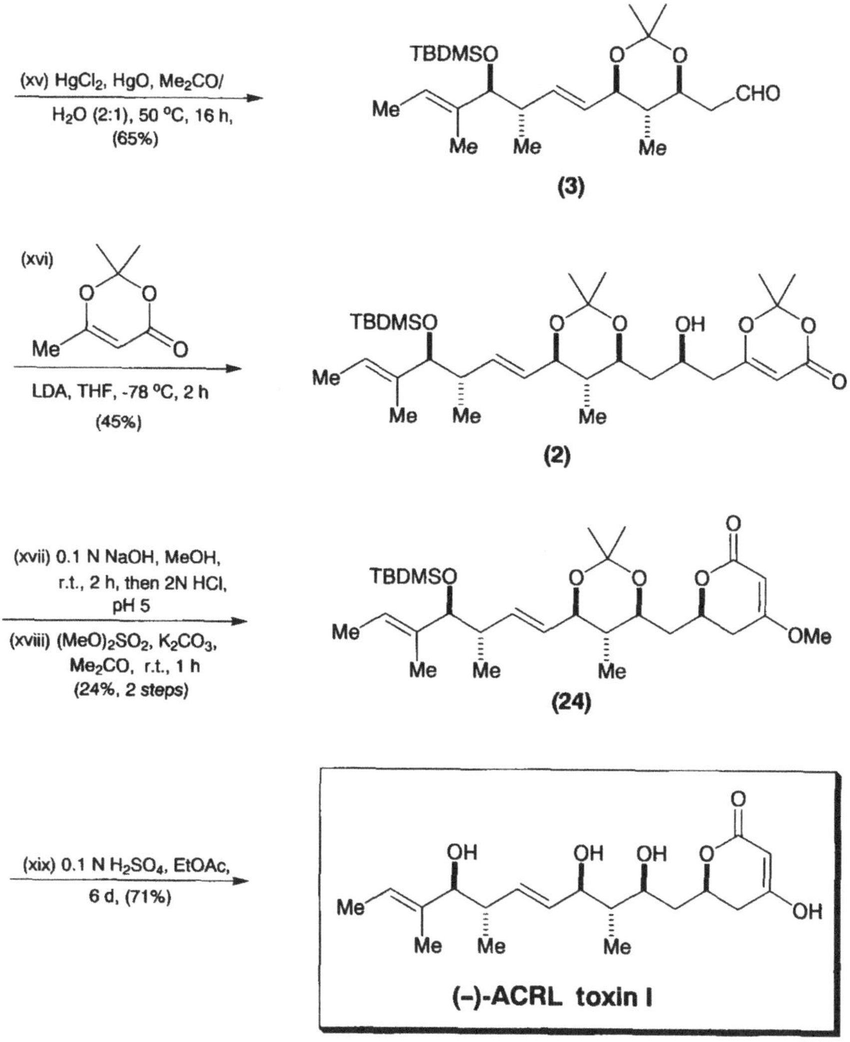

(xv) HgCl₂, HgO, Me₂CO/
H₂O (2:1), 50 °C, 16 h,
(65%)

(3)

(xvi)

LDA, THF, -78 °C, 2 h
(45%)

(2)

(xvii) 0.1 N NaOH, MeOH,
r.t., 2 h, then 2N HCl,
pH 5

(xviii) (MeO)₂SO₂, K₂CO₃,
Me₂CO, r.t., 1 h
(24%, 2 steps)

(24)

(xix) 0.1 N H₂SO₄, EtOAc,
6 d, (71%)

(–)-ACRL toxin I

Scheme 14.4   (Continued).

This paved the way for an S$_N$2 displacement with potassium thiophen-oxide, and sulfide oxidation to obtain phenylsulfone **5**. Metallation of **5** was accomplished at $-78°C$ with *n*-butyllithium in tetrahydrofuran (THF). The resulting α-phenylsulfonyl anion added readily to aldehyde **8** to afford a mixture of β-alkoxysulfones that underwent smooth *O*-acylation with acetic anhydride. The mixture of β-acetoxysulfones so formed was then reacted with 6% Na/Hg amalgam in methanol and ethyl acetate at −30°C to produce a 4:1 mixture of E/Z alkenes, from which the pure (E)-isomer **23** was separated in 70% yield.

After mercury(II)-assisted hydrolysis of the thioenol ether, aldehyde **3** was obtained. This was then subjected to the critical vinylogous aldol reaction needed to complete the carbon backbone of the natural product. The latter process furnished a 3.5:1 mixture of the γ to α addition products. The stereoselectivity observed in the installation of the C(5)-hydroxyl (natural product numbering) was only 2:1. Fortunately, the predominant isomer was the desired product **2**. In retrospect, it can be seen that the level of selectivity attained conformed to the predictions of the Still model.[4]

After purification, the dimethyl 1,3-dioxinone unit of **2** was hydrolysed with mild base to furnish the β-keto acid. Acidification then brought about lactonisation to the enolised β-keto lactone **1**. To facilitate work-up and isolation, the latter was converted to the vinylogous ester **24** with potassium carbonate and dimethylsulfate in acetone. Mild acid hydrolysis of **24** with 0.1 N H_2SO_4 in EtOAc eventually furnished (−)-ACRL toxin 1 in good yield after six days, completing this very elegant and beautifully crafted synthesis.

14.4   Items of mechanistic interest in the Lichtenhaler synthesis of the (−)-ACRL toxin I

The isomerisation of 25 to 19
Irradiation of phenyldisulfide cleaves the weak S—S bond to give a pair of thiophenyl radicals. One of these then adds to the less-substituted olefinic carbon of **25** to generate the tertiary alkyl radical **26** (Scheme 14.5). Bond rotation then ensues to give the more stable rotamer **27** in which there is minimal steric repulsion between the C(7)-methyl and the tetrahydrofuran framework. Elimination of the thiophenyl radical from **27** finalises the isomerisation to alkene **19**.

The Julia–Lythgoe–Kocienski olefination reaction[7]
between 5 and 8 to obtain alkene 23
The first step in this multistage reaction is the nucleophilic addition of sulfone anion **28** to aldehyde **8** (Scheme 14.6). This produces a β-alkoxysulfone intermediate **29** which is trapped with acetic anhydride. The resulting β acetoxysulfone mixture **22** is then subjected to a reductive elimination with Na/Hg amalgam to obtain alkene **23**. The tendency of Julia–Lythgoe–Kocienski olefinations to provide (E)-1,2-disubstituted alkenes can be rationalised if one assumes that an α-acyloxy anion is formed in the reduction step, and that this anion is sufficiently long-lived to allow the lowest energy conformation to be adopted. Clearly, this will

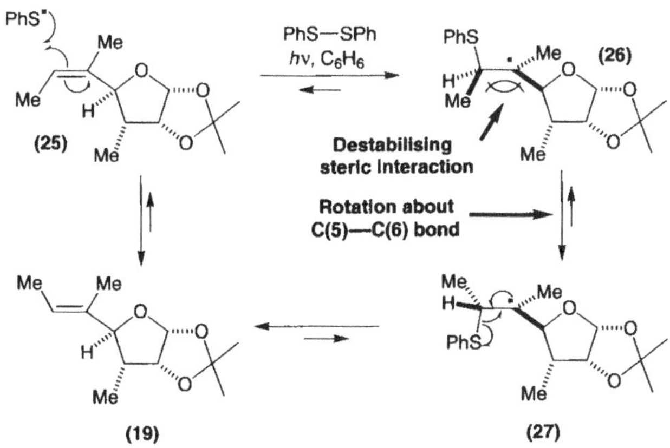

Scheme 14.5 Radical mediated photoisomerisation of an olefin.

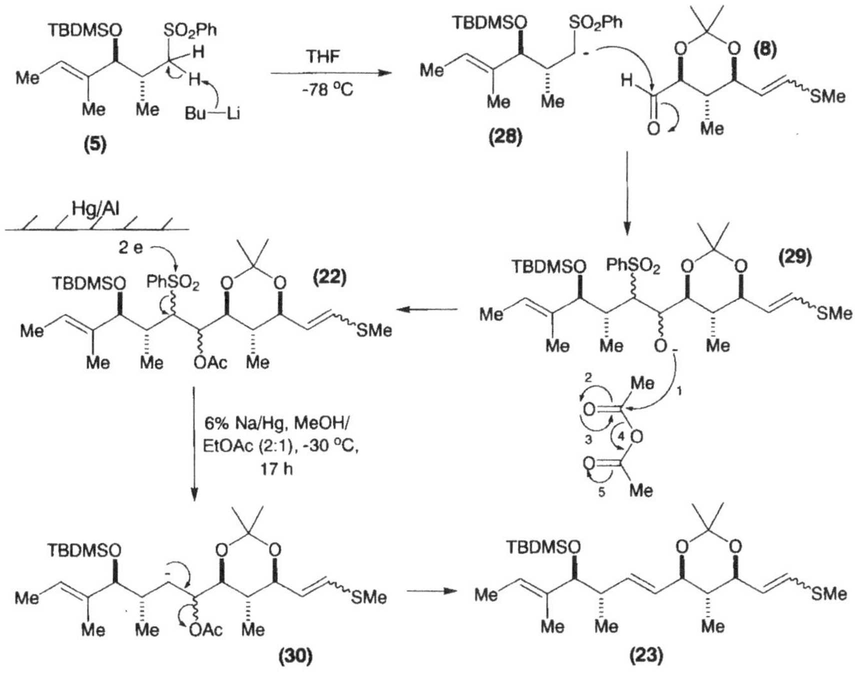

Scheme 14.6 The Julia–Lythgoe–Kocienski olefination method.

place the bulkiest R groups *trans* to one another, to minimise steric repulsions. Such an intermediate can then undergo E1*cb* elimination to give the desired (*E*)-1,2-disubstituted alkene as the predominant product.

14.5   Epilogue

Lichtenhaler's synthesis of the (−)-ACRL toxin I shows good levels of stereocontrol except for the installation of one stereocentre, the C(5)-hydroxyl. Its longest linear sequence of only 20 steps, for installing six asymmetric centres and two stereodefined double bonds, corresponds to 2.5 steps per stereogenic unit; a benchmark that many of us would be satisfied with when completing the *first* total synthesis of a natural product as complex as this one. Lichtenhaler's route provides yet another enchanting illustration of how one can cleverly exploit the topography of a monosaccharide framework to predictably solve difficult issues of regioselective functionalisation and stereochemical control.

References

1. (a) J.M. Gardner, Y. Kono, J.H. Tatum, Y. Suzuki and S. Takeuchi, 1985, *Agric. Biol. Chem.*, 49, 1235; (b) J.M. Gardner, Y. Kono, J.H. Tatum, Y. Suzuki and S. Takeuchi, 1985, *Phytochemistry*, 24, 2861; (c) Y. Kono, J.M. Gardner, J.H. Tatum, K. Kobayashi, Y. Suzuki, S. Takeuchi and T. Sakurai, 1986, *Phytochemistry*, 25, 69.
2. F.W. Lichtenhaler, J. Dinges and Y. Fukuda, 1991, *Angew. Chem. Int. Ed. Engl.*, 30, 1339.
3. (a) H. Hagiwara, K. Kimura and H. Uda, 1986, *J. Chem. Soc., Chem. Commun.*, 860; (b) W.S. Johnson, A.B. Kelson and J.D. Elliot, 1988, *Tetrahedron Lett.*, 29, 3757.
4. W.C. Still and J.A. Schneider, 1980, *Tetrahedron Lett.*, 21, 1035.
5. J. De Brabander, K. Vanhessche and M. Vandewalle, 1991, *Tetrahedron Lett.*, 32, 2821.
6. J.E. Baldwin, 1990, in *Challenges in Synthetic Organic Chemistry* by T. Mukaiyama, Oxford University Press, Oxford.
7. (a) M. Julia and J.M. Paris, 1973, *Tetrahedron Lett.*, 4833; (b) P.J. Kocienski, B. Lythgoe and S. Ruston, 1978, *J. Chem. Soc. Perkin Trans. 1*, 829; (c) P.J. Kocienski, B. Lythgoe and I. Waterhouse, 1980, *J. Chem. Soc. Perkin Trans. 1*, 1045; (d) Review: N.S. Simpkins, 1993, *Sulphones in Organic Synthesis*, Pergamon Press, Oxford, Chapter 7, 254; (e) Review: P.J. Kocienski, 1985, *Phosphorus and Sulfur*, 24, 97 and P.J. Kocienski, 1981, *Chem. Ind. (London)*, 548.
8. A.C. Richardson, 1969, *Carbohydr. Res.*, 10, 395.
9. A. Rosenthal and M. Sprinzl, 1969, *Can. J. Chem.*, 47, 3941.
10. Y.S. Cheng, W.L. Liu and S. Chen, 1980, *Synthesis*, 223.

15 (+)-Gabosine E

K.J. Hale

15.1 Introduction

(+)-Gabosine E belongs to an interesting family of pseudo-sugars that weakly inhibit the biosynthesis of cholesterol.[1] (+)-Gabosine E has recently been synthesised by Barry Lygo's group at Salford University, by a route which utilises D-ribose as the chiral starting material.[2]

15.2 The Lygo retrosynthetic analysis of (+)-gabosine E

The fact that (+)-gabosine E possesses three contiguous hydroxy stereocentres, buttressed within a cyclohexenone framework, suggests that there might be some tactical advantage gained if its synthesis was attempted from a monosaccharide precursor. For example, L-lyxopyranose has a perfect chirality match with (+)-gabosine E at its C(2), C(3) and C(4) hydroxyls, and, on paper at any rate, looks an excellent chiral building block from which to commence a total synthesis (Figure 15.1). Unfortunately, however, L-lyxose is an exceedingly rare sugar in nature, it being prohibitively expensive to buy. This makes its use unattractive for such a significant synthetic undertaking. Overlaying the structures of D-gulopyranose and (+)-gabosine E, likewise, reveals a good stereochemical correlation between both molecules. Now, however, it is not sugar availability that is the issue, it is the problem of finding a reasonably short pathway to the target from that starting sugar. Indeed, all the paper syntheses that we devised for obtaining (+)-gabosine E from readily-available 1,4-D-gulonolactone proved excessively lengthy, primarily

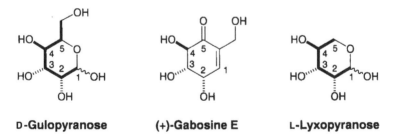

D-Gulopyranose   **(+)-Gabosine E**   **L-Lyxopyranose**

Figure 15.1 D-Gulopyranose, (+)-gabosine E and L-lyxopyranose.

because extensive protecting group adjustments were required. Another sugar that shows good stereochemical overlap with (+)-gabosine E is D-ribose. Its C(2) and C(3) hydroxyls both have appropriate absolute stereochemistry for this target, but its C(4) hydroxyl needs to be inverted. Because a ketone sits at C(5) in (+)-gabosine E (sugar numbering), there is the possibility of adjusting the C(4) stereochemistry of a D-ribose-derived precursor by base-mediated epimerisation. An awareness of this possibility led Lygo to select enone **1** as a subtarget (Scheme 15.1). A simple E1*cb* elimination and an epimerisation were considered appropriate for generating **1** from **2**. In turn, the presence of a β-hydroxy ketone within **2** opened up the possibility of introducing this unit by an intramolecular nitrile oxide–alkene cycloaddition reaction.[3] In order to apply this retrosynthetic transform, one has to conceptually convert the target β-hydroxy ketone to a β-hydroxyimine (Scheme 15.2). A bond then needs to be established between the imine nitrogen and the β-hydroxyl to form a cyclic oxime. This then fashions the isoxazole that will emerge from the required intramolecular nitrile oxide cycloaddition reaction. Implementation of a 1,3-dipolar cycloreversion then procures the appropriate nitrile oxide and alkene needed for the forward synthesis. The former should, in turn, be derivable from an aldehyde oxime.

Lygo's application of this transform intramolecularly to **2** led to the selection of alkenyl oxime **4** as an intermediate (Scheme 15.1).

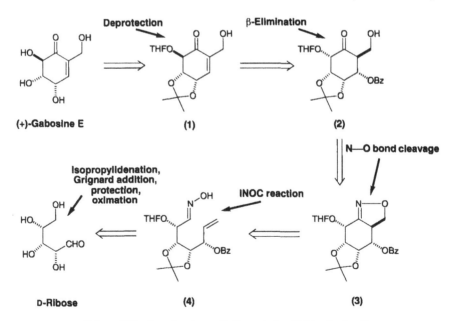

Scheme 15.1   Lygo's retrosynthetic analysis of (+)-gabosine E.

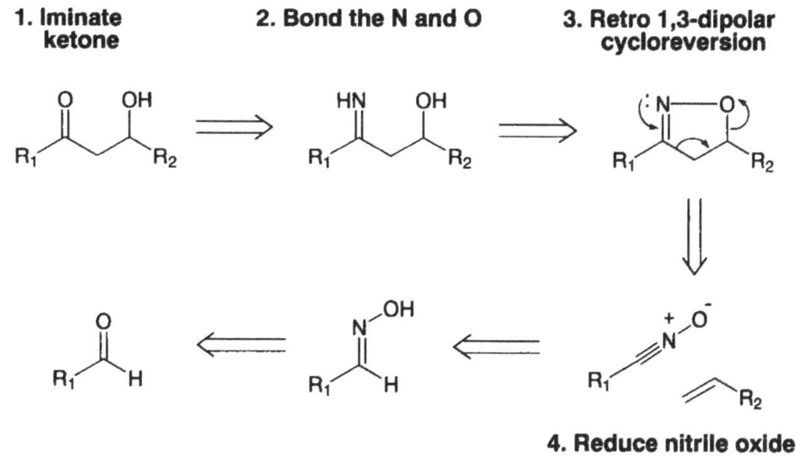

1. Iminate ketone   **2. Bond the N and O**   **3. Retro 1,3-dipolar cycloreversion**

4. Reduce nitrile oxide

Scheme 15.2   The intramolecular nitrile oxide cycloaddition retrosynthetic pathway to adducts.

Conceptually, the key steps needed for converting D-ribose into **4** are protection of the C(2) and C(3) hydroxyls with a base-stable blocking group, a two-carbon homologation of C(1) with a suitable vinyl carbanion, and oxime formation at C(5). In the coming section, we will show how Lygo put all this planning into practice.

15.3   The Lygo synthesis of (+)-gabosine E

Lygo's synthesis begins with the selective O-isopropylidenation of D-ribose[4] with dimethoxypropane and camphorsulfonic acid in acetone (Scheme 15.3). This afforded the mono-isopropylidene derivative **5**. Treatment of **5** with excess vinylmagnesium bromide initially caused hemiacetal ring-opening to the open-chain alkoxyaldehyde, which then underwent chelation-controlled Grignard addition[4] to provide **6**. The less-hindered primary alcohol of **6** was now blocked selectively as a $tert$-butyldimethyl silyl (TBDMS) ether, and the more nucleophilic allylic alcohol selectively O-benzoylated with benzoyl chloride and pyridine. The remaining secondary alcohol was then protected as an acid-labile O-tetrahydrofuranyl ether. Tetra-n-butylammonium fluoride deprotected the primary alcohol of **7** to permit Swern oxidation to the aldehyde, and oxime formation. Oxime **4** was converted to the reactive nitrile oxide by oxidation with sodium hypochlorite.[3] The internal 1,3-dipolar cyclo-addition delivered **3** as a single reaction product in 60% yield. Raney nickel then successfully brought about N—O bond cleavage[5] to give a

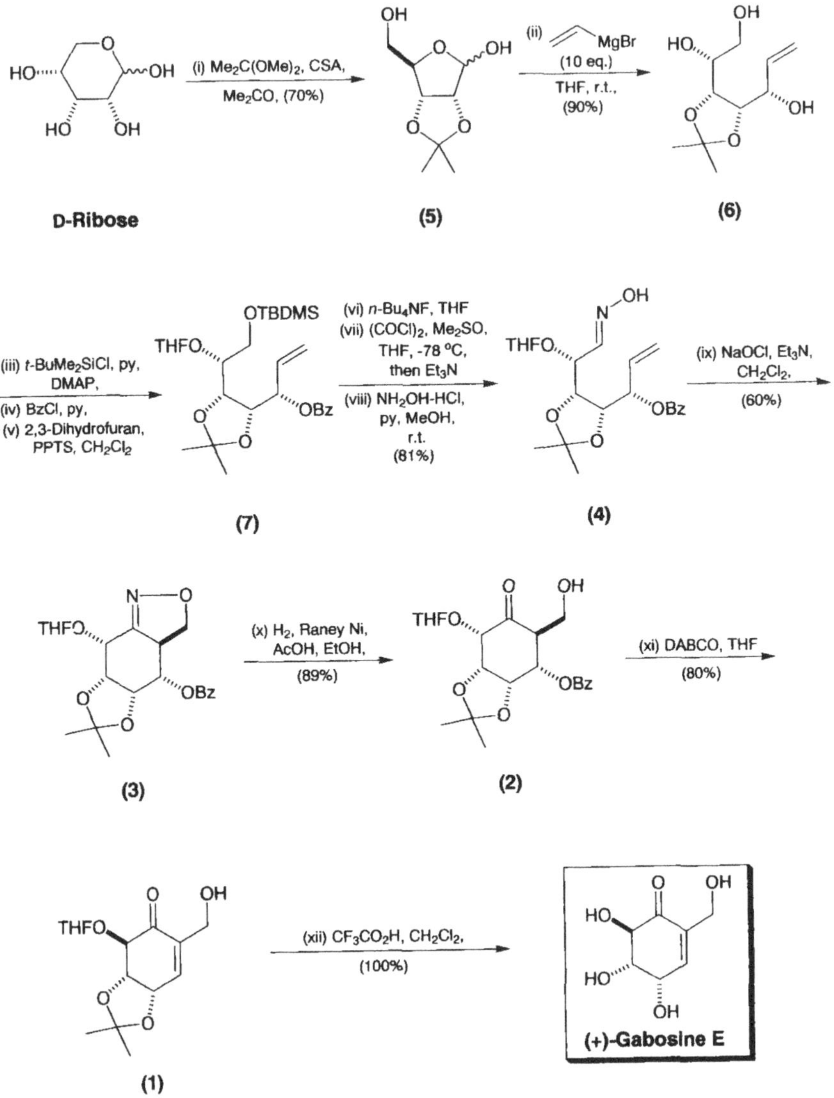

Scheme 15.3 Lygo's synthesis of (+)-gabosine E.

β-hydroxy imine, which subsequently lost ammonia, to provide the β-hydroxy ketone **2** in excellent yield. The final stages of the synthesis were a base-mediated epimerisation α to the carbonyl to obtain the correct hydroxyl group stereochemistry at C(4), and an *in situ* β elimination of

the benzoate group from C(1). It was essential not to expose the desired
product **1** to the 1,4-diazobicyclo[2,2,2]octane (DABCO) for more than
24 h, otherwise the C(4)-centre would re-epimerise to give a 2:1 equilibrium mixture of enone epimers enriched in **1**. Deprotection of pure **1**
with trifluoroacetic acid finally afforded (+)-gabosine E in excellent yield.

15.4   Points of mechanistic interest
in the Lygo (+)-gabosine E synthesis

The reaction sequence of most interest in this route is the conversion of **4**
into **3** (Scheme 15.4). Presumably, the NaOCl that is present in the
reaction mixture chlorinates the deprotonated oxime nitrogen to create
an *N*-chloro-nitrone **9** which then eliminates HCl to generate the alkenyl
nitrile oxide **10**. The latter then undergoes the anticipated 1,3-dipolar
cycloaddition reaction to form **3**.

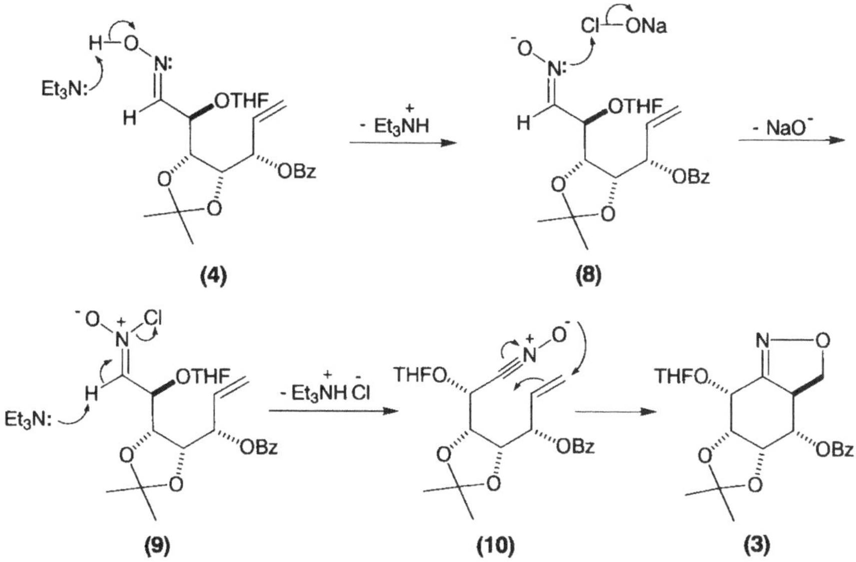

Scheme 15.4.

The conversion of 3 to 2
This is believed to involve a free radical cleavage of the isoxazoline to the
β-hydroxy imine **11**, which then undergoes hydrolysis (Scheme 15.5).[5]

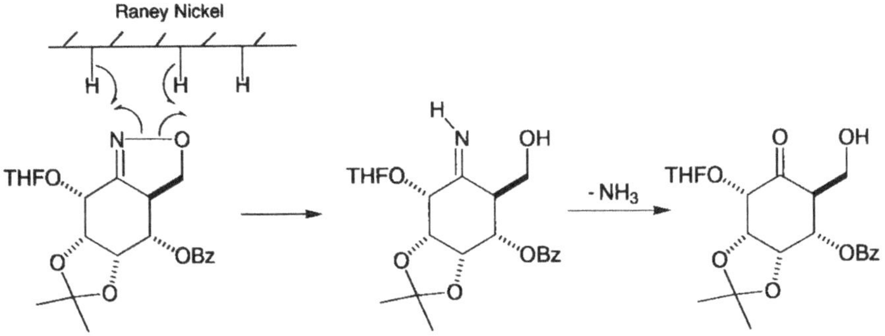

Scheme 15.5.

References

1. G. Bach, S. Breiding-Mack, S. Grabley, P. Hammann, K. Hutter, R. Thiericke, H. Uhr, J. Wink and A. Zeeck, 1993, *Ann. Chem.*, 241.
2. B. Lygo, M. Swiatyj, H. Trabsa and M. Voyle, 1994, *Tetrahedron Lett.*, 35, 4197.
3. (a) B.H. Kim, P.B. Jacobs, R.L. Elliot and D.P. Curran, 1988, *Tetrahedron*, 44, 3079; (b) Review: A.P. Kozikowski, 1984, *Acc. Chem. Res.*, 17, 410.
4. W.C. Still and J.H. McDonald, III, 1980, *Tetrahedron Lett.*, 21, 1031.
5. (a) D.P. Curran, 1982, *J. Amer. Chem. Soc.*, 104, 4024; (b) D.P. Curran, 1983, *J. Amer. Chem. Soc.*, 105, 5826.

16   (−)-Augustamine and (−)-amabiline

K.J. Hale

16.1   Introduction

Alkaloids of the *Amaryllidaceae* family have long captured the imagination of organic chemists, not only for their unusual structures, but also for their intriguing pharmacological effects.[1] Two representative members of this class are (−)-augustamine and (−)-amabaline. Their syntheses[2] from a common monosaccharide precursor are now described.

16.2   The Pearson retrosynthetic analysis of (−)-augustamine

After detailed evaluation of the (−)-augustamine problem, Pearson[2] considered that the C—C bond connecting the aryl ring to the acetal was the bond most prime for initial retrosynthetic disjunction (Scheme 16.1). This was because heterolysis of this bond produced the doubly stabilised dioxonium ion **1**, in which the activated aryl ring was perfectly positioned for nucleophilic attack upon the cationic centre. The desired electrophilic substitution pathway would probably be the one most favoured since this would lead to the product with least unfavourable steric interactions between the two dioxolane units. Pearson next considered how cation **1** could be generated in a forward synthesis. One possible progenitor could be the cyclic orthoester **2**, since this could readily eject methanol if treated with acid. Pearson recognised that **2** could potentially be derived from acetal **3** by protecting group interchange. The retrosynthetic manoeuvre that followed next was an exceedingly bold, and enormously simplifying, transform. Pearson heterolytically disconnected the *N*-methyl bond of **3** to generate a pyrrolidino anion. He then migrated this pair of electrons into the ring to induce rupture of the two α,β-bonds adjacent to the ring nitrogen and create the stabilised 2-azaallyl anion **4**. In electronic terms, he performed what would tantamount to a 1,3-anionic-cycloreversion reaction! When one considers what would now be required in the forward direction, it would be an intramolecular [π4s + π2s] 2-azaallyl-olefin cycloaddition reaction[3] on **4**. It transpires that the cycloaddition of 2-azaallyl anions to alkenes can, under appropriate circumstances, be a very good method for making pyrrolidines,[4] as the examples in Scheme 16.2 show.

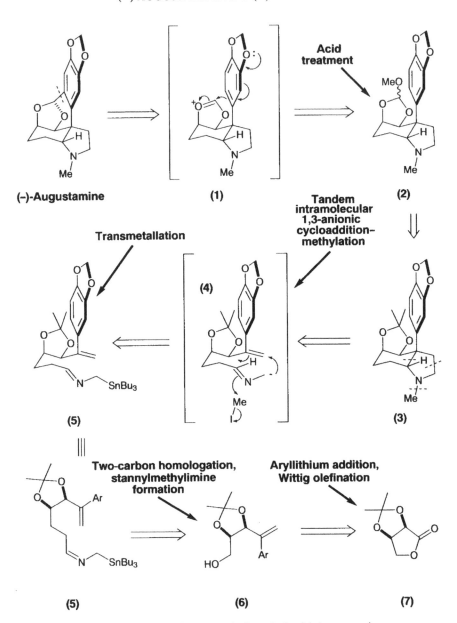

Scheme 16.1 Pearson's retrosynthetic analysis of (−)-augustamine.

In more recent times, 2-azaallyl anions have been much more conveniently generated by the transmetallation of 1-aminoalkylstannanes. Thus for the forward synthesis of (−)-augustamine, this would mean that the 2-azaallylstannane **5** would be the key intermediate needed.

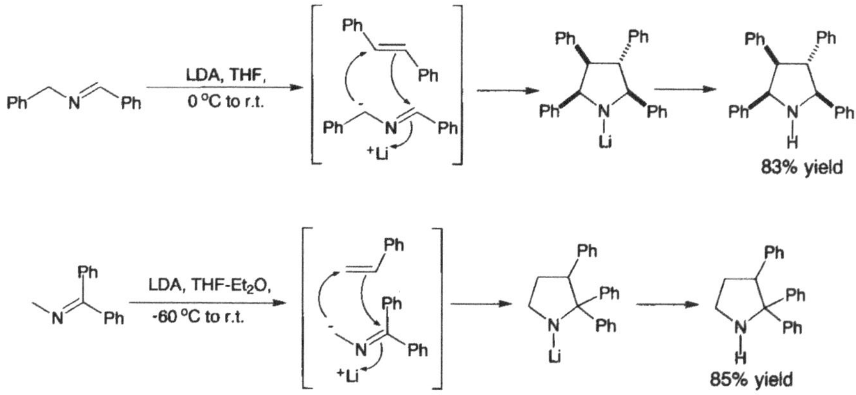

Scheme 16.2   Examples of cycloaddition of 2-azaallyl anions to alkenes.

One of the primary dividends resulting from this disconnection was its exposure of a possible relationship between the target and a mono-saccharide precursor. After systematic analysis, Pearson saw that the stereochemistry of **5** correlated perfectly with lactone **7**. Furthermore, lactone **7** had a structure appropriate for easy synthetic elaboration into **6** since it was functionally differentiated at C(1) and C(4). It therefore appeared fully capable of being selectively modified in the manner required to obtain **5**.

16.3   The Pearson total synthesis of (−)-augustamine

Lactone **7** (derived from D-isoascorbic acid) reacted readily with the aryllithium formed from bromide **8** to produce lactol **9** (Scheme 16.3). The latter underwent a facile ring-opening and Wittig olefination with methylenetriphenylphosphorane to give **6** in excellent overall yield. After O-triflation, a two-carbon chain extension was performed on **10** with the azaenolate derived from N-cyclohexylacetaldimine **11** and lithium diisopropylamine (LDA). After acid hydrolysis of the product imine, aldehyde **12** was isolated in 83% yield for the two steps. The (2-azaallyl)stannane **5** was prepared from aldehyde **12** in quantitative yield by treatment with (aminomethyl)tri-n-butylstannane.

The intramolecular anionic cycloaddition proceeded rapidly at −78°C after transmetallation with n-BuLi. The resulting N-lithiopyrrolidine was N-alkylated with methyl iodide to give a 1:5 mixture of **13** and **14** in which **14** predominated. Without purification, the isopropylidene group was cleaved from the unpurified mixture of **13** and **14**, and the resulting

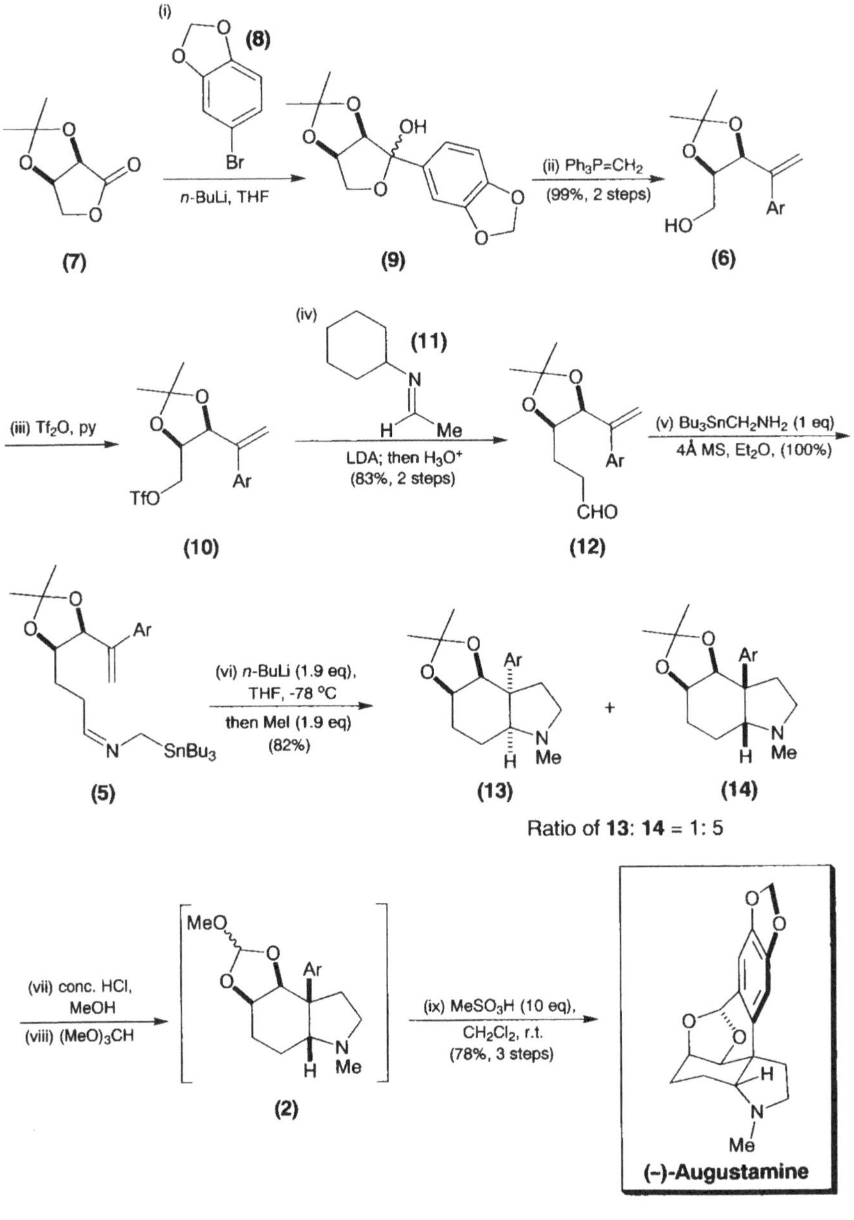

Scheme 16.3   Pearson's (−)-augustamine synthesis.

hydrochlorides mixed with trimethyl orthoformate to obtain **2**. Treatment of **2** with excess methanesulfonic acid in CH_2Cl_2 finally afforded (−)-augustamine in 44% overall yield from lactone **7**.

16.4   Pearson's synthesis of (−)-amabiline

Interestingly, when the aforementioned *N*-lithiopyrrolidine was trapped with water rather than methyl iodide, a mixture of **15** and **16** was obtained in 74% yield (Scheme 16.4). This gave Pearson the opportunity to make (−)-amabiline. For this, **16** was reacted with Eschenmoser's salt to produce the reactive iminium ion **17** which then underwent electrophilic aromatic substitution to give (−)-amabiline in excellent yield after acid-mediated transketalisation of the *O*-isopropylidene group.

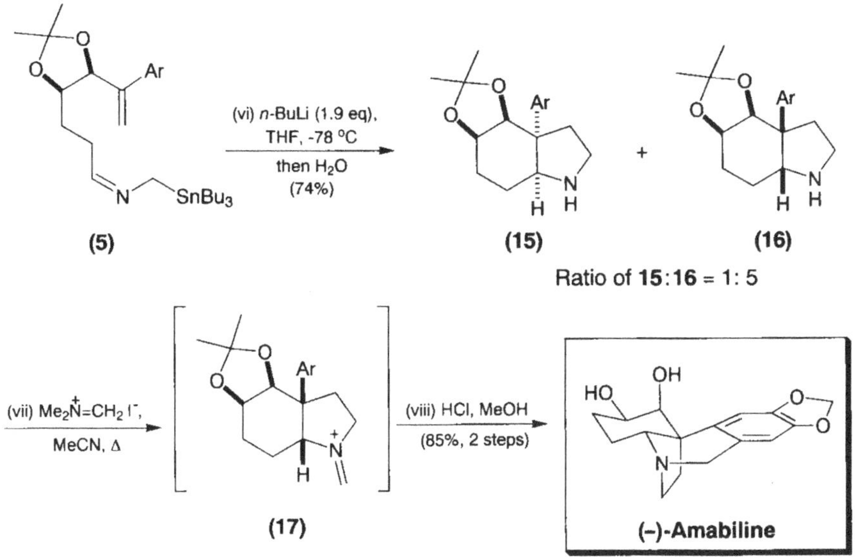

Scheme 16.4   Pearson's route to (−)-amabiline.

The Pearson syntheses of (−)-augustamine and (−)-amabiline by a common strategy are very noteworthy. They brilliantly show how the awesome power of intramolecular 2-azaallyl anion–olefin cycloadditions can be marshalled for the rapid assembly of complex pyrrolidine alkaloids with excellent efficiency and atom economy.

16.5   Mechanistic analysis of the (−)-augustamine and (−)-amabiline syntheses

The complete mechanism of the conversion of **2** into (−)-augustamine is shown in Scheme 16.5.

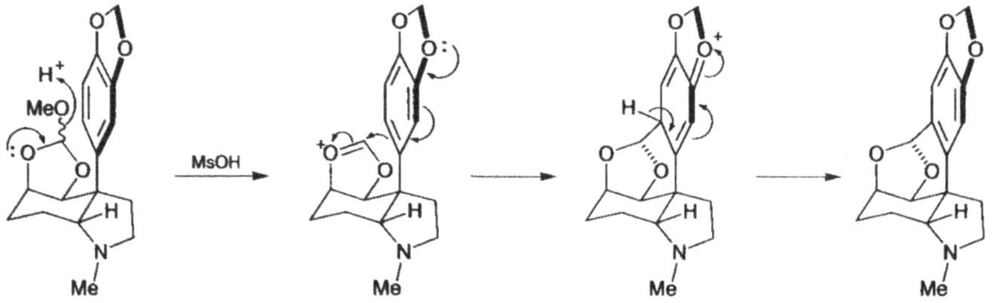

Scheme 16.5.

References

1. (a) (−)-Augustamine: A.A. Ali, H. Kating, A.W. Frahm, A.M. El-Moghazi and M.A. Ramadan, 1981, *Phytochemistry*, 20, 1121; A.A. Ali, H. Hambloch and A.W. Frahm, 1983, *Phytochemistry*, 22, 283; (b) (−)-Amabiline: K. Likhitwitayawuid, C.K. Angerhofer, H. Chai, J.M. Pezzuto, G.A. Cordell and N. Ruangrungsi, 1993, *J. Nat. Prod.*, 56, 1331; (c) Review on *Amaryllidaceae* alkaloids: S.F. Martin, 1987, *The Alkaloids* (A. Brossi, ed.), Academic Press, London, 30, 251.

2. (a) W.H. Pearson and F.E. Lovering, 1995, *J. Amer. Chem. Soc.*, 117, 12336; (b) W.H. Pearson and F.E. Lovering, 1998, *J. Org. Chem.*, 63, 3607.

3. (a) W.H. Pearson, M.A. Walters and K.D. Oswell, 1986, *J. Amer. Chem. Soc.*, 108, 2769; (b) W.H. Pearson, D.P. Szura and W.G. Harter, 1988, *Tetrahedron Lett.*, 29, 761; (c) W.H. Pearson, D.P. Szura and M.J. Postich, 1992, *J. Amer. Chem. Soc.*, 114, 1329; (d) W.H. Pearson and M.J. Postich, 1992, *J. Org. Chem.*, 57, 6354; (e) W.H. Pearson and E.P. Stevens, 1994, *Tetrahedron Lett.*, 35, 2641; (f) W.H. Pearson and V.A. Jacobs, 1994, *Tetrahedron Lett.*, 35, 7001.

4. (a) T. Kauffmann, H. Berg and E. Koppelmann, 1970, *Angew. Chem. Int. Ed. Engl.*, 9, 380; (b) T. Kauffmann and E. Koppelmann, 1972, *Angew. Chem. Int. Ed. Engl.*, 11, 290; (c) Review: T. Kauffmann, 1974, *Angew. Chem. Int. Ed. Engl.*, 13, 627.

17 (−)-FK506

K.J. Hale

17.1 Introduction

In 1987, Tanaka and associates at the Fujisawa Pharmaceutical Company in Japan reported their discovery of the powerful immunosuppressant (−)-FK506 in the culture filtrates of *Streptomyces tsukubaensis*.[1] (−)-FK506 is of great medical interest on account of its ability to inhibit interleukin 2 production, the mixed lymphocyte culture response, and cytotoxic T-cell generation at 100 times lower concentration than cyclosporin A, the most effective immunosuppressant available previously.[2] These powerful immunosuppressive properties later manifested themselves in several very promising human organ-transplantation trials.[3] To date, four total syntheses[4] and three formal total syntheses[5] of (−)-FK506 have been accomplished, and numerous synthetic approaches to various subunits of (−)-FK506 have been recorded.[6, 47] In this chapter, we will give an overview of the formal total syntheses reported by Danishefsky[5a] and Smith,[5b] since both exploit monosaccharides as sources of chirality. Both routes also construct advanced intermediates which have previously been converted to (−)-FK506 by the Merck Process Team during their first total synthesis[4a] of this molecule.

17.2 The Danishefsky retrosynthetic analysis of (−)-FK506

For their formal total synthesis of (−)-FK506, Danishefsky and co-workers[5a] made the Merck advanced intermediate **34** (see Scheme 17.8) their primary synthetic goal. This had previously been transformed into (−)-FK506 by T.K. Jones *et al.*[4a] in twelve steps (see Section 17.4). Danishefsky's plan for the assemblage of (−)-FK506 called for the implementation of two stereoselective Julia–Lythgoe–Kocienski olefinations[7] to fashion the C(19)—C(20) and C(27)—C(28) (*E*)-trisubstituted alkenes in **2** (Scheme 17.1). The first Julia transform allowed **3** and **4** to be selected as retrosynthetic reaction partners, the second enabled **7** and **9** to be chosen (Scheme 17.2). Danishefsky envisioned accessing phenylsulfone **7** from a regioselective functionalisation sequence on chiral acid **8**.

Although D-talose might initially be adjudged to be an appropriate precursor of **9**, because of its chirality overlap with the C(24) and C(26)

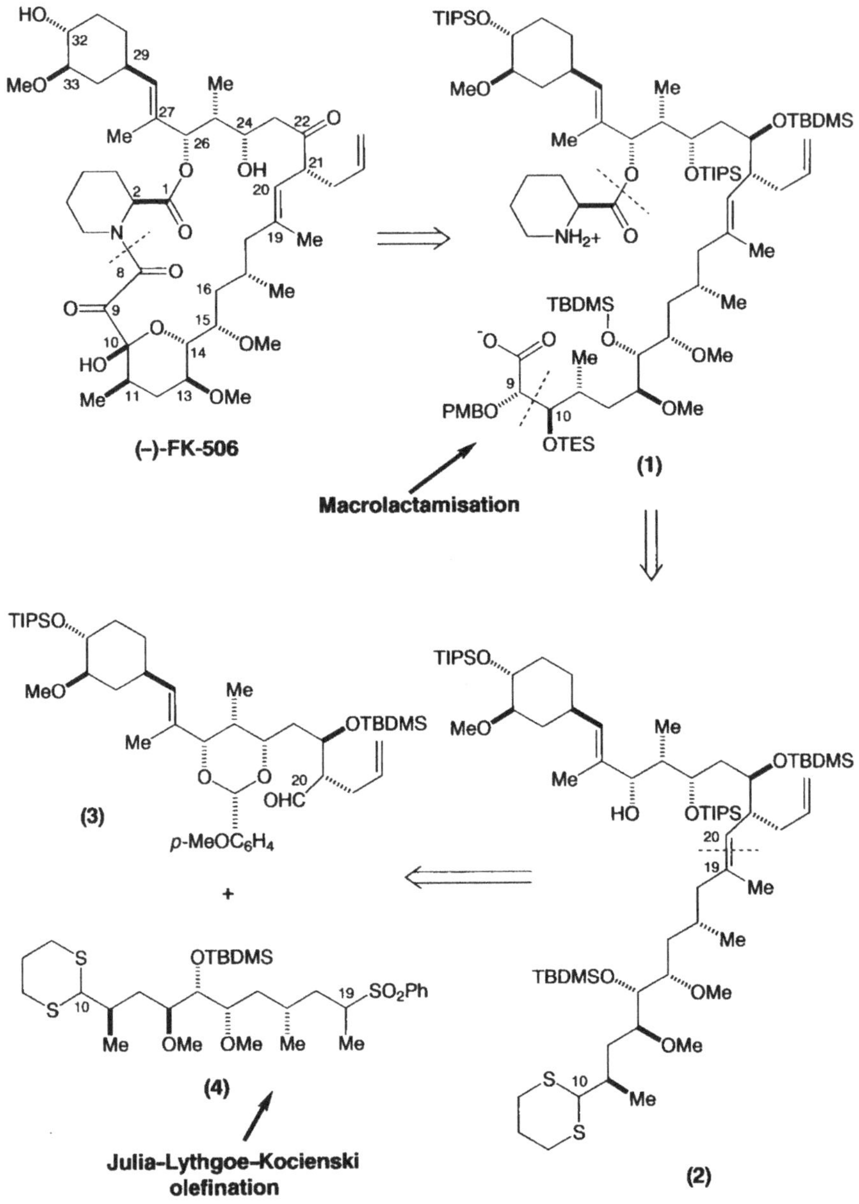

Scheme 17.1 Danishefsky's retrosynthetic analysis of (−)-FK 506.

hydroxyls of the target, the added necessity for a stereospecific methylation at C(3) in this starting sugar suggested that a D-talo epoxide such as **11** might be a more synthetically appropriate intermediate for the

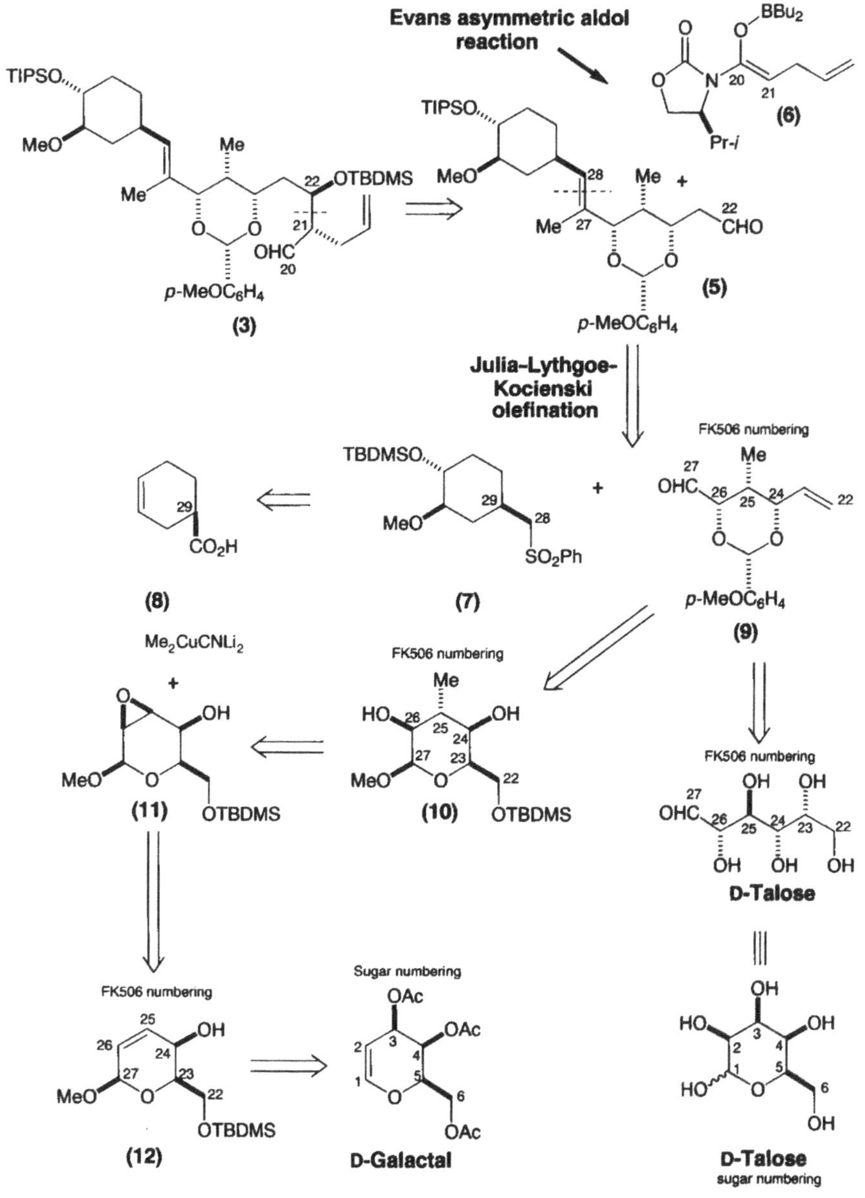

Scheme 17.2 Danishefsky's retrosynthetic analysis of aldehyde **3**.

task at hand (Scheme 17.2). Accordingly, Danishefsky proposed to ring-open epoxide **11** in *trans* diaxial fashion with the Lipshutz dilithium dimethylcyanocuprate reagent.[8] This would nicely fashion the stereotriad present within **10**. Danishefsky next envisioned constructing **11** through a hydroxyl-directed Henbest epoxidation[9] on alkene **12** with *m*-chloroperoxybenzoic acid (*m*-CPBA). Alkene **12** would, in turn, be derivable from D-galactal by a Lewis-acid-mediated Ferrier rearrangement[10] with methanol.

If one now considers what would probably be needed in the synthetic direction once aldehyde **5** had been reached, a *syn*-selective Evans asymmetric aldol reaction[11] with boron enolate **6** could potentially set the C(21) and C(22) stereocentres in **3**. All that would be required subsequently would be product liberation from the auxiliary by reduction, and oxidation at C(20). With compound **3** in hand the stage would then be set for implementation of the second Julia olefination tactic.

In like fashion, by careful stereochemical correlation of the three masked hydroxy stereocentres in **4** [i.e. C(13), C(14) and C(15) using FK506 numbering] with various monosaccharide derivatives (Scheme 17.3), Danishefsky was able to establish an appropriate chirality match with the C(2), C(3) and C(4) stereocentres of D-galactose. Methyl β-D-galactopyranoside was therefore selected as the starting material for his synthesis of **4**.

17.3   The Danishefsky formal total synthesis of (−)-FK506

Scheme 17.4 illustrates the synthetic pathway that was developed to secure the C(10)–C(19) sector **4**. The route commenced with a selective *O*-silylation of the C(6) hydroxyl in **18** with *t*-butyldimethylsilyl chloride. This was followed by a highly regioselective benzylation of OH(3) using *O*-stannylene acetal technology.[12] The remaining hydroxyls were then *O*-methylated with sodium hydride and iodomethane in tetrahydrofuran (THF) to obtain **19**, and a cleavage of the *O*-silyl ether and methyl glycoside simultaneously performed with aqueous acid. Undoubtedly, the β-configuration of this glycoside aided this step by conferring good acid lability on the acetal. Reduction of lactol **17** next afforded the ring-opened triol, whose terminal 1,2-diol was protected as an acetonide. The remaining alcohol was then oxidised to the aldehyde. Compound **15** readily participated in a fully stereocontrolled Wittig olefination with **16** to obtain enoate **14**. After reduction to the allylic alcohol, chlorination was effected with triflyl chloride and lithium chloride in dimethylformamide (DMF). This was the prelude to cyanide displacement for obtention of the nitrile **20**. The latter was isolated in 69% yield, along with a 22% yield of the α,β unsaturated nitrile. Multiple reduction of the cyano group in **20** furnished alcohol **21**, which was converted to the corresponding

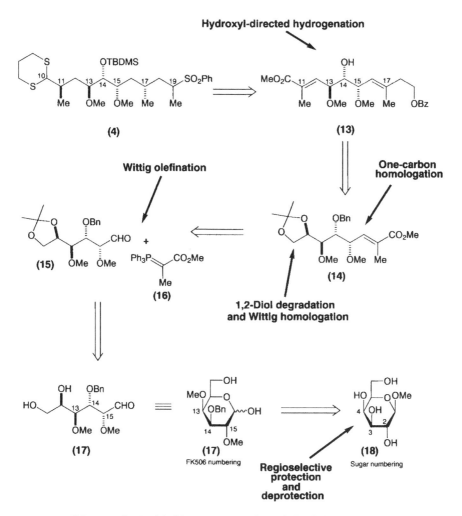

Scheme 17.3   Danishefsky's retrosynthetic analysis of phenylsulfone **4**.

tetraol by sequential *O*-debenzylation and acetonide hydrolysis. Periodate cleavage of the terminal 1,2-diol next revealed an aldehyde that was converted to enoate **13** by olefination and selective *O*-benzoylation.

At this juncture, Danishefsky decided to investigate the possibility of setting the C(11) and C(17) methyl stereocentres of **23** by hydroxyl-directed hydrogenation.[13] Some years earlier, Evans had shown that cationic rhodium– and iridium–phosphine complexes can mediate highly diastereoselective reductions of trisubstituted homoallylic alcohols. However, for excellent stereoselectivities to generally be observed, it

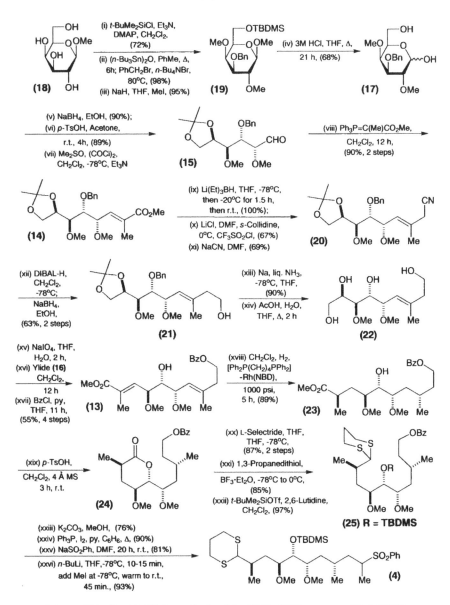

Scheme 17.4 Danishefsky's synthesis of the C(10)–C(19) segment of (−)-FK506.

was usually necessary for the alkene component to have a preexisting allylic stereocentre.[13] For alkene **13**, this criterion was fulfilled on two counts. Accordingly, when Danishefsky attempted the Rh-catalysed hydrogenation of **13**, it was not surprising to find that a 19:2.1:1 mixture of three diastereoisomers was formed in which **23** predominated. After separation of this mixture, mild acid treatment of **23** was sufficient to cause lactonisation. Reduction of **24** next led to the lactol, which was ring-opened to **25** by thioketalisation and O-silylation. A further four steps were necessary to convert **25** into the methylated phenylsulfone **4**.

The route eventually formulated for reaching **3** is depicted in Schemes 17.5 and 17.6. It sets off with a stannic-chloride-mediated Ferrier rearrangement[10] on D-galactal with methanol (Scheme 17.5). O-Deacetylation and selective O-silylation of the primary hydroxyl then followed to yield the allylic alcohol **12**. A Henbest hydroxyl-directed *syn*-epoxidation[9] on **12** nicely negotiated the issue of having stereospecifically to form an epoxide with D-*talo* stereochemistry. Significantly, epoxy alcohol **11** then underwent a highly regioselective *trans*-diaxial ring opening at C(3) with the Lipschutz higher-order cuprate reagent Li_2Me_2CuCN[11] to give **10** in excellent yield. Diol **10** was next protected as a *p*-methoxybenzylidene acetal. O-Desilylation and iodination subsequently afforded iodide **26**. A two-step sequence involving reductive elimination with zinc dust, and acetal equilibration with mild acid, allowed aldehyde **9** to be obtained. This readily condensed with the lithio anion derived from phenylsulfone **12** to give a mixture of two β ketosulfones **27** after Dess–Martin oxidation. Happily, both these products underwent smooth desulfonylation with lithium napthalenide, to give a single ketone that reacted readily with methyl magnesium bromide. The pair of tertiary methyl carbinols that emerged were now dehydrated with the Burgess reagent.[14] This led to a 6:1.5:1 mixture of three olefins in which the desired C(27)—C(28) (*E*)-isomer predominated. After protecting group interchange in the cyclohexyl ring to obtain **28**, a hydroboration and Dess–Martin oxidation secured the desired (*E*)-aldehyde **5**. The latter participated in a *syn*-selective Evans aldol reaction with the (*Z*)-boron enolate formed from chiral oxazolidinone **29** to give **30** after O-silylation. This protection step permitted efficient cleavage of the auxiliary with lithium benzyloxide, and reduction of the benzyl ester to the primary alcohol. Swern oxidation completed subtarget **3** (Scheme 17.6).

The anion derived from sulfone **4** coupled readily to aldehyde **3** (Scheme 17.7) to give a mixture of β hydroxysulfones that could be converted to a 2.5:1 mixture of the olefins **2** and **32**, after acylation and reductive desulfonylation. Unfortunately, problems were subsequently encountered during the attempted removal of the acetal unit from **2** and **32** with pyridinium *p*-toluenesulfonate in propanol and acetonitrile. Apparently, silyl ether cleavage competed with the desired deacetalation process. Notwithstanding this, the C(24),C(26) diol was isolated pure in 33% yield, along with 28% yield of recovered starting material. This diol

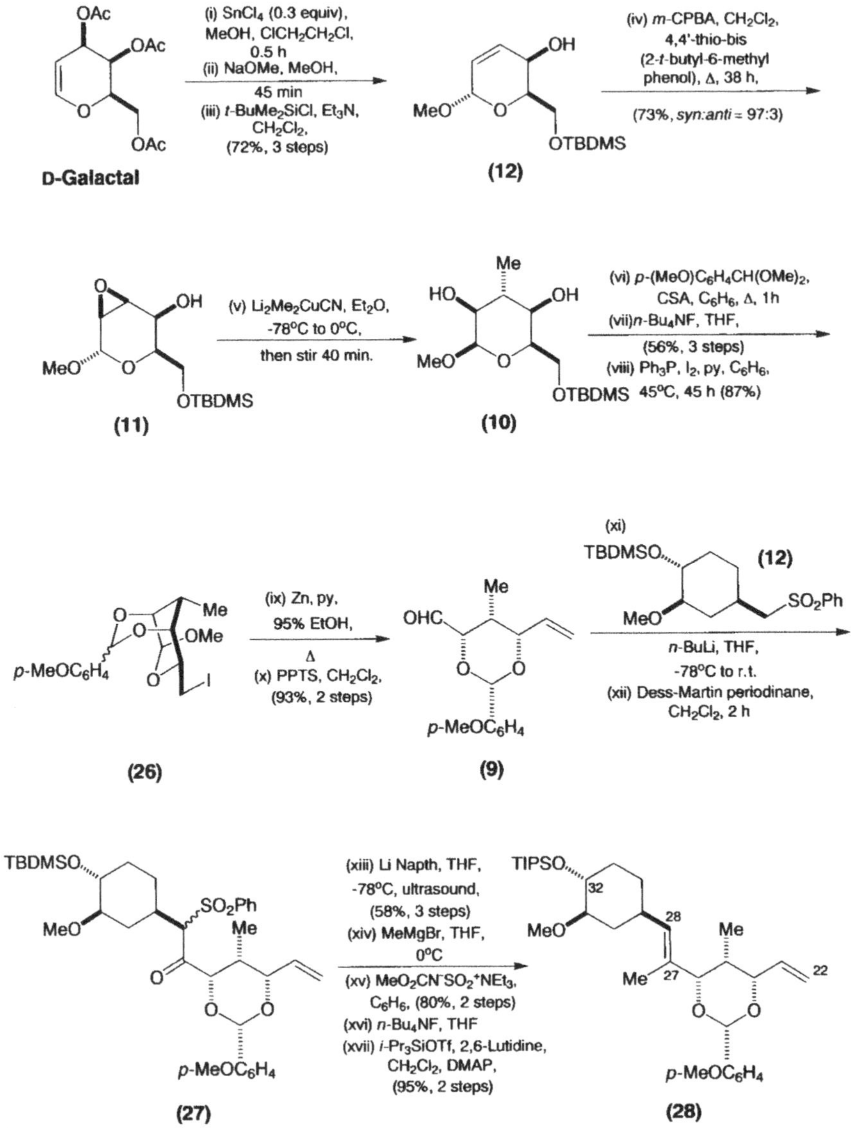

Scheme 17.5 Danishefsky's route to the C(22)–C(34) sector **28** of (−)-FK 506.

was then selectively monosilylated at C(24) with excess triisopropylsilyl triflate to give **33** (Scheme 17.8). This operation allowed the protected pipecolic acid moiety to be cleanly introduced at C(26). Danishefsky finally arrived at the Merck aldehyde **34** after transketalisation of the

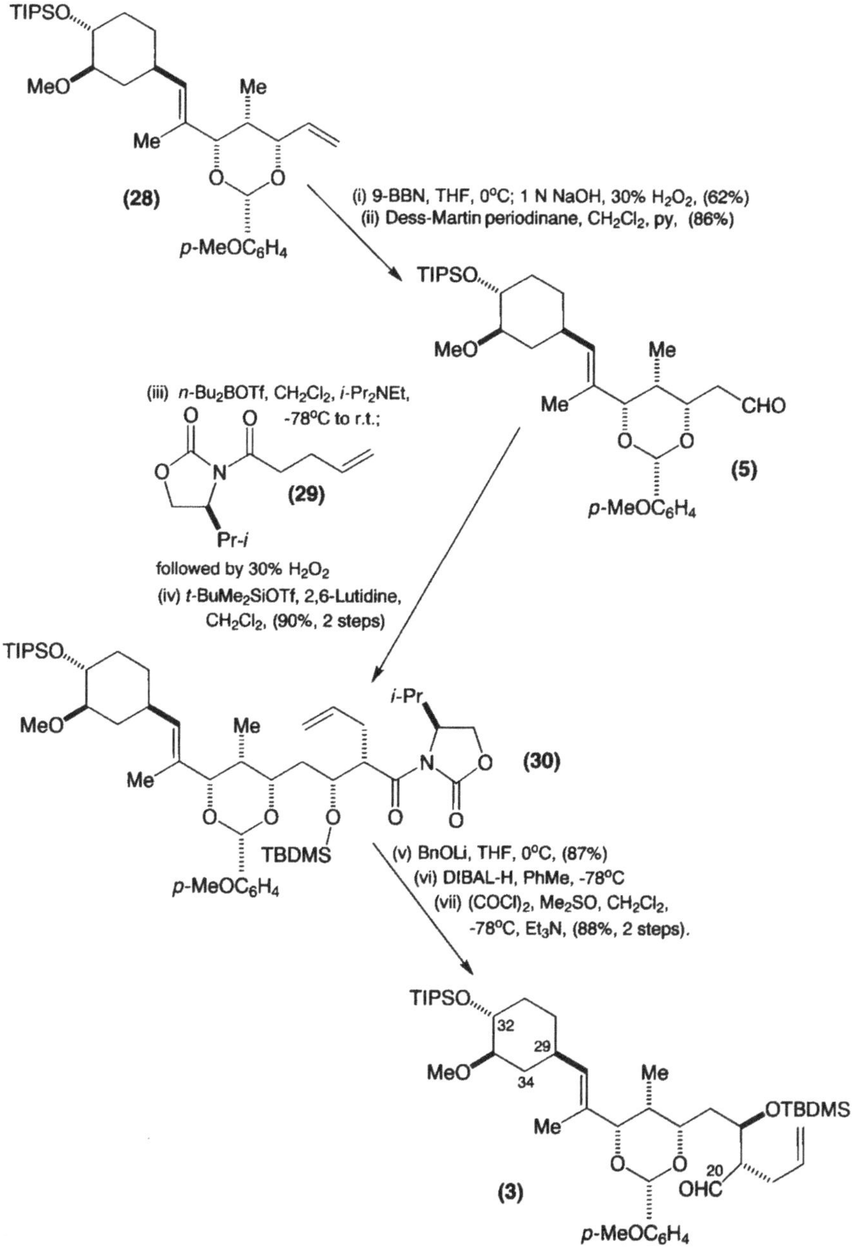

Scheme 17.6   Danishefsky's route to the C(20)–C(34) sector **3** of (−)-FK506.

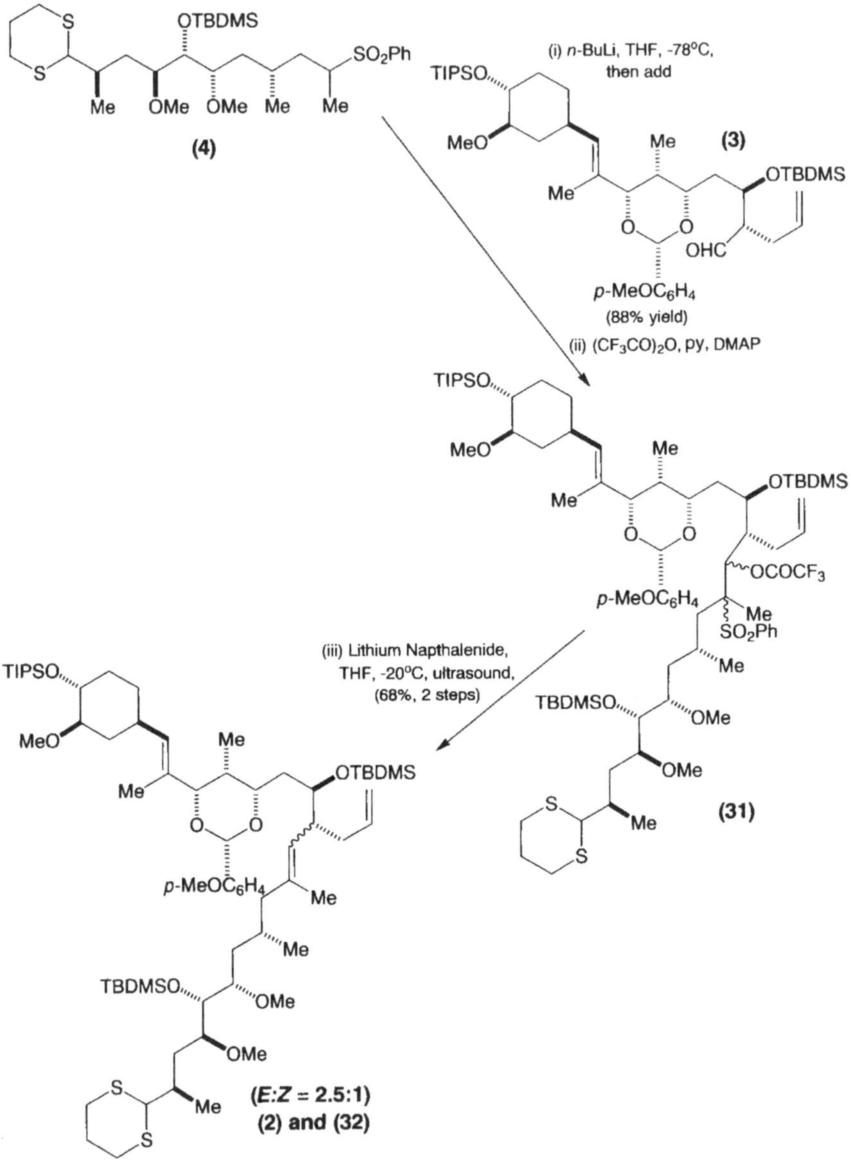

Scheme 17.7 Danishefsky's route to alkene **2** and its (*Z*)-isomer **32**.

1,3-dithiane with methanol, and hydrolysis of the dimethylacetal with pyridinium *p*-toluenesulfonate in dichloromethane. This completed a very elegant formal total synthesis of (−)-FK506.

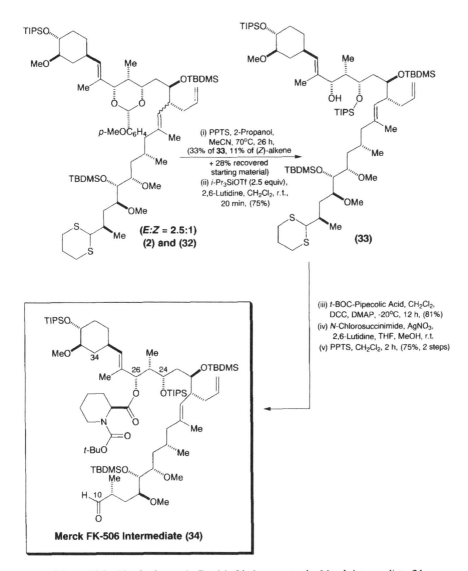

Scheme 17.8 The final steps in Danishefsky's route to the Merck intermediate **34**.

17.4   The Merck endgame used in the first total synthesis of (−)-FK506

T.K. Jones and co-workers had already converted the Danishefsky intermediate **34** into (−)-FK-506 using the chemistry outlined in Schemes

17.9 and 17.10.[4a] The C(8) and C(9) carbons were appended by a *syn*-selective Evans asymmetric aldol reaction. Lithium hydroperoxide freed the aldol adduct from the chiral auxiliary with high efficiency. The *t*-Boc protecting group on nitrogen was then removed by sequential treatment with triethylsilyl triflate and silica gel. The latter hydrolysed the intermediary silyl carbamate to liberate the desired amino acid **38**. This

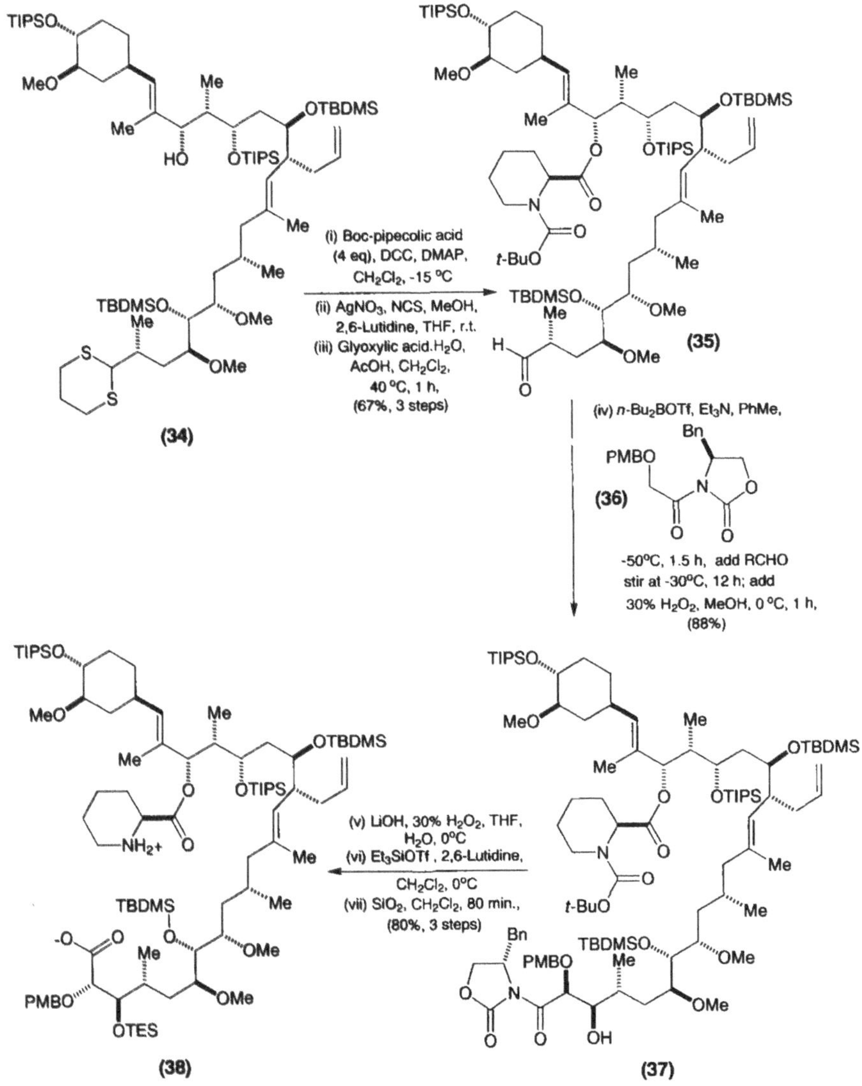

Scheme 17.9 The Merck synthetic endgame for (−)-FK506.

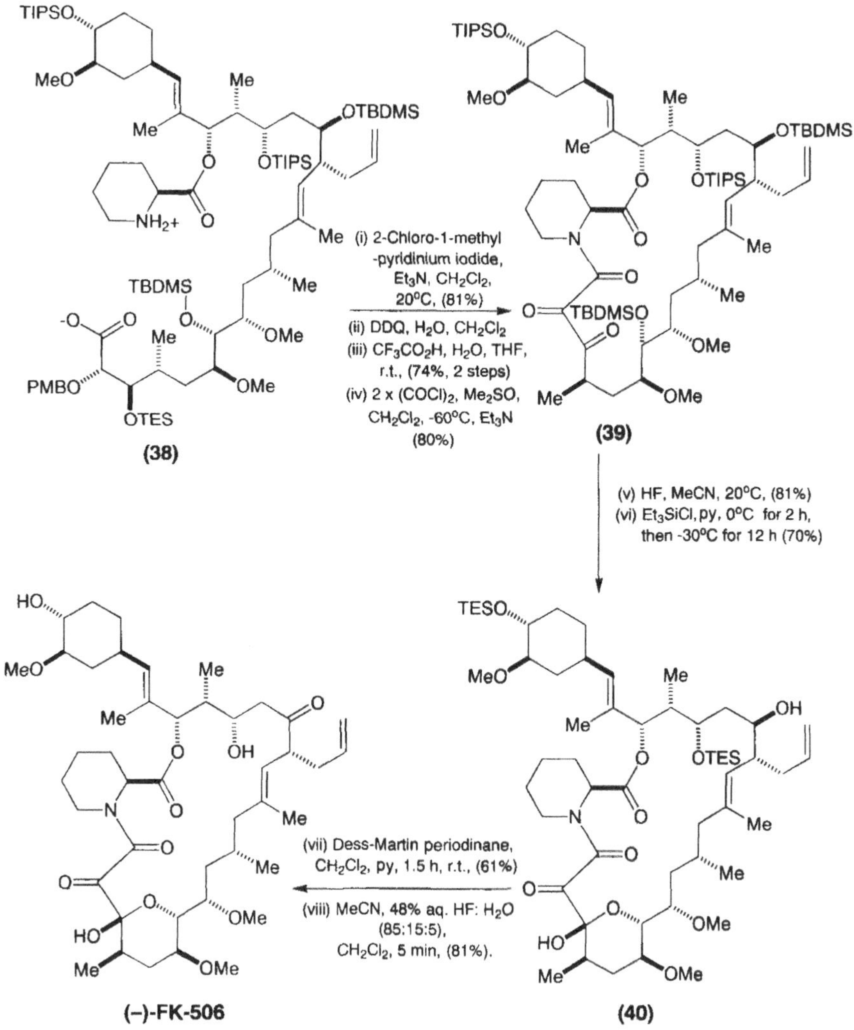

Scheme 17.10 The Merck synthetic endgame for (−)-FK506.

then underwent a highly efficient macrolactamisation reaction after the carboxylic acid group was activated with 2-chloro-1-methylpyridinium iodide.

The final stages of the Merck synthesis involved differential oxidation of the C(9), the C(10) and the C(22) secondary hydroxyl groups. Initially, the C(9)-p-methoxybenzyl group was oxidatively cleaved with 2,3-di-chloro-5,6-dicyano-1,4-benzoquinone (DDQ), and the C(10)-triethylsilyl ether hydrolysed with aqueous trifluoroacetic acid. This supplied the diol needed for the Swern oxidation to diketoamide **39**. Aqueous hydrofluoric acid in acetonitrile then detached the remaining silyl protecting groups from **39** to provide 22-dihydro-FK-506. Fortunately, it proved possible selectively to *O*-silylate the hydroxyls at C(24) and C(32) with triethylsilyl chloride to obtain **40**. This permitted oxidation of the C(22) alcohol with the Dess–Martin reagent. Deprotection with hydrofluoric acid in acetonitrile finally provided (−)-FK-506 in 81% yield.

17.5   Smith's retrosynthetic analysis of (−)-FK506

A.B. Smith and co-workers[5b] have also devised a synthetic strategy for (−)-FK506 which again intersects with an advanced intermediate **41** on the Merck pathway (Scheme 17.11). Smith's retrosynthetic plan for **41** called for the intermediacy of β-hydroxy ketone **42** as a subtarget, and proposed a chelation-controlled *syn* reduction[15] for stereospecific installation of the C(24) hydroxyl. Retrosynthetic conversion of the C(24) ketone in **42** to a 1,3-dithiane[16] subsequently primed the C(23)—C(24) bond for reverse anionic cleavage into dithiane **43** and iodide **44**.

Retrosynthetic simplification of the C(21)-allyl unit now appeared timely, when accompanied by reverse transketalisation and *O*-benzylation (Scheme 17.12). These combined manoeuvres ultimately afforded **45** as an intermediate. The existence of a 1,3-*anti*-relationship between the two methyl ether stereocentres in **45** lured Smith into considering a hydroxyl-directed *anti*-reduction[17] on **46** for creation of the C(15) stereocentre. Smith was attracted to the idea of using **46** as an intermediate since it looked amenable to construction from phenylsulfone **47** and methyl ester **48**.[18] The latter, in turn, would be accessible from alkene **52** by stereoselective hydrogenation on the less-hindered β face of the double bond,[19] opposite to the axial methoxy group at C(1). There were two main advantages associated with selecting **52** as an intermediate. The first derived from the fact that it had previously been prepared in six steps by Inch and Lewis[20] from methyl α-D-glucopyranoside. The second dividend arose from its guaranteed stereochemical integrity for the C(13) and C(14) stereocentres in **48**.

Although there are numerous sites in **47** for potential retrosynthetic disassembly, Smith considered the σ-bond linking the C(20) and C(21) carbons the one most strategic for breakage, since this offered the opportunity for simultaneously controlling the stereochemistry of the C(19)–C(20) alkene and the C(21) stereocentre through a vinylalanate

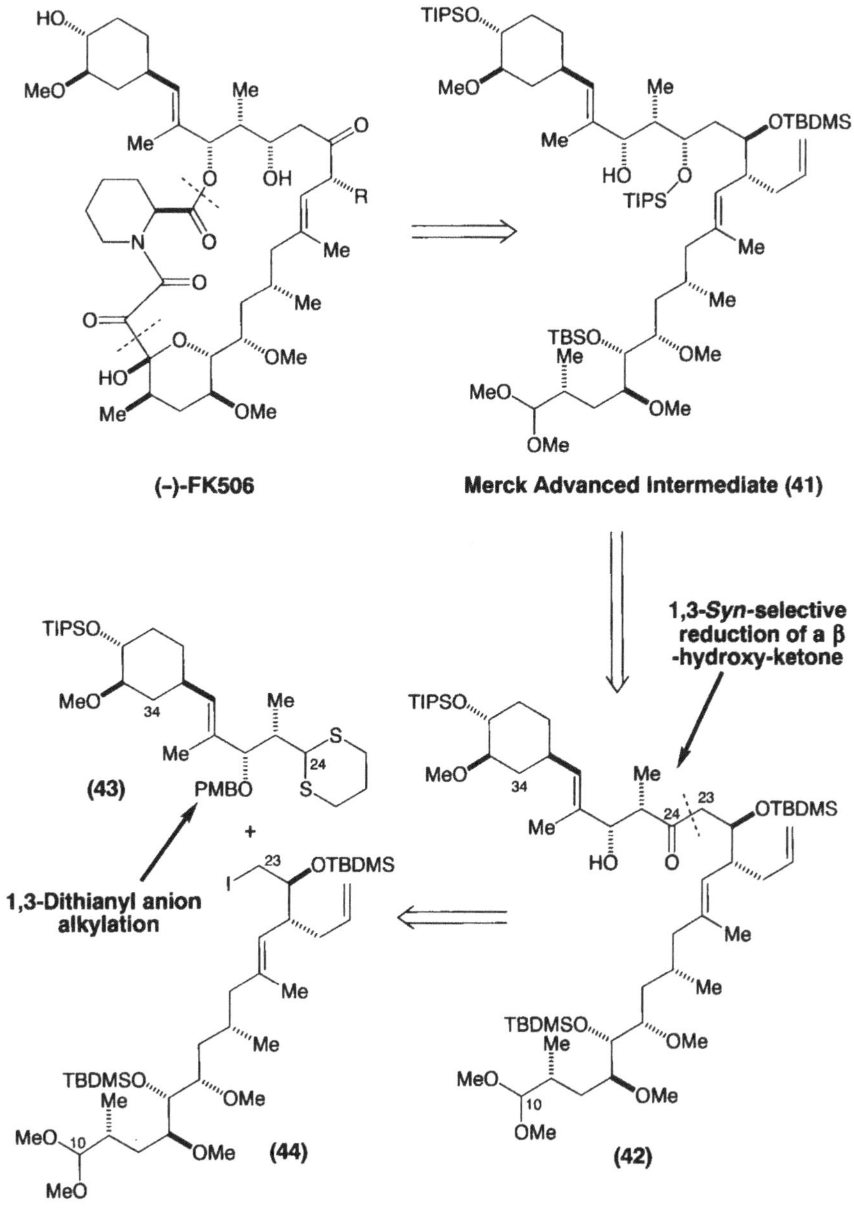

Scheme 17.11 Smith's retrosynthetic analysis of (−)-FK506.

disubstituted epoxide ring-opening between **49** and epoxide **50**. The use of a C2-symmetrical epoxide for this coupling would sidestep any issues associated with site regioselectivity during epoxide ring-opening. Epoxide

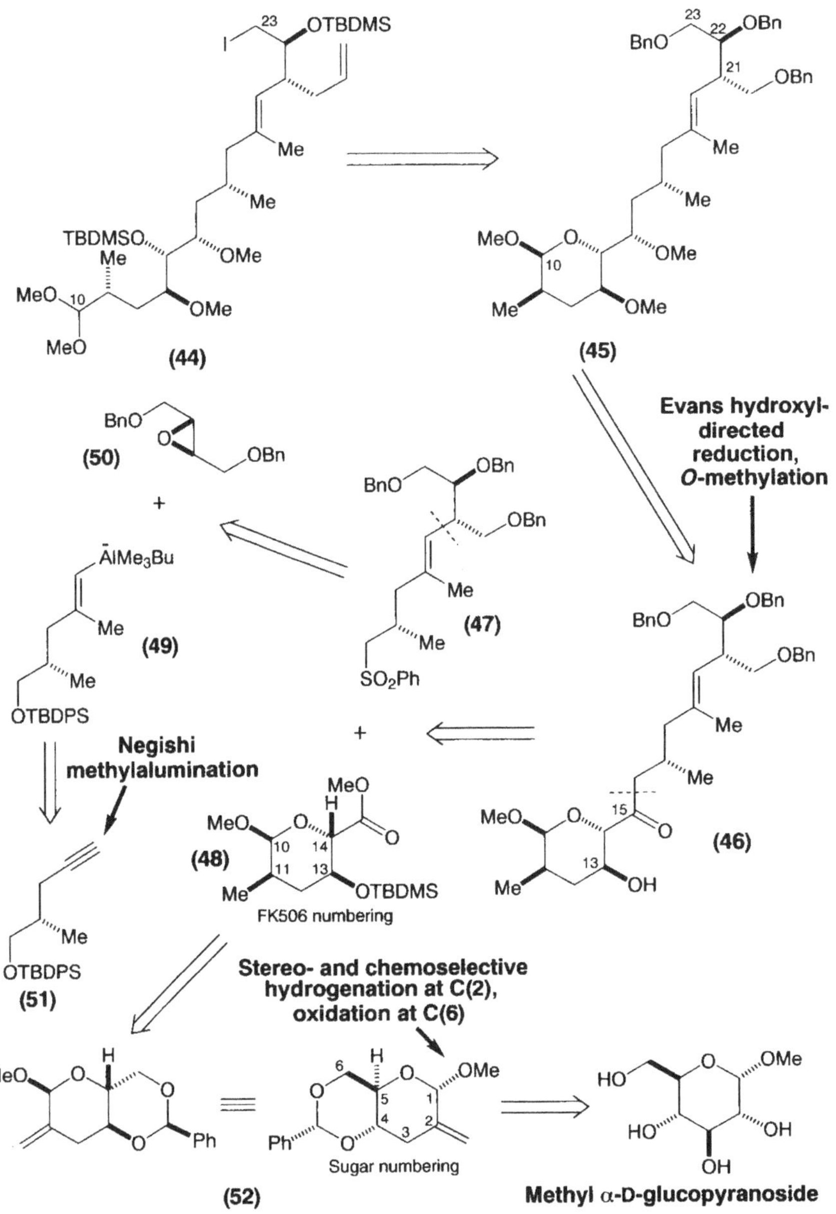

Scheme 17.12 Smith's retrosynthetic analysis of the C(10)–C(23) sector of (−)-FK 506.

50 had added attraction, in as much as it had previously been prepared by Nicolaou *et al.* during their synthesis of the ionophore antibiotic X-14547A.[21]

Smith's stratagem for the controlled introduction of the trisubstituted alkene in dithiane **43** rested upon the successful creation of enol triflate **53** under kinetically controlled enolisation conditions. It would then be employed for a McMurry–Scott cross-coupling reaction[22] with lithium dimethylcuprate (Scheme 17.13). Molecular models of ketone **55** suggested that the bulky side-chains emanating from either side of the carbonyl would probably prefer to orientate themselves *trans* to one another to minimise destabilising steric interactions. As a consequence, it was postulated that kinetic deprotonation of **55** at low temperature would markedly favour the stereoselective formation of Z-enolate **54**, which would give **53** after O-triflation. Since cuprate cross-coupling reactions usually proceed with retention of olefin geometry in the parent enol triflate, Smith felt satisfied that this technology would successfully control the (E)-geometry of the C(27)–C(28) trisubstituted olefin. Further functionalisation and disconnection of **56** pointed to sulfone **57** and aldehyde **58** as appropriate progenitors.

17.6   The Smith formal total synthesis of (−)-FK506

Smith's route to **48** was founded on the use of olefin **52** for assembly of the C(10)–C(15) subunit of (−)-FK506. The latter was available in six steps from methyl α-D-glucopyranoside using chemistry developed by Inch and Lewis[20] (Scheme 17.14). Stereoselective hydrogenation of **52** with palladium on charcoal in ethanol produced a 3.5:1 mixture of **66:65** from which **66** was isolated in pure condition in 67% yield after recrystallisation. A second hydrogenation with palladium hydroxide next allowed the benzylidene acetal to be cleanly detached from **66**. The most satisfactory protocol for selectively benzoylating[23] the primary position of this diol exploited the Mitsunobu reaction. The remaining hydroxyl in **67** was then blocked as a *tert*-butyldimethylsilyl ether. In order to prevent silyl migration from O(4) to the sterically less crowded O(6) position, the aforementioned benzoate was cleaved with diisobutylaluminium hydride to obtain the primary alcohol **68**. This underwent Sharpless oxidation[24] to the carboxylic acid with catalytic ruthenium tetraoxide. Esterification was then effected with methyl iodide and potassium carbonate in N,N-dimethylformamide.

Sulfone **47** was prepared in nine steps from known iodide **69** by the sequence shown in Scheme 17.15. After initial conversion of **69** into aldehyde **70** by cyanide displacement and reduction with diisobutylaluminium hydride (DIBAL), a Corey–Fuchs alkynylation[25] yielded alkyne **51**, the key substrate needed for the proposed carboalumination-epoxide ring-opening step. Carboalumination[26] was achieved by treating

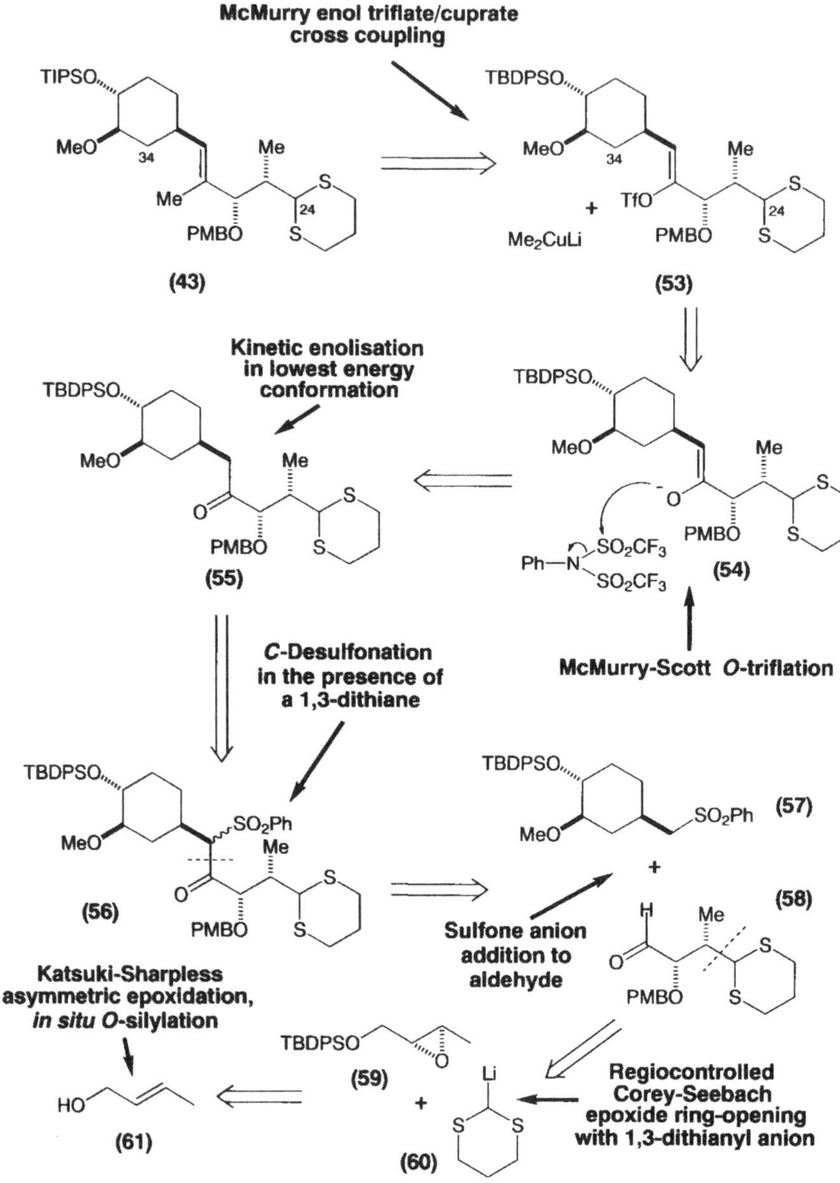

Scheme 17.13 Smith's retrosynthetic analysis of C(24)–C(34) sector of (−)-FK506.

51 with two equivalents of trimethylaluminium and a catalytic quantity (0.1 equiv.) of zirconocene dichloride in 1,2-dichloroethane. After 2 h stirring at room temperature, excess trimethylaluminium and solvents

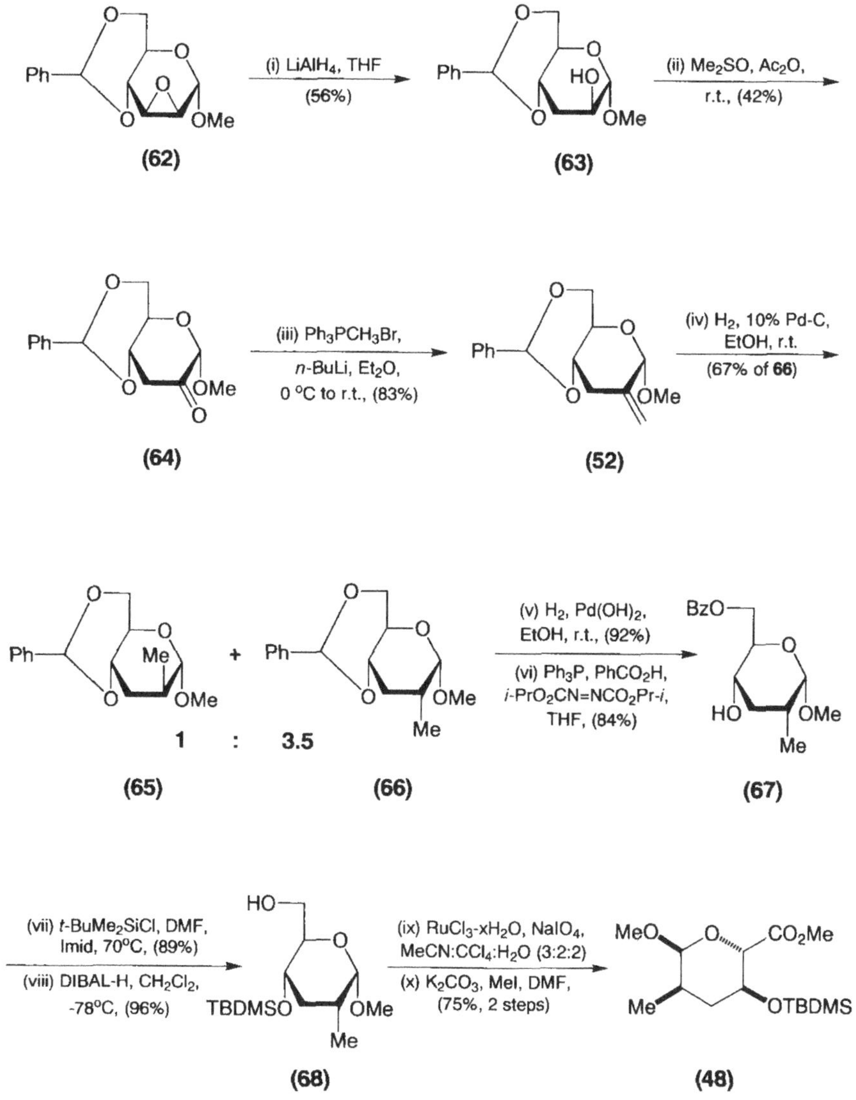

Scheme 17.14 Smith's route to the FK506 C(10)–C(15) pyranoside **48**.

were removed by vacuum distillation, and the putative vinylalane intermediate converted to its aluminate complex by treatment with *n*-butyllithium in hexanes at 0°C. The latter coupled smoothly to homochiral epoxide **50**[21] to give alkene **71** after aqueous work-up. Alkene **71** was subsequently *O*-benzylated, *O*-desilylated, converted to the

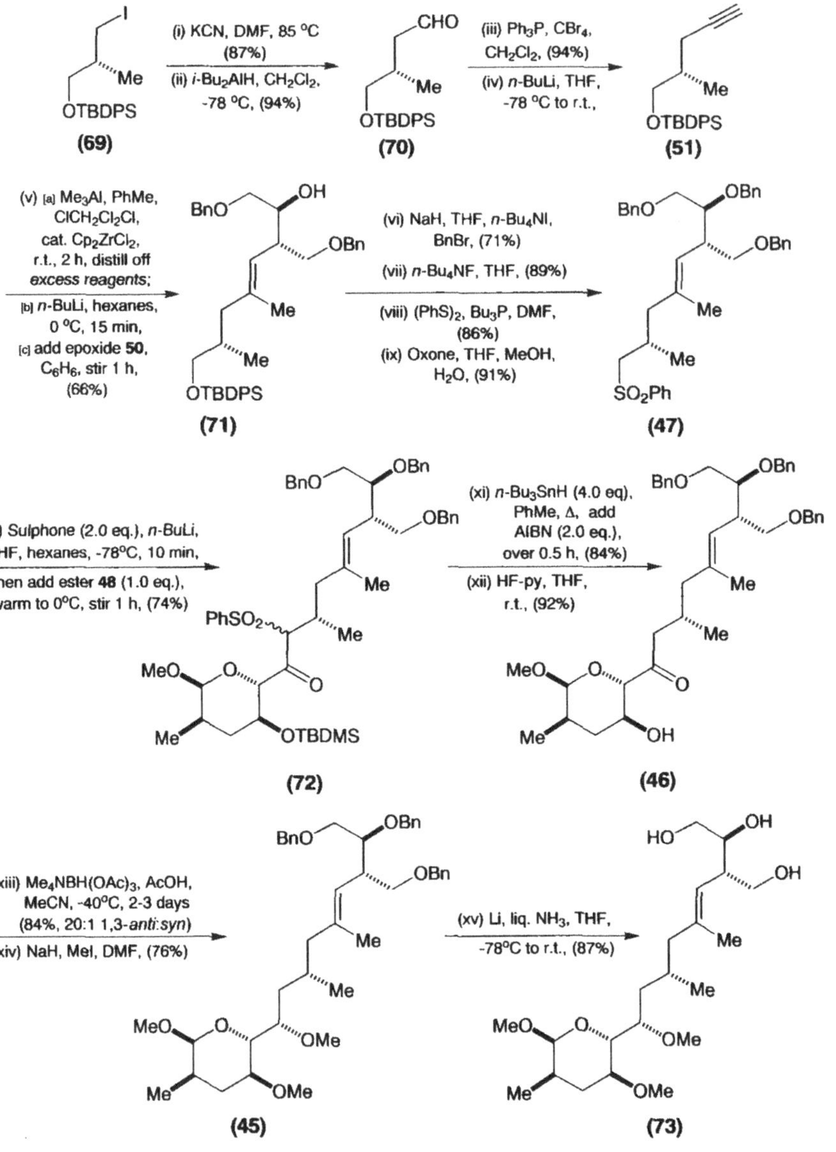

Scheme 17.15 Smith's route to the $(-)$-FK 506 C(10)–C(23) triol **73**.

thiophenyl ether, and oxidised under the Trost–Curran[27] conditions to obtain phenylsulfone **47**.

Two equivalents[18] of the anion derived from sulfone **47** were needed to obtain a satisfactory yield of the β-ketosulfone mixture **72** (Scheme 17.15) from the coupling reaction with ester **48**. The use of excess sulfone anion was necessary for this reaction because of the acidity of the two β-ketosulfone adducts, which partially quench the parent sulfone anion. Owing to **72** resisting desulfonylation by standard literature procedures, Smith developed a new reaction for this purpose.[28] This entailed heating **72** with excess tri-*n*-butyltin hydride and azobisisobutyronitrile (AIBN) in toluene at reflux. It accomplished the desired transformation to **98** (see Section 17.7) very cleanly and rapidly in 84% yield.

O-Desilylation of this ketone with HF–pyridine complex next furnished the β-hydroxy ketone **46**, which was subjected to Evans hydroxyl-directed reduction with tetra-*n*-methylammonium triacetoxyborohydride in acetic acid and acetonitrile at −40°C.[17] This reaction afforded a 20:1 mixture of *anti:syn* 1,3-diols which proved readily separable by flash chromatography. The major diol was *O*-methylated under standard conditions to obtain **45**, and *O*-debenzylation effected by dissolving metal reduction. In order to protect the terminal 1,2-diol of **73** regioselectively, to allow further elaboration at C(21), Smith reacted **73** with 2,4-dimethyl-3-pentanone dimethyl ketal and *p*-TsOH in THF (Scheme 17.16). The desired acetal was isolated as a single isomer in 90% yield. Tosylation next led to **74** which underwent nucleophilic displacement with lithium acetylide–ethylenediaminetetra-acetic acid (EDTA) complex in dimethylsulfoxide (DMSO) to provide the expected alkyne. Its structure was verified by X-ray crystallography. A standard Lindlar reduction[29] transformed this alkyne into **75** in high yield. Interestingly, when the Lipshutz higher-order vinyl cyanocuprate reagent[30] (CuCN, vinyllithium, Et₂O, 0°C) was used to elaborate **74** into **75**, the desired product was formed only in a rather meagre 37% yield, the major product being the desulfonylated alcohol which was obtained in 44% yield.

Five steps were required to convert **75** into the iodide coupling partner **44** needed for union with dithiane **43** (Scheme 17.16). Thioacetal formation[31] and concomitant deketalisation were instigated by reacting **75** with 1,3-propanedithiol under Lewis acid conditions. Triol **76** was isolated in 65% yield. The less-hindered primary hydroxyl of **76** was then selectively *O*-tosylated, and the remaining hydroxyls masked as *tert*-butyldimethyl silyl (TBDMS) ethers. Iodide displacement on **77** with sodium iodide and copper bronze,[32] and transketalisation with *N*-chlorosuccinimide (NCS)/silver nitrate[33] finally secured **44**.

Smith exploited a Lewis-acid-mediated asymmetric Diels–Alder reaction[34] between 1,3-butadiene and the Oppolzer acryloyl camphorsultam **78** to set the remote C(29) stereocentre in phenylsulfone **57** (Scheme 17.17). This tactic procured acid **8** in 93% enantiomeric excess (ee), after base-promoted hydrolysis of the chiral auxiliary. Utilising a procedure published by Martin and co-workers,[35] Smith iodolactonised acid **8**, and

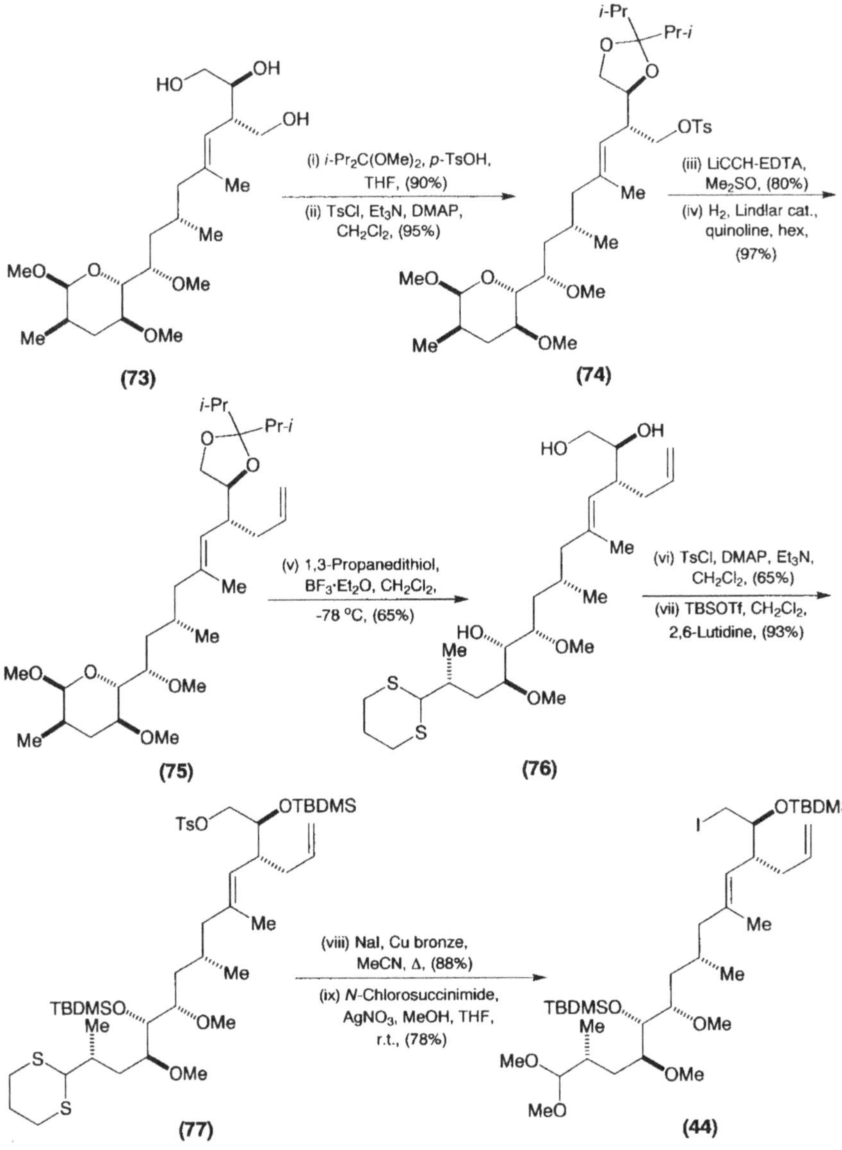

Scheme 17.16 Smith's synthetic pathway to iodide **44**.

eliminated the resulting iodide with 1,8-diazobicyclo [5.4.0] undec-7-ene (DBU) to access alkene **80**. The strained lactone in **80** was readily reduced to the corresponding diol with lithium aluminium hydride, and this

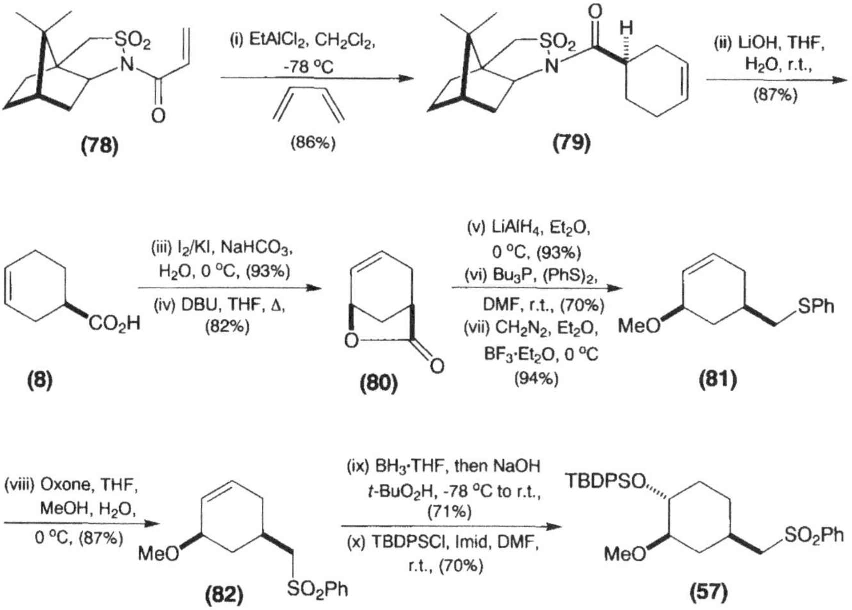

Scheme 17.17   Smith's asymmetric Diels–Alder strategy to phenylsulfone **57**.

chemoselectively thioetherified with tri-*n*-butylphosphine and phenyldisulfide[36] at its less-hindered primary hydroxyl. *O*-Methylation with ethereal diazomethane and BF₃-etherate[37] provided **81**. A Trost–Curran oxidation on **81** with oxone finally afforded the alkenylphenylsulfone **82**.

Previous work by Pasto[38a] and by Schreiber[38b] had already shown that the hydroboration and oxidation of 3-alkoxycyclohexenes generally results in the installation of a hydroxyl vicinal and *trans* to the 3-alkoxy group. It was not that surprising, therefore, to find that when **82** was hydroborated and oxidised, a 17.5:1.7:1.0 mixture of three products was obtained, enriched in the desired isomer. The latter was isolated in 71% yield after flash chromatography. *O*-Silylation with *t*-butyldiphenylsilyl chloride blocked this hydroxyl group and produced the crystalline derivative **57**, whose structure was verified by X-ray crystallography.

Let us turn now to the other coupling partner, aldehyde **58**; it was synthesised by the pathway shown in Scheme 17.18. The first step was a Sharpless catalytic asymmetric epoxidation on (*E*)-crotyl alcohol with the oxidant derived from (−)-diethyl tartrate. An *in situ* derivatisation with *t*-butyldiphenylsilylchloride was then performed. The desired epoxide **59** was readily isolated in 76% overall yield after chromatography. Treatment of **59** with 2-lithio-1,3-dithiane in THF and 1,3-dimethyl-3,4,5,6-tetrahydro-2(1*H*)-pyrimidinone (DMPU; also known as *N*,*N*-

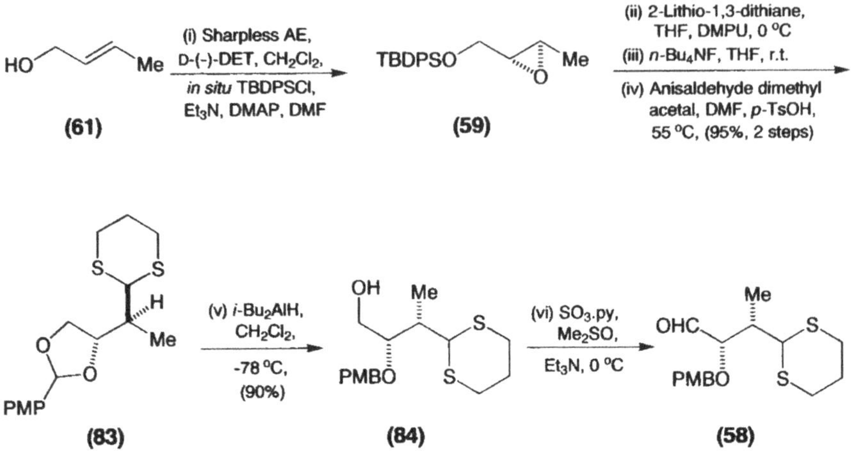

Scheme 17.18 Smith's synthesis of aldehyde **58**.

dimethylpropyleneurea) at 0°C afforded the C(3) ring-opened product exclusively, owing to the C(3) position being considerably less sterically crowded than C(2). Note how this epoxide-opening nicely controlled the *syn*-stereochemical relationship between the C(25) and C(26) stereo-centres. Given that a direct *p*-methoxybenzylation of the product alcohol was not achievable under all the conditions tested, a three-step protocol was eventually devised for positioning this protecting group at O(26). The sequence involved *O*-desilylation, *p*-methoxybenzylidenation of the terminal 1,2-diol, and regioselective acetal reduction[39] of **83** with DIBAL. This last reaction afforded a 5:1 mixture of regioisomers enriched in the desired alcohol **84**. Although the majority of oxidants investigated caused significant epimerisation at the C(26) stereocentre in **58**, this side-reaction could be avoided if the SO₃-pyridine/DMSO oxidant of Parikh and von Doehring[40] was used.

Aldehyde **58** reacted efficiently with the anion derived from phenylsulfone **57** to provide a diastereomeric mixture of β-hydroxysulfones **85** in 89% yield Scheme 17.19. Oxidation of **85** to the corresponding β-ketosulfones was best accomplished with excess trifluoroacetic anhydride and DMSO[41] at low temperature. Desulfonylation could be realised with Al/Hg amalgam[42] in aqueous THF under the Corey–Chaykovsky conditions. Having secured a viable pathway to ketone **55**, effort was now concentrated on stereoselectively transforming it into the (Z)-enol triflate **53**. Significantly, when DMPU[43] was added to the reaction mixture, it not only promoted *O*-triflation of the (Z)-enolate generated from ketone **55**, but it also improved the stereoselectivity of *O*-sulfonylation

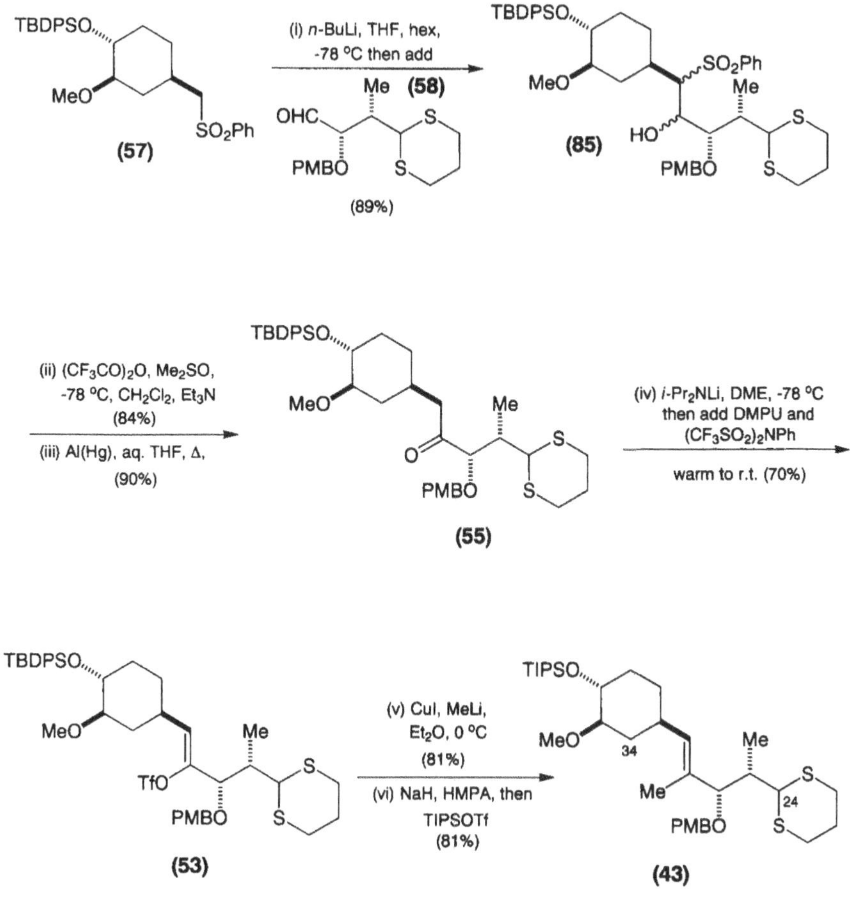

Scheme 17.19 Smith's pathway to the C(24)–C(34) sector of (–)-FK 506.

from 9:1 in favour of **53** to exclusively this product. Treatment of **53** with excess lithium dimethylcuprate in Et$_2$O at 0°C gave rise to two products in 7:1 ratio according to high-performance liquid chromatography (HPLC) analysis. The major product was the desired (*E*)-olefin isomer which was isolated in 73% yield after flash chromatography. The other product was not the (*Z*)-alkene, but another by-product, possibly the alkyne. A one-step protecting group interchange was now implemented to convert the newly-formed (*E*)-alkene into **43**.

Having successfully developed syntheses of the two key fragments needed to access the Merck advanced intermediate **41**, Smith next set about joining these two subunits together. The best conditions for efficiently generating the desired 1,3-dithianyl anion deprotonated **43** with two equivalents of *t*-BuLi in THF with 10% hexamethylphosphoric triamide (HMPA) as a cosolvent. Fortunately, the resulting anion proved highly reactive, it readily engaging in an efficient *C*-alkylation reaction with primary iodide **44**. Gratifyingly, this trapping process was complete within half an hour at −78°C, the desired product **86** being isolable in a noteworthy 74% yield (Scheme 17.20). After deprotection of the dithiane, the *p*-methoxybenzyl (PMB) ether was removed with DDQ[44] to unveil the requisite β-hydroxy ketone **42**. A low-temperature hydroxyl-directed *syn* reduction[15] with lithium aluminium hydride and lithium iodide now furnished the desired *syn*-1,3-diol as a single diastereoisomer. The final step of Smith's formal synthesis was a selective protection of the C(24) hydroxyl as a triisopropylsilyl ether. Compound **41** was obtained in 55% overall yield from **87**. The Merck group had previously converted **41** into the natural product.

17.7   Items of interest in the Danishefsky and Smith total syntheses of (−)-FK506

The conversion of 5 to 88

This is an example of the Evans asymmetric aldol reaction.[11] The *syn* selectivity observed in this addition almost certainly derives from the (*Z*)-enolate geometry of **6** (Scheme 17.21), and from the great desire of these enolates to engage in cyclic chair-like transition states which minimise repulsive steric interactions between the groups of both reaction partners. It is the vacant p orbital associated with the trivalent boron atom which allows chelation to the aldehyde carbonyl. In such highly compacted transition states, unfavourable 1,3-diaxial interactions tend to be magnified by the very short length (1.36–1.47 Å) of the B—O bond. The transition state arrangement most readily adopted is the one which reduces the number of 1,3-diaxial interactions between the groups of both participants. The absolute stereochemical outcome of these reactions is controlled by the bulky isopropyl group on the oxazolidinone, which permits approach of the aldehyde to the enolate from only one direction. For enolate **6** and aldehyde **5**, the most favoured transition state assembly is shown in Scheme 17.22.

For comparison purposes, a disfavoured chair transition state (TS) is also depicted (Scheme 17.23). This would lead to the *anti*-aldol product **91**. It will soon be realised that this transition state is plagued by a seriously destabilising 1,3-diaxial interaction between the oxazolidinone unit and the bulky R-group protruding from the aldehyde. A similar interaction can be seen for the B-Bu and R groups:

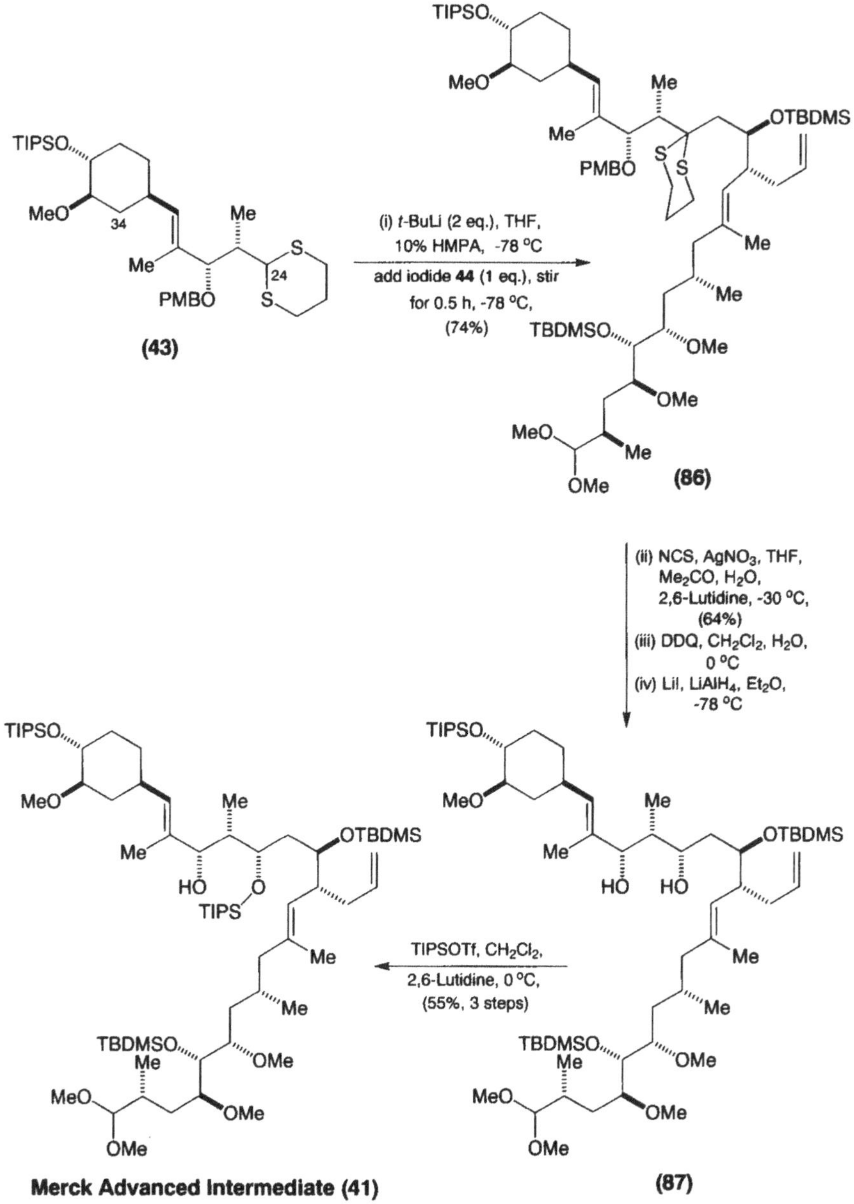

Scheme 17.20 The final stages of Smith's formal synthesis of (−)-FK506.

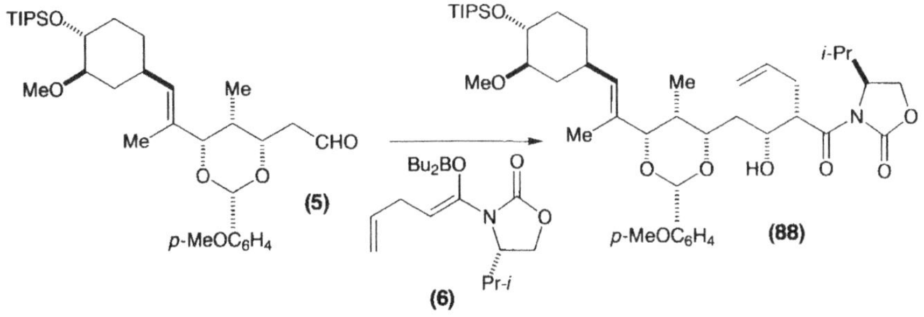

Scheme 17.21.

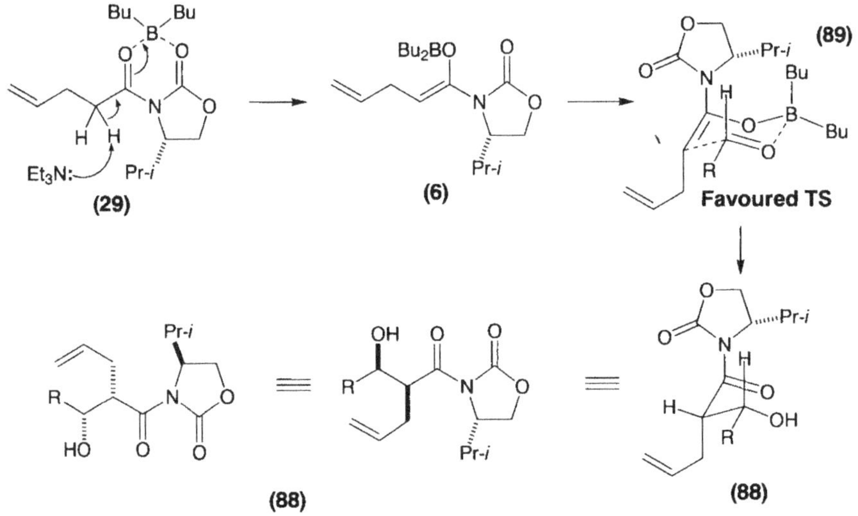

Scheme 17.22.

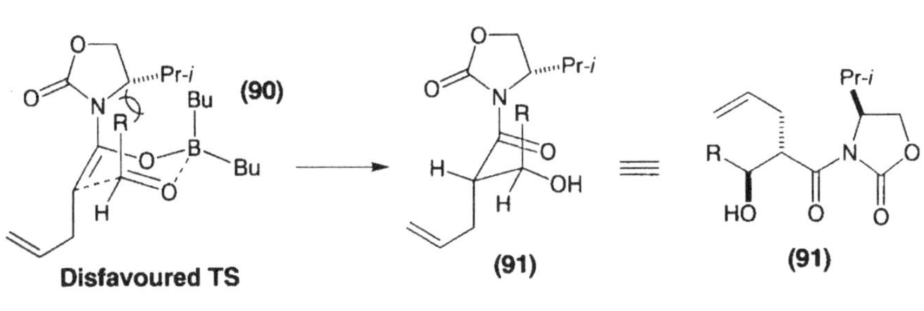

Scheme 17.23.

The conversion of 51 into 71

The first event in this multistage reaction is a Negishi[26] zirconium-catalysed carboalumination on alkyne **51** to obtain vinylalane **96** (Scheme 17.24), a protocol that proceeds with excellent stereo- and regiocontrol.

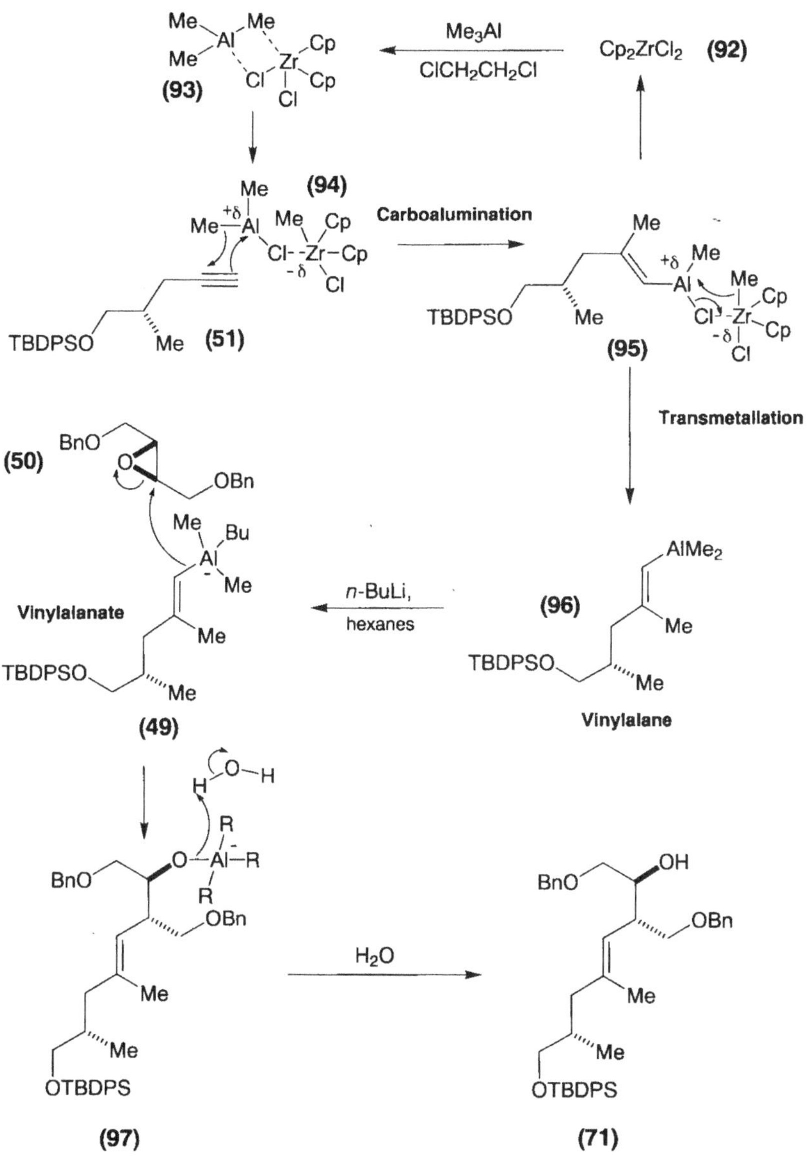

Scheme 17.24.

Mechanistically,[45] this reaction can be thought of as operating through a concerted four-centre transition state in which one of the Me—Al bonds adds *cis* to the alkyne unit, by a pathway that is not far removed from that in hydroboration. In essence, the zirconocene dichloride enhances the electrophilicity of trimethylaluminium and makes it a much more willing participant in the carbometallation process. Addition of *n*-butyllithium later fashions the ate complex **49** from **96**. The former is a much more potent nucleophile than **96**, it rapidly opening epoxide **50** by a traditional S_N2 pathway. Significantly, this was the first example of a trisubstituted vinylalanate complex being used to open a disubstituted epoxide. Thus, in this powerful new tandem process, an (*E*)-trisubstituted olefin and a new allylic asymmetric centre were stereo- and regiospecifically constructed in a simple one-pot operation. Versatile chemistry indeed, and a significant augmentation of earlier work by Negishi.

It is also worthwhile noting, in passing, that Peter Wipf and Sungtaek Lim[46] at Pittsburgh have discovered that Cp_2ZrCl_2–Me_3Al carboalumination reactions can be dramatically accelerated by the addition of a stoichiometric amount of water. A mechanistic rationale of this rather curious behaviour was presented in their report.

The conversion of β-ketosulfone 72 into ketone 98

Two possible reaction pathways can be visualised for this reaction (Scheme 17.25). One invokes an attack by a tributyltin radical on the

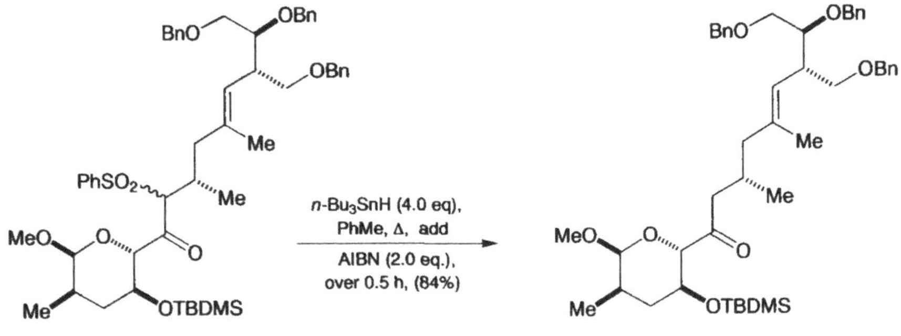

n-Bu₃SnH (4.0 eq),
PhMe, Δ, add
AIBN (2.0 eq.),
over 0.5 h, (84%)

(72)   (98)

Scheme 17.25.

phenylsulfone oxygen atom, the other an attack on its sulfur atom (Scheme 17.26). Both processes can lead to a stabilised α-keto radical that can abstract hydrogen from the alkylstannane to give **98**.

Although one can equally well envision a mechanism whereby a tin radical adds to the carbonyl oxygen to furnish a ketyl radical, this intermediate would have to eliminate a phenylsulfonyl radical and form

Mechanism:

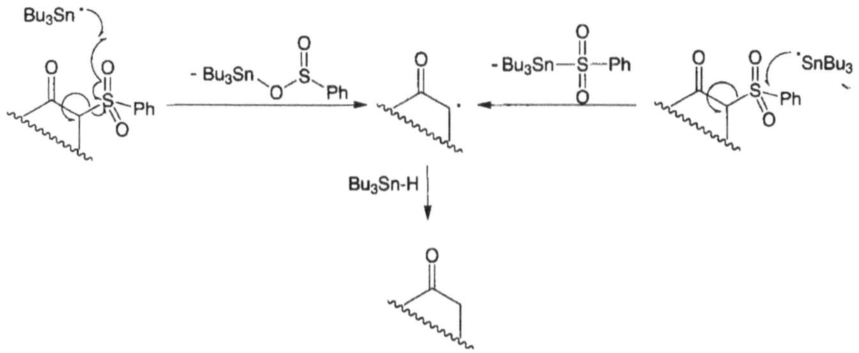

Scheme 17.26.

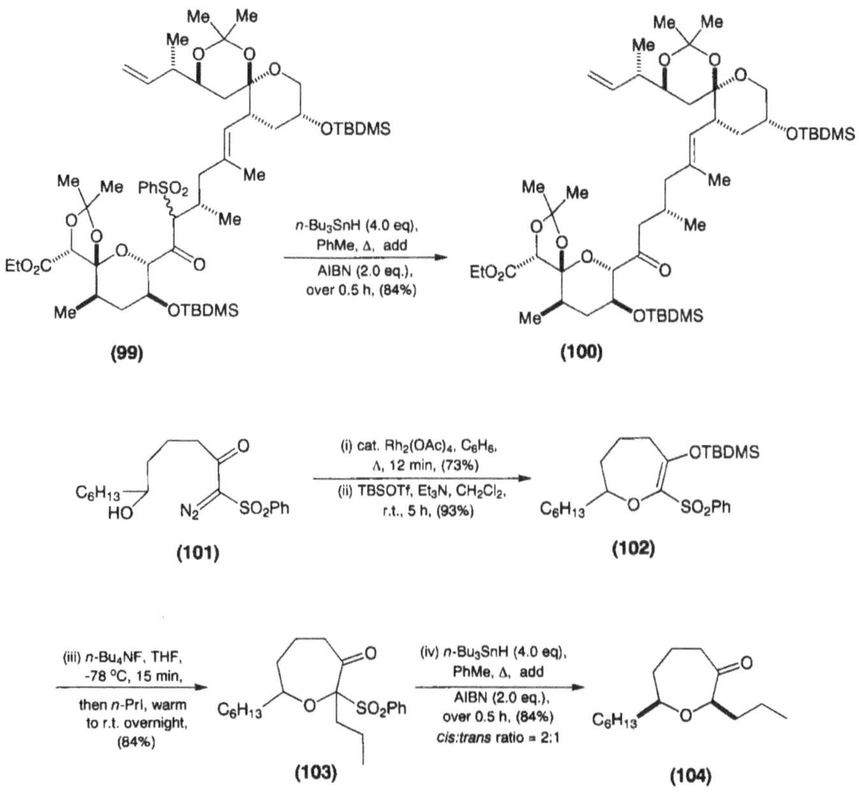

Scheme 17.27.

an enolate to arrive at the product. Because Smith was unable to intercept a putative tributylstannyl enolate, this mechanism was considered far less likely. Likewise a range of electron-transfer mechanisms that can also generate enolate intermediates have been discounted.

As the following examples show, the Smith–Hale desulfonylation method is now being used increasingly in natural product synthesis. The first example in Scheme 17.27 depicts a desulfonylation reported in 1992 by Ireland et al.[47] on an even more elaborate FK 506 intermediate **99**. The second sequence in Scheme 17.27 shows how this method was applied by Davies et al.[48] for the synthesis of a cis-2,7-disubstituted oxepane related to the marine natural product isolaurepinnacin.

The transformation of 46 into 105

Evans has proposed[17] that the anti-selective reduction of β-hydroxy ketones with Me₄NBH(OAc)₃ proceeds through an alkoxydiacetoxy kerning borohydride intermediate, which internally delivers hydride to the proximal ketone by a chair-like transition state, in which repulsive 1,3-diaxial interactions are minimised. For compound **46** (Scheme 17.28), two possible chair transition states can be envisaged. However, only the assembly that leads to the anti-1,3-diol (i.e. **106**) negates the creation of a destabilising 1,3-diaxial interaction between the ketone R group and the axial acetoxy group on boron. As a consequence, this transition state is believed to be considerably more favoured than the one (**107**) that would produce the 1,3-syn-diol **108**.

It is important to recognise that it is only through alkoxy/acetoxy interchange around the parent triacetoxyborohydride that a sufficiently reactive hydride donor can be generated that can perform the ketone reduction at an acceptable rate. Ordinarily, Me₄NBH(OAc)₃ will reduce ketones exceedingly slowly.

The conversion of 78 to 79

Oppolzer and co-workers[34] have attributed the high topological bias observed in this [4 + 2]-cycloaddition to the chelated endo transition state shown in **109**, wherein the diene adds to the less-hindered α face of the rigidly-held acrylimide (Scheme 17.29). This orientation for the alkene component appears to minimise steric repulsions between it and the camphorsultam framework.

The conversion of 53 into 43

The McMurry–Scott cross coupling reaction[22] between enol triflates and organocuprates generally proceeds with retention of the configuration in the parent enol triflate component. In these reactions it is widely assumed that a planar alkenylcopper(III) intermediate is formed by oxidative

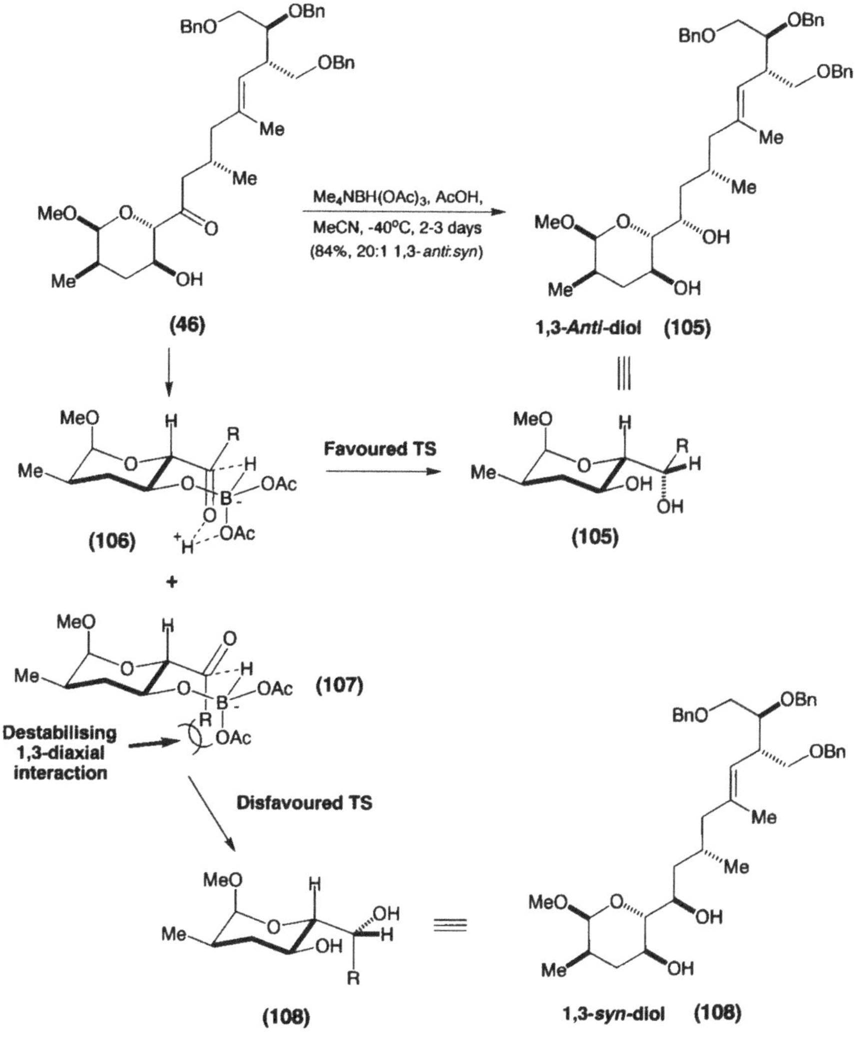

Scheme 17.28.

addition of the d^{10} cuprate to the enol triflate. In the present system, this intermediate would probably have a structure resembling **110** (Scheme 17.30). The latter would then undergo a rapid reductive *cis*-elimination to furnish the coupled product **43**.

The conversion of 42 into 87
Suzuki and co-workers[15] have found that β-alkoxy and β-hydroxy ketones can be reduced with high *syn*-stereoselectivity using the LiAlH₄–

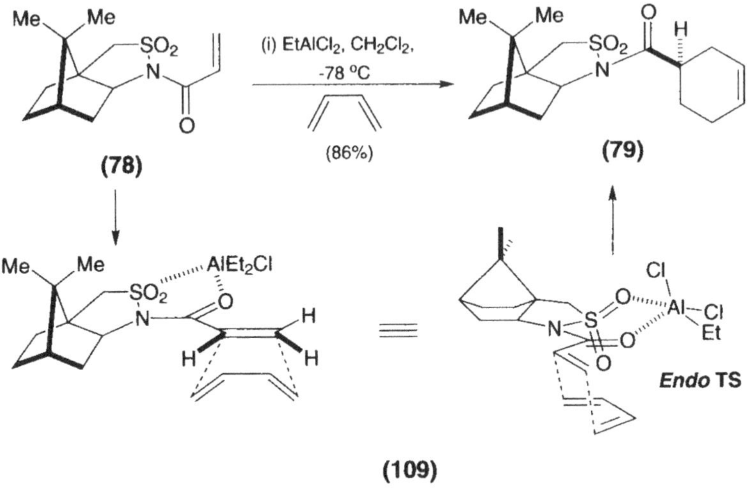

Scheme 17.29.

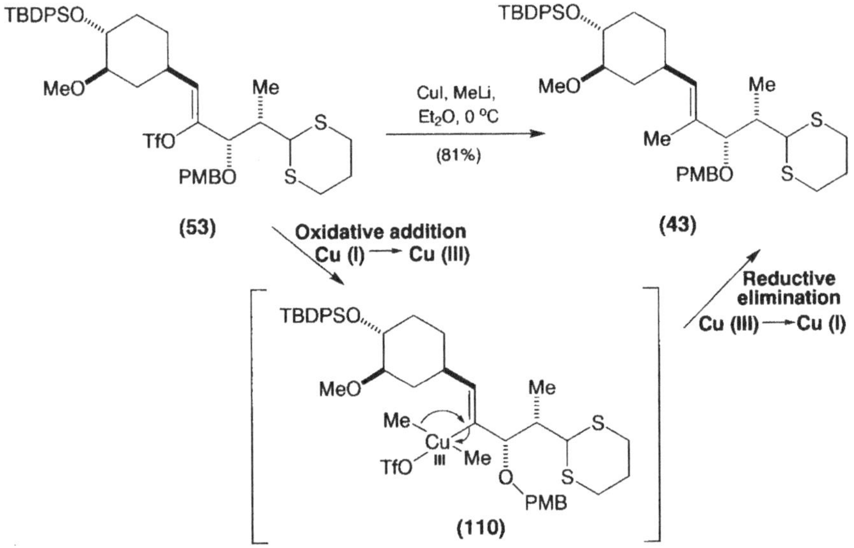

Scheme 17.30.

LiI reagent system. It is widely believed that the high *syn*-selectivity originates from a chelate formed between the β-hydroxy ketone and the lithium cation. This locks the ketone into a rigid conformation that is preferentially reduced from the least-hindered direction. For compound

42, the less sterically-encumbered face of the chelate is the α side (Scheme 17.31). This is the face opposite to the pseudoaxial C(25) methyl substituent. Such a transition state nicely accounts for the product stereochemistry observed in **87**.

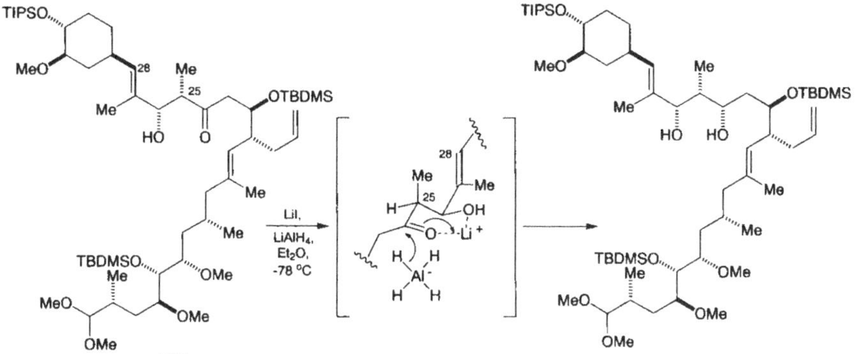

Scheme 17.31.

References

1. Isolation and structure elucidation: H. Tanaka, A. Kuroda, H. Marusawa, H. Hatanaka, T. Kino, T. Goto, M. Hashimoto and T. Taga, 1987, *J. Amer. Chem. Soc.*, 109, 5031.
2. (a) T. Ochiai, M. Nagata, K. Nakajima, T. Suzuki, K. Sakamoto, K. Enomoto, Y. Gunji, T. Uematsa, T. Goto, S. Hori, T. Kenmochi, T. Nakagouri, T. Asano, K. Isono, K. Hamaguchi, H. Tsuchida, K. Nakahara, N. Imamura and T. Goto, 1987, *Transplantation*, 44, 729; (b) T. Ochiai, K. Nakajima, M. Nagata, S. Hori, T. Asano and K. Isono, 1987, *Transplantation*, 44, 734.
3. A.W. Thomson, 1989, *Immunol. Today*, 10, 6.
4. Total syntheses: (a) T.K. Jones, S.G. Mills, R.A. Reamer, D. Askin, R. Desmond, R.P. Volante and I. Shinkai, 1989, *J. Amer. Chem. Soc.*, 111, 1157; D. Askin, D. Joe, R.A. Reamer, R.P. Volante and I. Shinkai, 1990, *J. Org. Chem.*, 55, 5451; (b) M. Nakatsuka, J.A. Ragan, T. Sammakia, D.B. Smith, D.E. Uehling and S.L. Schreiber, 1990, *J. Amer. Chem. Soc.*, 112, 5583; (c) R.E. Ireland, J.L. Gleason, L.D. Gegnas and T.D. Highsmith, 1996, *J. Org. Chem.*, 61, 6856; (d) R.E. Ireland, L. Liu and T.D. Roper, 1997, *Tetrahedron*, 53, 13221; R.E. Ireland, L. Liu, T.D. Roper and J.L. Gleason, 1997, *Tetrahedron*, 53, 13257.
5. Formal total syntheses: (a) R.G. Linde, II, M. Egbertson, R.S. Coleman, A.B. Jones and S.J. Danishefsky, 1990, *J. Org. Chem.*, 55, 2771; A. Villabolos and S.J. Danishefsky, 1990, *J. Org. Chem.*, 55, 2776; A.B. Jones, A. Villabolos, R.G. Linde, II and S.J. Danishefsky, 1990, *J. Org. Chem.*, 55, 2786; (b) A.B. Smith, III and K.J. Hale, 1989, *Tetrahedron Lett.*, 30, 1037; A.B. Smith, III, K.J. Hale, L.M. Laakso, K. Chen and A. Riera, 1989, *Tetrahedron Lett.*, 6939; A.B. Smith, III, K. Chen, D.J. Robinson, L.M. Laakso and K.J. Hale, 1994, *Tetrahedron Lett.*, 35, 4271; (c) R. Gu and C. Sih, 1990, *Tetrahedron Lett.*, 31, 3283 and 3287.

6. Synthetic approaches: (a) D.R. Williams and J.W. Benbow, 1988, *J. Org. Chem.*, 53, 4643; (b) P. Kocienski, M. Stocks, D. Donald, M. Cooper and A. Manners, 1988, *Tetrahedron Lett.*, 29, 4481; (c) R.E. Ireland and P. Wipf, 1989, *Tetrahedron Lett.*, 30, 919; (d) H.H. Wasserman, V.M. Rotello, D.R. Williams and J.W. Benbow, 1989, *J. Org. Chem.*, 54, 2785; (e) E.J. Corey and H.C. Huang, 1989, *Tetrahedron Lett.*, 30, 5235; (f) Z. Wang, 1989, *Tetrahedron Lett.*, 30, 6611; (g) A.V.R. Rao, T.K. Chakraborty and K.L. Reddy, 1990, *Tetrahedron Lett.*, 31, 1439; (h) A.V.R. Rao, T.K. Chakraborty and A. Purandare, 1990, *Tetrahedron Lett.*, 31, 1443; (i) M. Stocks and P. Kocienski, 1990, *Tetrahedron Lett.*, 31, 1637; (j) R.E. Ireland, P. Wipf and T.D. Roper, 1990, *J. Org. Chem.*, 55, 2284; (k) M.E. Maier and B. Schoeffling, 1991, *Tetrahedron Lett.*, 32, 53; (l) K. Maruoka, S. Saito, T. Ooi and H. Yamamoto, 1991, *Synlett*, 579; (m) Y. Morimoto, A. Mikami, S. Kuwabe and H. Shirahama, 1991, *Tetrahedron Lett.*, 32, 2909; (n) Z. Wang, 1991, *Tetrahedron Lett.*, 32, 4631; (o) M. Chini, P. Crotti, F. Machia and L.A. Flippin, 1992, *Tetrahedron*, 48, 539; (p) J.D. White, S.G. Toske and T. Yakura, 1994, *Synlett*, 591; (q) K. Jarowicki, P. Kocienski, S. Norris, M. O'Shea and M. Stocks, 1995, *Synthesis*, 195; (r) J.A. Marshall and S. Xie, 1995, *J. Org. Chem.*, 60, 7230.

7. (a) M. Julia and J.M. Paris, 1973, *Tetrahedron Lett.*, 4833; (b) P.J. Kocienski, B. Lythgoe and S. Ruston, 1978, *J. Chem. Soc. Perkin Trans. 1*, 829; (c) P.J. Kocienski, B. Lythgoe and I. Waterhouse, 1980, *J. Chem. Soc. Perkin Trans. 1*, 1045; (d) Review: N.S. Simpkins, 1993, *Sulphones in Organic Synthesis*, Pergamon Press, Oxford, Chapter 7, 254; (e) Reviews: P.J. Kocienski, 1985, *Phosphorus and Sulfur*, 24, 97 and P.J. Kocienski, 1981, *Chem. Ind. (London)*, 548.

8. B.H. Lipshutz, J. Kozlowski and R.S. Wilhelm, 1982, *J. Amer. Chem. Soc.*, 104, 2305; B.H. Lipshutz, R.S. Wilhelm, J.A. Kozlowski and D. Parker, 1984, *J. Org. Chem.*, 49, 3928.

9. H.B. Henbest and R.A.L. Wilson, 1957, *J. Chem. Soc.*, 1958.

10. G. Grynkiewicz, W. Priebe and A. Zamojski, 1979, *Carbohydr. Res.*, 68, 33.

11. D.A. Evans, J. Bartroli and T.L. Shih, 1981, *J. Am. Chem. Soc.*, 103, 2127.

12. Review: S. David and S. Hanessian, 1985, *Tetrahedron*, 41, 643.

13. D.A. Evans and M.M. Morrissey, 1984, *J. Amer. Chem. Soc.*, 106, 3866; D.A. Evans, M.M. Morrissey and R.C. Dow, 1985, *Tetrahedron Lett.*, 26, 6005; D.A. Evans, R.L. Dow, T.L. Shih, J.M. Takacs and R. Zahler, 1990, *J. Amer. Chem. Soc.*, 112, 5290.

14. E.M. Burgess and H.R. Penton, 1977, *Org. Synth.*, 56, 40; E.M. Burgess, H.R. Penton and E.A. Taylor, 1973, *J. Org. Chem.*, 38, 26.

15. Y. Mori, Y. Kohchi and M. Suzuki, 1991, *J. Org. Chem.*, 56, 631; Y. Mori, M. Kuhara, A. Takeuchi and M. Suzuki, 1988, *Tetrahedron Lett.*, 29, 5419; Y. Mori, A. Takeuchi, H. Kageyama and M. Suzuki, 1988, *Tetrahedron Lett.*, 29, 5423.

16. (a) D. Seebach and E.J. Corey, 1975, *J. Org. Chem.*, 40, 231; (b) Review: B.T. Grobel and D. Seebach, 1977, *Synthesis*, 357.

17. D.A. Evans, K.T. Chapman and E.M. Carreira, 1988, *J. Amer. Chem. Soc.*, 110, 3561.

18. For an earlier coupling of a sulfone anion to a methyl ester, see: B.M. Trost, J. Lynch, P. Renaut and D.H. Steinman, 1986, *J. Amer. Chem. Soc.*, 108, 284.

19. For related reductions see: (a) B.J. Fitzsimmons, D.E. Plaumann and B. Fraser-Reid, 1979, *Tetrahedron Lett.*, 3925; (b) M. Miljkovic and D. Glisin, 1975, *J. Org. Chem.*, 40, 3357.

20. (a) T.D. Inch and G.J. Lewis, 1972, *Carbohydr. Res.*, 22, 91; (b) S. Jarosz, D.R. Hicks and B. Fraser-Reid, 1982, *J. Org. Chem.*, 1982, 935.

21. K.C. Nicolaou, D.P. Papahatjis, D.A. Claremon, R.A. Magolda and R.E. Dolle, 1985, *J. Org. Chem.*, 50, 1440.

22. J.E. McMurry and W.J. Scott, 1980, *Tetrahedron Lett.*, 21, 4313.

23. I.D. Jenkins and S. Bottle, 1984, *J. Chem. Soc. Chem. Comm.*, 385.

24. (a) T. Katsuki and K.B. Sharpless, 1980, *J. Amer. Chem. Soc.*, 102, 5974; (b) B.E. Rossiter, T. Katsuki and K.B. Sharpless, *J. Amer. Chem. Soc.*, 1981, 103, 464; (c) Y. Gao, R.M. Hanson, J.M. Klunder, S.Y. Ko, H. Masamune and K.B. Sharpless, 1987, *J. Amer. Chem. Soc.*, 109, 5765.

25. E.J. Corey and P.L. Fuchs, 1972, *Tetrahedron Lett.*, 14, 3769.
26. (a) D.E. Van Horn and E. Negishi, 1978, *J. Amer. Chem. Soc.*, 100, 2252; (b) T. Yoshida and E. Negishi, 1981, *J. Amer. Chem. Soc.*, 103, 4985; (c) Review: E. Negishi and D. Choueiry, 1995, in *Encyclopedia of Organic Reagents* (L.A. Paquette, ed.), John Wiley and Sons, Chichester, 5195.
27. B.M. Trost and D.P. Curran, 1981, *Tetrahedron Lett.*, 22, 1287.
28. A.B. Smith, III, K.J. Hale and J.P. McCauley, Jr., 1989, *Tetrahedron Lett.*, 30, 5579.
29. E.N. Marvell and T. Li, 1973, *Synthesis*, 8, 457.
30. Reviews: B.H. Lipshutz, R.S. Wilhelm and J.A. Kozlowski, 1984, *Tetrahedron*, 40, 5005; B.H. Lipshutz, 1987, *Synthesis*, 325; B.H. Lipshutz, 1994, *Organometallics in Synthesis*, Ed. M. Schlosser, Chapter 4, 282.
31. T. Nagata, S. Nagao, N. Mori and T. Oishi, 1985, *Tetrahedron Lett.*, 26, 6461 and 6465.
32. D.F. Taber, E.M. Petty and K. Raman, 1985, *J. Amer. Chem. Soc.*, 107, 196.
33. E.J. Corey and B.W. Erickson, 1971, *J. Org. Chem.*, 36, 3553.
34. W. Oppolzer, C. Chapuis and G. Bernadinelli, 1984, *Helv. Chim. Acta*, 67, 1397.
35. S.F. Martin, M.S. Dappen, B. Dupre, C.J. Muphy and J.A. Colapret, 1989, *J. Org. Chem.*, 54, 2209.
36. I. Nakagawa, K. Aki and T. Hata, 1983, *J. Chem. Soc. Perkin Trans. 1*, 1315.
37. M.C. Caserio, J.D. Roberts, M. Neeman and W.S. Johnson, 1985, *J. Amer. Chem. Soc.*, 80, 2584.
38. D.J. Pasto and J. Hickman, 1968, *J. Amer. Chem. Soc.*, 90, 4445; S.L. Schreiber and D.B. Smith, 1989, *J. Org. Chem.*, 54, 9.
39. S. Takano, M. Akiyama, S. Sato and K. Ogasawara, 1983, *Chem. Lett.*, 1593.
40. J.R. Parikh and W.E. von Doering, 1967, *J. Amer. Chem. Soc.*, 89, 5505.
41. K. Omura, A.K. Sharma and D. Swern, 1976, *J. Org. Chem.*, 41, 957.
42. E.J. Corey and M.J. Chaykovsky, 1965, *J. Amer. Chem. Soc.*, 87, 1345.
43. T. Mukhopadhyay and D. Seebach, 1982, *Helv. Chim. Acta.*, 65, 385.
44. K. Horita, T. Yoshioka, T. Tanaka, Y. Oikawa and O. Yonemitsu, 1986, *Tetrahedron*, 42, 3021.
45. E. Negishi, D.E. Van Horn and T. Yoshida, 1985, *J. Amer. Chem. Soc.*, 107, 6639.
46. P. Wipf and S. Lim, 1993, *Angew. Chem. Int. Ed. Engl.*, 32, 1068.
47. R.E. Ireland, T.K. Highsmith, L.D. Gegnas and J.L. Gleason, 1992, *J. Org. Chem.*, 57, 5071.
48. M.J. Davies, C.J. Moody and R.J. Taylor, 1991, *J. Chem. Soc. Perkin Trans.1*, 1; M.J. Davies and C.J. Moody, 1991, *J. Chem. Soc. Perkin Trans.1*, 1991, 9.

18 (3S,5S)-5-Hydroxypiperazic acid

K.J. Hale

18.1 Introduction

(3S,5S)-5-Hydroxypiperazic acid is a rare hexahydropyridazine encountered in the monamycin family of antifungal antibiotics. These are
architecturally novel cyclodepsipeptides first isolated and characterised
by Hassall and co-workers in the early 1970s.[1,2] Quite recently, (3S,5S)-5-
hydroxypiperazic acid has been synthesised in optically pure form for the
first time by the Hale group,[3] as part of their total synthesis programme
on monamycin D_1. Their retrosynthetic planning and synthetic strategy
for this molecule are presented below.

18.2 The Hale retrosynthetic analysis
of (3S,5S)-5-hydroxypiperazic acid

Although the five-carbon framework, and (5S)-alcohol stereochemistry,
of this target both suggest that it could potentially be derived from a D-
pentose precursor, a careful analysis of possible routes from such starting
materials soon reveals that many are tactically compromised by their
significant length. Hale and co-workers therefore evaluated the possibilities for deriving this target from an appropriately functionalised D-
glyceraldehyde synthon. Conceptually, this would require a two-carbon
chain extension being performed at one terminus of such a starting sugar,
and a hydrazination at the other terminal carbon. The (5S)-hydroxyl
stereochemistry would emanate from the starting sugar. A Wittig reaction
with a suitable $Ph_3P=CHCO_2R$ synthon could be envisaged for installing the appropriate two-carbon fragment, if it were combined with a
hydrogenation reaction to remove the resulting double bond. However,
this would still leave the issue of stereospecific hydrazination and
formation of the hexahydropyridazine ring system. Since Evans and
Vederas had both introduced good technology for the asymmetric
electrophilic hydrazination of N-acyloxazolidinone enolates,[4,5] thought
was given to the possibility of introducing the (3S)-hydrazino grouping
using such a reaction. Because these reactions invariably lead to the
formation of a terminal hydrazino anion, it was felt that it might be
possible to trap this anion in a tandem S_N2 ring-closure process, provided
a suitable leaving group was positioned at the terminal carbon of an

appropriate valeryl enolate. The summary of this retrosynthetic thinking is presented in Scheme 18.1. A key element of this strategy was its
proposed usage of a phosphonoacetate that already incorporated the
Evans oxazolidinone chiral auxiliary. It would be condensed with the
known triose **7** to give Wittig product **5**, which could then be saturated
and brominated at its terminal carbon to provide the key cyclisation

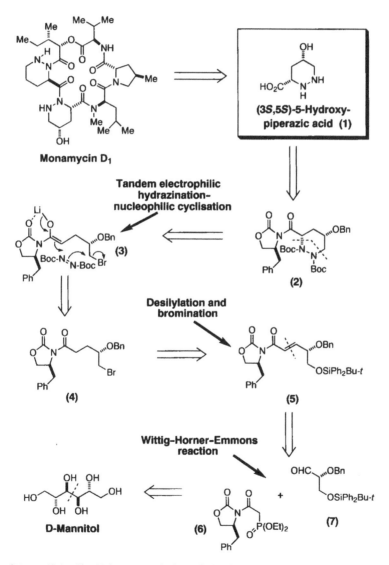

Scheme 18.1   The Hale retrosynthetic analysis of (3S,5S)-5-hydroxypiperazic acid.

precursor **4**. The latter would then be subjected to the tandem electrophilic hydrazination–S_N2 ring closure tactic needed to arrive at **2**.

18.3   The Hale total synthesis of (3*S*,5*S*)-5-hydroxypiperazic acid

Triose **7** had previously been prepared by Welzel *et al.*[6] in five steps from D-mannitol. When treated with known phosphonate **6**,[7] under the Roush–Masamune conditions,[8] aldehyde **7** underwent a facile Wittig–Horner olefination to produce the crystalline alkene **5** as a single geometrical isomer (Scheme 18.2). The silyl group was detached from **5** by exposing it to 40% aq. HF in tetrahydrofuran (THF)/MeCN. This gave the alcohol **9** in 85% overall yield from **8**. The double bond of **9** was then chemoselectively hydrogenated, without disturbing the *O*-benzyl ether, to produce the desired alcohol in good yield. The latter underwent bromination with excess Ph$_3$P and carbon tetrabromide in dry THF at room temperature to furnish **4** as an oil in 83% yield from **9**. This prepared the substrate needed for the key tandem electrophilic hydrazination–nucleophilic cyclisation process.[9]

Low-temperature enolisation of bromide **4** with lithium diisopropylamide (LDA) gave the chiral bromovaleryl enolate **3** which underwent instantaneous hydrazination[4,5] with di-*tert*-butylazodicarboxylate (DBAD) with high diastereocontrol. For the desired tandem cyclisation to proceed efficiently, it was found necessary to add dry 1,3-dimethyl-3,4,5,6-tetrahydro-2(1*H*)-pyrimidinone (DMPU) to the reaction mixture, and then to warm it to room temperature for *ca.* 50 min. After chromatographic purification, the desired product **2** was typically isolated in 50–65% yield. It transpired that the best protocol for removing the chiral auxiliary from **2** was to react it with sodium methoxide in CH$_2$Cl$_2$ and methanol at −30°C for 15 min. This regime generally delivered **10** as an oil in 93% yield after chromatography. The following set of reactions proved most satisfactory for completing the synthesis of **1**. First, the *O*-benzyl ether was cleaved by catalytic hydrogenolysis over Pd(OH)$_2$ in methanol. The product alcohol was then *O*-acetylated to obtain **11**. A clean and high yielding deprotection of the two Boc groups was next effected with trifluoroacetic acid. Crude **12** was finally reacted with excess LiOH in THF and H$_2$O for 1 h at 0°C to obtain **1**. The latter was purified by reverse-phase C$_{18}$ chromatography.

This synthesis of **1** is noteworthy for its seminal demonstration of the utility of tandem electrophilic hydrazination–nucleophilic cyclisation[9] for preparing *functionalised* homochiral cyclic α hydrazino acids.[10] It highlights, yet again, the great utility of functionalised glyceraldehyde synthons for the installation of remote hydroxy stereocentres in natural product target molecules.

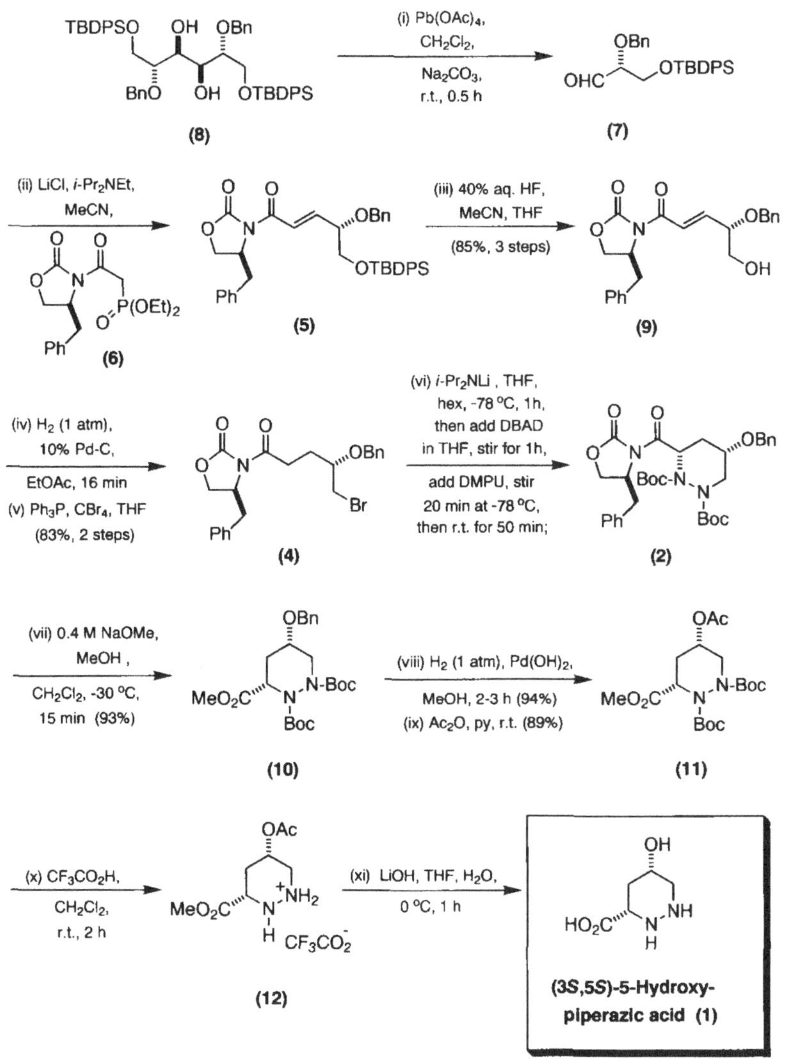

Scheme 18.2 Hale's enantiospecific synthesis of (3S,5S)-5-hydroxypiperazic acid.

18.4 Points of mechanistic interest in the Hale (3S,5S)-5-hydroxypiperazic acid synthesis

The stereochemical outcome of the conversion of 4 into 2

The stereochemical outcome of this hydrazination can be explained if one assumes that the (Z)-lithium enolate **3** is internally chelated to the oxazolidinone carbonyl at low temperature. This would hold the enolate

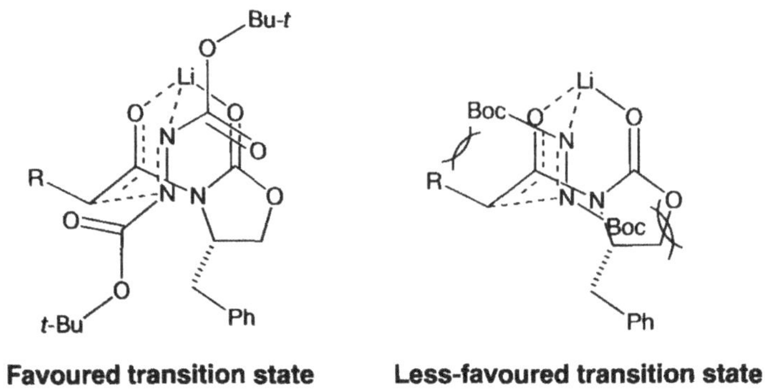

Favoured transition state **Less-favoured transition state**

Scheme 18.3.

in a fairly rigid conformation in which one side of the enolate double bond is sterically shielded by the phenylmethyl group of the auxiliary. Evans[4] has suggested that there are two possible chair-like six-centred transition states that can in principle be adopted during such hydrazinations. However, steric considerations strongly suggest that only one of these is adopted in practice. The favoured transition state is that shown above. Note how this invokes the azodicarboxylate approaching the enolate from the face that is opposite to the phenylmethyl group.

References

1. (a) K. Bevan, J.S. Davies, C.H. Hassall, R.B. Morton and D.A.S. Phillips, 1971, *J. Chem. Soc. (C)*, 514; (b) C.H. Hassall, Y. Ogihara and W.A. Thomas, 1971, *J. Chem. Soc. (C)*, 522; (c) C.H. Hassall, R.B. Morton, Y. Ogihara and D.A.S. Phillips, 1971, *J. Chem. Soc. (C)*, 526; (d) C.H. Hassall, W.A. Thomas and (in part) M.C. Moschidis, 1977, *J.C.S. Perkin Trans. 1*, 2369.

2. (3S,5S)-5-Hydroxypiperazic acid has also been obtained by optical resolution of the racemate with quinine, see: C.H. Hassall and K.L. Ramachandran, 1977, *Heterocycles*, 7, 119.

3. K.J. Hale, N. Jogiya and S. Manaviazar, 1998, *Tetrahedron Lett.*, 39, 7163.

4. (a) D.A. Evans, T.C. Britton, R.L. Dorow and J.F. Dellaria, 1988, *Tetrahedron*, 44, 5525; (b) D.A. Evans, T.C. Britton, R.L. Dorow and J.F. Dellaria, 1986, *J. Amer. Chem. Soc.*, 108, 6395.

5. L.A. Trimble and J.C. Vederas, 1986, *J. Amer. Chem. Soc.*, 108, 6397.

6. U. Peters, W. Bankova and P. Welzel, 1987, *Tetrahedron*, 43, 3803.

7. C.A. Broka and J. Ehler, 1991, *Tetrahedron Lett.*, 32, 5907.

8. M.A. Blanchette, W. Choy, J.T. Davis, A.P. Essenfield, S. Masamune, W.R. Roush and T. Sakai, 1984, *Tetrahedron Lett.*, 25, 2183.

9. K.J. Hale, J. Cai, V. Delisser, S. Manaviazar, S.A. Peak, G.S. Bhatia, T.C. Collins and (in part) N. Jogiya, 1996, *Tetrahedron*, 52, 1047.

10. For another, conceptually different, and very elegant total synthesis of a protected (3R,5R)-hydroxypiperazic acid derivative, which again exploits Evans–Vederas hydrazination technology, see: T.M. Kamenecka and S.J. Danishefsky, 1998, *Angew. Chem. Int. Ed. Engl.*, 37, 2995.

Index

Printed and bound by CPI Group (UK) Ltd, Croydon, CR0 4YY

10/06/2024

14512954-0002